Proceedings in Life Sciences

Population Biology and Evolution

Edited by
K. Wöhrmann and V. Loeschcke

With 74 Figures

Springer-Verlag
Berlin Heidelberg New York Tokyo 1984

Professor Dr. KLAUS WÖHRMANN
Lehrstuhl für Populationsgenetik
Institut für Biologie II der Universität Tübingen
Auf der Morgenstelle 28
7400 Tübingen, FRG

Dr. VOLKER LOESCHCKE
Institute of Ecology and Genetics
University of Aarhus
Ny Munkegade 550
8000 Aarhus C, Denmark

ISBN-13:978-3-642-69648-0 e-ISBN-13:978-3-642-69646-6
DOI: 10.1007/978-3-642-69646-6

Library of Congress Cataloging in Publication Data. Main entry under title: Population biology and evolution. (Proceedings in life sciences) Papers presented at a symposium held May 15–19, at the guest house of the University of Tübingen at Oberjoch and sponsored by the Deutsche Forschungsgemeinschaft. Includes index. 1. Population biology – Congresses. 2. Evolution – Congresses. I. Wöhrmann, K. (Klaus), 1928–. II. Loeschcke, V. (Volker), 1950–. III. Deutsche Forschungsgemeinschaft. IV. Series. QA352.P573 1984 575.1'5 84-5329

Softcover reprint of the hardcover 1st edition 1984

2131/3130-543210

Preface

This volume contains the papers presented at a symposium on population biology sponsored by the Deutsche Forschungsgemeinschaft. It was held at the guest house of the University of Tübingen at Oberjoch on May 15–19, 1983.

Prior to this conference a small group of European biologists had met in Berlin (June 1981) and Pavia (September 1982) to discuss research problems on the borderline between population genetics and evolutionary ecology.

From the contributions and discussions at these meetings it became evident that the unification of approaches to evolutionary problems in population genetics and evolutionary ecology has not yet been successful and requires further efforts. It was the consensus that a larger symposium with international participation would be helpful to confront and discuss the different approaches to population biology in order to assess "where we are now" and "where we should be going."

As a result an organizational committee was formed (F. Christiansen, S. Jayakar, V. Loeschcke, W. Scharloo, and K. Wöhrmann) to identify topics that seemed, at least to them, to be fruitful in tackling problems in population biology. Consequently, a number of colleagues were asked to participate in the meeting.

We have divided this book into chapters corresponding to the eight topics chosen. The volume begins with the relation between genotype and phenotype and is followed by a chapter on quantitative genetics and selection in natural populations. Chapter 3 deals with theoretical aspects of density regulation and life histories followed by discussions of genetic heterogeneity and ecological factors. The next chapter concerns genetic structure and demography in plant populations and Chap. 6 covers population differentiation and asexual reproduction. These are followed by contributions on theoretical aspects of coevolution and concludes with some comments on models in population genetics and evolutionary ecology.

The papers are revised versions of those presented at Oberjoch. They were reviewed by participants of the meeting including colleagues who did not present papers. We wish to thank all the reviewers for their help and the members of the organizational committee for their support in arranging the conference.

Special thanks are due to the leaders of the evening discussions, W. Scharloo and P. Grant, and to the chairmen of individual sections. We are grateful to Mrs. M. Bullinger and Mrs. C. Rehm from the Institute of Biology II in Tübingen for assistance in organizing the conference and to Doth Anderson, Annie Jensen, Kirsten Petersen, and Marianne Szygenda from the Institute of Ecology and Genetics in Aarhus for editorial assistance.

Aarhus and Tübingen
April 1984

KLAUS WÖHRMANN
VOLKER LOESCHCKE

Contents

Theoretical Aspects of Density Regulation and Life Histories

Genetic Heterogeneity and Ecological Factors

Genetic Structure and Demography in Plants

Population Differentiation and Asexual Reproduction

Theoretical Aspects of Coevolution

Comments on Models

List of Contributors

You will find the addresses at the beginning of the respective contribution

Introduction: Genotype and Phenotype: Basic Concepts in Population Biology

V. LOESCHCKE[1] and K. WÖHRMANN[2]

An individual is characterized by its particular genotype. In sexually reproducing species two individuals rarely share a genotype. The phenotype of an individual is the result of the interplay between its genotype and its environment: of single gene effects, the interaction between its genes, and the interaction between the genes and the environment. The phenotype includes all aspects of the individual's morphology, physiology, behavior, as well as its ecological relationships to other individuals.

In practice, we use the concepts of genotype and phenotype in a more restricted sense by referring to a subset of the genotype and to a partial phenotype at some stage of the life history. In this restricted sense individuals of the same species commonly share genotypes and phenotypes. If we consider a specific partial phenotype, then generally we will find that several genes are relevant in the formation of particular characters. These genes are located on chromosomes and are therefore more or less linked to groups of other genes. Genes contributing to the determination of particular characters may well have pleiotropic effects on other characters, which in turn are affected by other genes. At the molecular level, they are coding for products that may be interacting functionally with other gene products.

It is the individual that survives and reproduces. Its success in doing so under specific environmental conditions is characterized by the fitness. Thus, fitness is a function of the phenotype and the environment. The differential success of distinct phenotypes in contributing to the next generation induces a change in the relative proportions of genes transmitted to the next generation (Fig. 1). Thus, essential features of evolutionary processes occur at two distinct levels. Selection is acting on phenotypes, but only genes record the history of the selective processes.

The processes that link genotype and phenotype are only known in a few cases and will hopefully be better understood in the future. So far, evolution has been considered by evolutionary ecologists mainly as changes in life history characters and other dynamically interesting phenotypes in relation to environmental conditions. On the other hand, population geneticists view evolution as changes in heritable phenotypic variation that can be traced back to genetic variation and as changes in the corresponding gene frequencies. Both disciplines have tried to characterize the evolutionary role of variation at each level in relation to environmental conditions. One of the major tasks of contemporary population biologists is to relate the processes that occur at both levels and, in addition, to differentiate concepts of the environment when

1 Institute of Ecology and Genetics, University of Aarhus, 8000 Aarhus C, Denmark
2 Institute of Biology II, University of Tübingen, 7400 Tübingen, FRG

Population Biology and Evolution
Ed. by K. Wöhrmann and V. Loeschcke

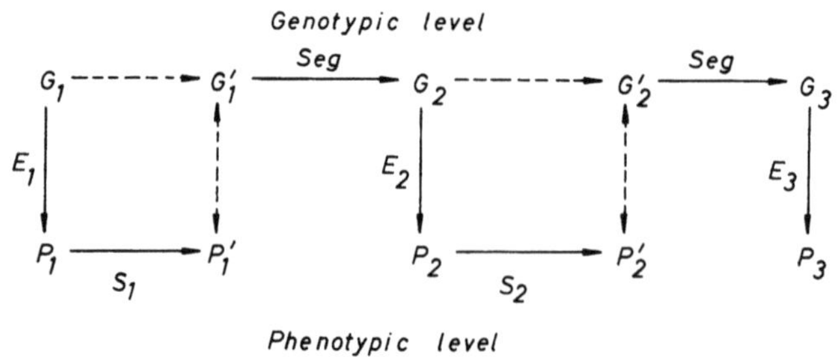

Fig. 1. Some basic processes within and between the genotypic and phenotypic levels. The phenotypes, P_1, that correspond to the genotypes, G_1, are influenced by the environment E_1. Selection, S_1 is operating on the phenotypes, P_1, and thereby conditions a change in the phenotypic composition of the population from P_1 to P'_1. Consequent upon this change there is a change in the genotypic composition from G_1 to G'_1. At the genotypic level segregation, *Seg*, occurs which determines the genotypic composition, G_2, at the beginning of generation 2. The process continues in similar fashion in subsequent generations

describing these processes. The contributions to the symposium are attempts toward this goal using a variety of approaches which hopefully will provide an impression of the diversity within contemporary population biology.

Genotype and Phenotype

Genetics of Adaptive Reactions

W. SCHARLOO[1]

Neo-Darwinism is a theory which is completely focused on genes: evolutionary change is considered to be a change of gene frequencies. This view has its origin in the theory of population genetics. The mathematical theory bestowed prestige on evolutionary genetics. As a consequence this prestigious mathematical framework became more and more identified with the whole field.

Nevertheless, at the end of the fifties and the early sixties there was still a place for quantitative genetics in evolutionary biology [e.g., see SEB Symposium on Evolution (Brown and Danielli 1953) and the Cold Spring Harbor Symposium on Population Genetics (Demerec 1955)]. Quantitative genetics is not so much a theory of genes as a theory of phenotypes. Selection operating in quantitative genetics is selection on phenotypes as it is, of course, almost always in nature. Genetic effects of selection are determined by the developmental and/or physiological processes which connect genes and phenotypes.

In the early sixties quantitative genetics started (1) to identify and localize genes involved in quantitative variation (Thoday 1961) and (2) to analyze physiological and developmental processes underlying quantitative variation (Robertson 1959; Spickett 1963).

Electrophoretic Variation

The discovery of electrophoretic variation by Harris (1966) and Lewontin and Hubby (1966) stopped this almost completely. Experimental population genetics became electrophoresis. After the International Congress of Genetics in Berkeley in 1973 a well-known quantitative geneticist wrote to me: "Quantitative genetics is dead."

By a dogmatic inverted application of the dictum "one gene-one polypeptide" stained bands became equalled with genes, often without proper genetic analysis. Electrophoresis strengthened the gene frequency approach in evolutionary genetics. A positive interaction of the prestige of the mathematical theory of population genetics and the esteem for a molecular approach made this gene frequency approach to evolution the dominant tenet in evolutionary biology without much regard for what is beyond the genome.

1 Department of Population and Evolutionary Biology, University of Utrecht, 3508 TB Utrecht, The Netherlands

Population Biology and Evolution
Ed. by K. Wöhrmann and V. Loeschcke

Revolutions

Since then, two revolutions have tended to restore the balance between genotype and phenotype in evolutionary theory. Paleontologists and ecologists were the revolutionaries. The paleontologists could not relate morphological evolution as it is revealed in the paleontological record with frequency changes of electrophoretic variants. Ecologists could not recognize change of frequencies of electrophoretic variants as underlying adaptive change of characters with ecological importance, e.g., rates and timing of reproduction.

These revolutions are agenetic and sometimes even antigenetic.

Paleontologists proclaimed the theory of punctuated equilibria, indulged in coquetry with saltational events and promoted species selection as the directing force in evolution instead of individual selection in populations (Gould 1980, Stanley 1979).

Thereby they try to remove population genetics from its central position in the explanation of evolution. Some people see Marxism behind this revolution.

In contrast, ecologists staged a counterrevolution (see Stearns 1976) and went straight back to Darwin's gradualism. The optimalization of life history characters implies that the right selection will find suitable genetic variability to produce smooth gradual adaptation (perhaps some philosopher of science will relate this to Reagonomics or Thatcherism).

Phenotypic Aspects

Whatever their further merits, these developments have contributed much to a renewed interest for the phenotypic aspects of evolutionary genetics, in particular for phenotypic characters which have ecological significance. Such a genetics of ecology leads inevitably to a renewed interest in quantitative genetics (Van Noordwijk et al. 1980). This approach is complementary to what I would like to call the ecology of genes in which the ecological significance of variation on individual gene loci, e.g., alcohol dehydrogenase or amylase is investigated (cf. De Jong and Scharloo 1976, Scharloo et al. 1977).

However, simple quantitative genetics, i.e., quantitative genetics which deals only with statistical parameters is not enough. We have to go again into the developmental and physiological processes underlying quantitative characters. This was the message which I tried to convey when introducing what I called the Magic Triangle (M.T.) of evolutionary genetics (Scharloo et al. 1977).

The Magic Triangle

The angles represent three levels of observation on which genetic variation can be observed.

M: molecular variation, i.e., variation in single genes as observed by electrophoresis and by analysis of amino acid sequences in proteins or base sequencies in DNA (review Lewontin 1974).

P: phenotypic variation, variation in physiological, morphological, or behavioral characters under control of a large number of genes. Its genetic basis is revealed by artificial selection (see Falconer 1960).

F: variation in fitness, the ultimate parameter in population genetics, one could almost say, the parameter of the last judgement which determines the selective forces acting on phenotypes and genes. This type of variation has been revealed when chromosomes of *Drosophila* were made homozygous and by inbreeding (see Dobzhansky 1970).

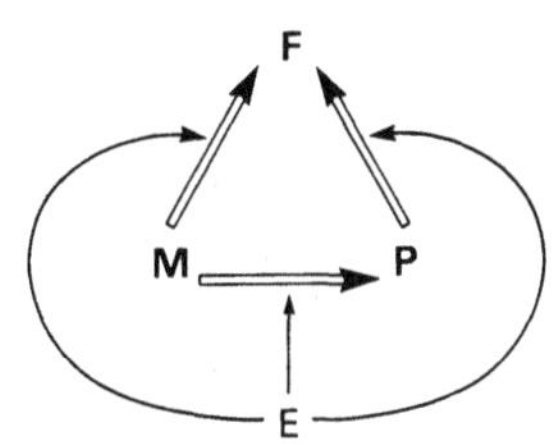

Fig. 1. The Magic Triangle of evolutionary genetics. *M*: molecular variation; *P*: phenotypic variation; *F*: variation in fitness; *E*: environmental effects. Further explanation in text

On all levels of observation we have abundant evidence for the presence of genetic variation. However, evidence on the connection between the variation on the three levels is limited. This is even the case for the connection M → F, the relation between properties of proteins and fitness (see Koehn et al. 1983), let alone for the route M → P → F where both arrows represent mainly ignorance. Because genetic effects of selection depend on these connections, the essential problems of evolutionary genetics are the sides of our Magic Triangle which form the links between genes and fitness, between genes and phenotypes, and between phenotypes and fitness.

The M.T. concentrates on what happens within organisms. However, fitness is dependent on environmental factors as well as on properties of the organism.

Environmental Interference

The M.T. is not a closed system. The environment interferes

1. When a character (either a molecule M produced by one gene or a phenotypic character P generated by the joint action of many genes) has obtained its final shape; fitness depends on the interaction of the character and the environment.

When there is no malaria, sickle cell heterozygotes have no advantage; when there are no thrush predators (as some suppose to be the case in France), there is no selection on snail color and banding.

2. However, the environment does not only affect fitness by its interaction with a character, morphological, physiological, or behavioral, when it is fully formed. The environment often already affects the making of a character, it interferes in the processes between genes and phenotype.

These environmental effects on the formation of characters are not necessarily random; they are often adaptive. The environmental factor causes a change in the character which promotes fitness under selection by the same environmental factor.

Adaptive Significance

A major problem is to decide whether an environmental effect can be considered to be an adaptive reaction. Schmalhausen (1949) in his book *Factors of Evolution* recognizes two types of reaction: (1) morphoses, phenotypic reactions on environmental factors which are not a normal part of the environment; (2) adaptive phenotypic reactions caused by environmental factors which regularly occur in the habitat.

Morphoses show considerable variation and are often monstrosities, phenotypes with considerably lower fitness. Adaptive reactions would be performed by all individuals of the population: the uniform reaction would produce phenotypes with improved fitness under the conditions which caused the reaction.

However, the uniformity criterion has its pitfalls: if the change caused by the environmental factor is a mechanical or physicochemical necessity, there is no reason to call it an adaptation.

Recently, Smith-Gill (1983) proposed a new criterion: adaptive reactions would involve the switching on of a coordinated set of genes. Morphoses would imply only the activation of one or a few uncoordinated genes. In our present state of ignorance concerning the processes of gene regulation and development, this criterion will be difficult to apply. Moreover, switching on a set of coordinated genes is just what happens when phenocopies of homeotic mutants are obtained by such abnormal environmental factors as heatshock or ether treatment (Garcia-Bellido 1977). Having a leg instead of an antenna is surely not an adaptive reaction.

Uniformity of Adaptive Reactions

The relative uniformity in reaction to environmental change is, of course, a necessity when the usual uniformity of wild phenotypes, the relative limited variation in normal phenotypes, is considered. This limitation of variation in wild populations is particularly striking when it is compared with the variety of phenotypes and reaction norms revealed in morphological mutants or in populations submitted to an abnormal environmental factor (see Waddington 1957, Scharloo 1962).

Origin of Adaptive Reactions

The Neo-Darwinian explanation of the uniform adaptive reactions is, of course, fixation of one type of adaptive reaction by natural selection acting on a variable array of reactions. A model for such a process can be found in the work of Waddington (review, 1961) and Bateman (1960).

They found in *Drosophila* that a heat shock in the first hours after puparium formation induces a variety of phenotypes, e.g., extra and missing cross veins, changes in wing shape, etc. Artificial selection for a specific phenotypic reaction rapidly produced lines which have almost exclusively the specific reaction type. This implies, of course, that there is genetic variation for reaction type and that selection can fix this variation. Further, artificial selection ended in what Waddington (1961) called genetic assimilation, i.e., genetic fixation of the phenotype previously only produced after the application of the environmental stimulus.

Artificial Selection

In wild populations with established reaction norms to their normal environment, genetic variation in these reaction norms will be of a quantitative nature. Also in these situations when genetic variation of reaction norms is quantitative, selection can be effective. Selection will then result in an increase or decrease of the intensity of thc reaction.

This was shown when artificial selection was applied on the sensitivity of the expression of mutants to temperature (e.g. Waddington 1960, Waddington and Robertson 1966 on Bar and Scharloo et al. 1972 on ci^{D-G}.

Natural Selection: Body Size and Anal Papillae

Waddington (1959) suggested that such processes could also occur with a naturally variable character under natural selection. This was shown by Anderson (1966) and by Powell (1974) for body size in *Drosophila*. Body size of *Drosophila* has an inverse relation with temperature of rearing. This is a uniform reaction, but we do not have direct evidence on its adaptive significance. However, it was shown that when populations are reared continuously at high or low temperature there is a genetic change of body size in the same direction as the phenotypic reaction. This suggests action of natural selection and the adaptive nature of the phenotypic reaction.

Waddington (1959) studied what he thought was the size of anal papillae in *Drosophila* larvae and resistance against the presence of sodium chloride in the food medium. The anal papillae were described by Gloor and Chen (1950) as organs consisting of groups of specialized epidermal cells which are involved in osmoregulation and could be stained by immersion of the larvae in an $AgNO_3$ solution.

Waddington took his measurements in the pupae where remnants of the anal organs can be observed which are not continuously changing in size and shape as in the larvae. This measurement increased when populations were reared on increasing salt concentrations. This was described as an adaptive response. Moreover, he reared populations during a number of generations on increasing salt concentrations chosen to kill approximately 50% of the population. When he tested the resulting selection lines on a media series with increasing salt concentrations he found (1) the sodium chloride resistance of all selection populations was higher than in the base populations; (2) the phenotypic reaction, i.e., the increase of the measurements with increasing salt concentration had become stronger; (3) without addition of salt to the medium, the mean measurements were higher in the lines submitted to selection than in the base populations, i.e., there was genetic assimilation. Taking the same measurements as Waddington, te Velde could confirm his results.

This seems to be the perfect experiment in that (1) the action of the selective factor is known; (2) an organ is known that is involved in the interaction of the individual with the selective factor; (3) there is a uniform phenotypic response of this organ to the selective forces; (4) genetic changes in resistance caused by natural selection are accompanied by strengthening the phenotypic response and by genetic assimilation.

However, similar organs occur in other diptera, e.g., *Chironomus* and *Aedes*, and Wigglesworth (1933) has described that the size of the anal papillae in *Culex* and *Aedes* larvae *decreases* when the sodium chloride concentration increases. Moreover, when related *Aedes* species live in habitats with different salt concentrations the species living at the higher concentration have smaller papillae (Beadle 1939).

In our laboratory it has been shown by measurement of papillae in larvae that also in *Drosophila* the size of the papillae is smaller when larvae are reared at higher concentrations (te Velde et al. in preparation). Moreover, it is very clear that what Waddington (1959) measured in pupae was not size of anal papillae, but the size of remnants of inactive parts of the epidermis *between* the anal papillae in larvae, a measurement which has only a remote relation to the size of the papillae.

Nevertheless, by measurements in larvae it could be shown that changes of anal papillae occur when populations adapt to sodium chloride. However, *Drosophila* populations can also adapt to sodium chloride without any detectable change of the papillae.

That anal papillae can be involved in adaptation to sodium chloride is shown by artificial selection. In a population on 2% NaCl two different types of anal papillae are found: retracted papillae (R) and stretched papillae (S). By artificial selection, lines were obtained which had either the S or the R type almost fixed. Chromosome exchange between the R and S lines showed that the difference between the lines is determined polygenically.

Moreover, when both lines are grown on 6% NaCl all larvae of the R line die, but in the S line more than 50% survive.

The different survival levels are a consequence of the different ability of the two lines to maintain the osmotic value of their hemolymph when feeding on media with extreme salt concentrations. The S-line does it better at high concentrations of NaCl, the R-line on strongly diluted media.

But not only the outward appearance of the papillae has changed. There are important structural changes in the papillae cells. These cells have a zone of apical membrane folding where they are in contact with the outside environment. With increasing sodium chloride concentrations the lamellar zone disappears. On diluted media the extent and length of the folds increase and the number of mitochondria in close contact with the lamellea increases suggesting an energy requiring intake of ions (Chen and Brugger 1973). This difference in structure which is a phenotypic reaction to differences in ionic strength of the medium is clearly genetically fixed in our selection lines which show a striking difference in adaptation to salt concentrations.

The adaptive reaction is not a simple change. In addition to changes in the size of the anal papillae, important structural changes occur in the cells composing this organ.

Polygenic determination and complex reactions can be expected in such characters. But what happens with simple characters, such as enzymes, which are so close to their structural gene?

Enzyme Variation

When we started to work on selective effects on electrophoretic variants we thought to study the effects of having different molecules on fitness, the side M-F of our Magic Triangle. Because electrophoretic variation is based on amino acid substitution, the change of enzyme structure would change enzyme action and thus fitness. Undoubtedly, we were strongly influenced by the sickle cell story. There would be no place for adaptive reactions in the path between the gene and the molecule on which selection acts directly.

Amylase in *Drosophila*

We tried to show that natural selection acts on electrophoretic variants which differed in total amylase activity. We showed first in cage populations that the frequency of the $Amy^{4,6}$ variant with high enzyme activity increased and the frequency of the variant Amy^{1} with low activity decreased on food media in which starch was an important component (De Jong et al. 1972, Scharloo et al. 1977).

In experiments on competition between $Amy^{4,6}$ larvae and Amy^{1} larvae it was confirmed that the relative fitness of the high activity variant increased when the concentration of starch in the food was increased. This occurred only when the food composition was such that addition of starch improved survival.

Comparison with controls in which instead of starch, equivalent amounts of the product of its degradation maltose were added, exclude the possibility that genetic variation beyond the amylase enzyme system was involved (De Jong and Scharloo 1976, Scharloo and De Jong 1980). But this did not mean that the fitness differences between the two genotypes were a consequence of differences in molecular structure between the enzyme variants. We found evidence that the regulatory aspects of gene activity were decisive (Hoorn and Scharloo 1980a). The differences in amylase activity between the two strains were at first only known to occur in adult flies. When studying the complete ontogeny we showed that such a difference is found in larvae, but occurs only in the last part of the second instar (Hoorn and Scharloo 1980b).

On optimal medium this difference in activity occurs after the pupation decision is made and the strains have identical rates of development. As a consequence, the difference in amylase activity cannot affect viability. But on the media used in the competition experiments, when starch addition contributes to survival, amylase activity is 4–5 times higher, the difference in activity appears already in the second instar, and the high activity variant $Amy^{4,6}$ has a far shorter developmental time. This explains the outcome of the earlier competition experiments of De Jong and Scharloo (1976) (Hoorn and Scharloo 1981).

Moreover, the activity profile in flies is also different and the amylase activity in the two stocks reacts differently to addition of food components (Hoorn and Scharloo 1978).

The possibility of involvement of regulatory genes in selection on amylase variants is underlined by the discovery by Abraham and Doane (1978) of genes determining in which part of the midgut amylase is produced. One such variable gene was located on a short distance (2–3 map units from the structural amylase locus (Abraham and Doane 1978, Klarenberg unpublished). There is evidence for other regulator genes tightly linked to the structural locus.

The action of selection on these regulators is indicated by the consistent and strong linkage disequilibrium between the MAP (Midgut Activity Pattern) regulator variants and the variants of the structural locus (Klarenberg and Scharloo 1982). Moreover, flies with the regulator variant in which a larger part of the midgut produces amylase survive longer when fed on starch than the variant with short amylase producing part of the gut (Klarenberg unpublished).

Moreover, the amylase production pattern in the gut reacts to environmental change: under conditions when starch is important for survival, amylase is produced in a larger part of the larval gut when the starch concentration is lower (Klarenberg et al. 1983). It seems reasonable to assume that the reaction in total amylase activity (Hoorn and Scharloo 1981) and in production pattern are adaptive, i.e., promoting a better utilization of a scarce resource.

What emerges here, for what we considered to be a simple genetic difference to be studied as a case of the ecology of a gene, is a complicated system of a structural gene with several regulatory loci which are influenced by environmental factors to form an amylase activity phenotype resembling more the P than the M in the Magic Triangle.

But not all environmental reactions of enzyme systems can be considered to be adaptive.

Alcohol Dehydrogenase in *Drosophila*

More effort has been put into alcohol dehydrogenase polymorphism than in any other system (reviews Van Delden 1982, Koehn et al. 1983). Nevertheless, definite conclusions on the selective forces acting on this polymorphism are still tentative, mainly because of the ignorance of its physiological function.

Even its precise role in the detoxification of ethanol is not yet completely understood. Degradation of ethanol by ADH was supposed to produce acetic aldehyde which cannot really be called detoxification because the product is more toxic than the substrate. Recently, it has been shown that ADH in fact catalyzes a two-step reaction in which not only is ethanol converted into acetic aldehyde, but acetic aldehyde is oxidized to the useful metabolic acetate (Heinstra et al. 1982, 1983). There is evidence that this occurs on one active site of the enzyme molecule and that the toxic intermediate product acetic aldehyde is not set free. This seems to be a perfect mechanism for alcohol tolerance (one would like to recommend genetical engineers to transfer it into human beings).

For ADH after electrophoresis five isozyme bands are found, ADH1 to ADH5 numbered according to their mobility.

It has been known for several years that the isozymes can be converted into each other by chemical treatment (Jacobson et al. 1972, Schwartz et al. 1979). Addition of acetone in the presence of NAD causes conversion of ADH5 into ADH3 and ADH1. The same effect is produced by feeding propan-2-ol (isopropanol) which is degraded to acetone by ADH. While primary alcohols can be oxidized to carbonic acids, ketones are end products of secondary alcohol oxidation. Both feeding of isopropanol and acetone causes, in addition to conversion of the isozymes, inactivation of ADH. This is described by Anderson and McDonald (1981) as an adaptive reaction because ketones would be more toxic than the corresponding alcohols.

However, Heinstra et al. (in preparation) have shown that there is an intimate relation between the isozyme conversion and the inactivation. The inactivation seems to be a consequence of the two-step mechanism of degradation of primary alcohols. On secondary alcohols only the first step can be performed and the product of the first step, the ketone, remains tightly bound to the active site forming a ternary complex of enzyme, first-step product, and NAD which causes the inhibition of ADH.

If this interpretation is correct, the inactivation of ADH by isopropanol feeding is a physicochemical necessity. It will occur in all flies – there is a uniform environmental reaction – but there is no need for an explanation in terms of adaptation.

Conclusion

Environmental effects modifying phenotypes before the action of natural selection are crucial for our understanding of evolutionary change. They will be often adaptive, but this cannot be taken for granted. These environmental effects are not only found in the classical characters of ecological importance, such as rates of reproduction and development and body size, but also in characters which seem on first sight closely connected to gene structure. Also when studying enzyme variation environmental reactions can be important.

More insight in the problems of environmental effects may be obtained by what can be called quantitative genetics of reaction norms for the classical type of quantitative characters and for enzyme activities. In these studies functional aspects of variation, both in their ecological and physiological context, must be paramount.

Acknowledgements. I thank my collaborators J. te Velde, A.J. Klarenberg, and P.W.H. Heinstra for unpublished results and discussion. Part of the investigations was supported by the Foundation for Fundamental Biological Research (BION), which is subsidized by the Netherlands Organization for Advancement of Pure Research Z.W.O.

References

Abraham I, Doane WW (1978) I. Genetic regulation of tissue specific expression of Amylase structural genes in *Drosophila melanogaster*. Proc Natl Acad Sci USA 75:4446–4450

Anderson SM, McDonald JF (1981) Effect of environmental alcohol on in vivo properties of *Drosophila* alcohol dehydrogenase. Biochem Genet 19:421–430

Anderson WW (1966) Genetic divergence in M. Vethukhiv's experimental populations of *Drosophila pseudoobscura*. 3. Divergence in body size. Genet Res 7:255–266

Bateman KG (1960) The genetic assimilation of four variation phenocopies. J Genet 56:443–473

Beadle LC (1939) Regulation of the haemolymph in the saline water mosquito larvae of *Aedes detritus*. J Exp Biol 16:346–362

Brown R, Danielli JF (eds) (1953) Symp Soc Exp Biol VII. Evolution, Cambridge

Chen PS, Brugger C (1973) An electronmicroscopic study of the anal organs of *Drosophila* larvae. Experientia 29:233–235

Delden Van W (1982) The alcoholdehydrogenase polymorphism in *Drosophila melanogaster*. Evol Biol 15:187–222

Demerec M (1955) Cold Spring Harbor Symp Quant Biol Vol XX. Population Genetics, New York

Dobzhansky Th (1970) Genetics of the evolutionary process. New York

Falconer DS (1960) Introduction to quantitative genetics. Edinburgh

Garcia-Bellido A (1977) Homeotic and atavistic mutations in insects. Am Zool 17:613–629

Gloor H, Chen PS (1950) Über ein Analorgan bei *Drosophila larven*. Rev Suisse Zool 57:751–756

Gould SJ (1980) Is a new and general theory of evolution emerging? Paleobiology 6:119–130

Harris H (1966) Enzyme polymorphisms in man. Proc R Soc London Ser B 164:298–310

Heinstra PWH, Eisses KTh, Schoonen WG, Aben W, Scharloo W, Thörig GEW (1982) A dual function of ADH in *Drosophila melanogaster*. Genetics 100:s29

Heinstra PWH, Eisses KTh, Schoonen GWEJ, Aben W, Winter De AJ, Horst Van Der DJ, Marrewijk Van WJA, Beenakkers AMTh, Scharloo W, Thörig GEW (1983) A dual function of alcohol-dehydrogenase in *Drosophila*. Genetica 60:129–137

Hoorn AJW, Scharloo W (1978) The functional significance of amylase polymorphism in *Drosophila melanogaster*. V. The effect of food components on amylase and alpha-glucosidase activity. Genetica 49:181–188

Hoorn AJW, Scharloo W (1980a) Selection on enzyme variants in *Drosophila*. Aquilo 20:41–48

Hoorn AJW, Scharloo W (1980b) The functional significance of amylase polymorphism in *Drosophila melanogaster*. III. Ontogeny of amylases and some alpha-glucosidases. Biochem Genet 18:51–63

Hoorn AJW, Scharloo W (1981) The functional significance of amylase polymorphism in *Drosophila melanogaster*. VI. Duration of development and amylase activity in larvae when starch is a limiting factor. Genetica 55:195–201

Jacobson KB, Murphy JB, Knopp AJ, Ortiz JR (1972) Multiple forms of *Drosophila* alcohol-dehydrogenase. III. Conversion of one form to another by NAD or acetone. Arch Biochem Biophys 149:22–35

Jong De G, Thörig GEW, Hoorn AJW, Scharloo W (1972) Amylase variants in *Drosophila melanogaster*. Nature 238:453–454

Jong De Ge, Thörig GEW, Hoorn AJW, Scharloo W (1972) Amylase variants in *Drosophila melanogaster*. Nature 238:453–454

Klarenberg AJ, Scharloo W (1982) Midgut activity pattern variation in alpha-amylase and other carbohydrases in *Drosophila melanogaster*. Genetics 100:s36–37

Klarenberg AJ, Scharloo W, Jacobs PJM, Vermeulen C (1983) Environmental and genetic control of larvae midgut pattern variation of α-amylase in *Drosophila melanogaster*. Genetics 104:s43
Koehn RK, Zera AJ, Hall JG (1983) Enzyme polymorphism and natural selection. In: Nei M, Koehn RK (eds) Evolution of genes and proteins (Sunderland, Mass.)
Lewontin RC (1974) The genetic basis of evolutionary change. (New York)
Lewontin RC, Hubby JC (1966) A molecular approach to the study of genic heterozygosity in natural populations. II. Amount of variation and degree of heterozygosity in natural populations of *Drosophila pseudoobscura*. Genetics 54:595–609
Noordwijk Van AJ, Balen Van JH, Scharloo W (1980) Heritability of ecologically important traits in the Great Tit. Ardea 68:193–203
Powell JR (1974) Temperature related genetic divergence in *Drosophila* body size. J Hered 65: 257–258
Robertson FW (1959) Studies in quantitative inheritance. XII. Cell size and number in relation to genetic and environmental variation of body size in *Drosophila*. Genetics 44:1113–1130
Scharloo W (1962) The influence of selection and temperature on a mutant character (ci^D) in *Drosophila melanogaster*. Arch Neerl Zool 14:431–512
Scharloo W, Jong De G (1980) Selection for amylase allozymes in *Drosophila melanogaster*. Evolution 34:608–610
Scharloo W, Zweep A, Schuitema KA, Wijnstra JG (1972) Stabilizing and disruptive selection in a mutant character in *Drosophila*. Genetics 71:551–566
Scharloo W, Dijken Van FR, Hoorn AJW, Jong De G, Thörig GEW (1977) Functional aspects of genetics variation. In: Christiansen FB, Fenchel T (eds) Measuring selection in natural populations. (Springer, Berlin Heidelberg New York), pp 131–148
Schmalhausen II (1949) Factors of evolution (Philadelphia)
Smith-Gill SJ (1983) Developmental plasticity, developmental conversion versus phenotypic modulation. Am Zool 23:47–55
Schwartz M, O'Donnell MJ, Sofer W (1979) Origin of multiple forms of alcoholdehydrogenase from *Drosophila melanogaster*. Arch Biochem Biophys 194:365–378
Spickett SG (1963) Genetic and developmental studies of a quantitative character. Nature (London) 199:870–873
Stanley SM (1979) Macroevolution. Pattern and process (San Francisco)
Stearns SC (1976) Life history tactics. A review of the ideas. Q Rev Biol 51:3–47
Thoday JM (1961) Location of polygenes. Nature (London) 191:368–370
Waddington CH (1957) The strategy of the genes (London)
Waddington CH (1959) Canalization of development and genetic assimilation of acquired characters. Nature (London) 183:1654–1655
Waddington CH (1960) Experiments on canalizing selection. Genet Res 1:140–150
Waddington CH (1961) Genetic assimilation. Adv Genet 10:257–293
Waddington CH, Robertson E (1966) Selection for developmental canalization. Genet Res 7: 303–312
Wigglesworth VB (1933) The effect of salt on the anal gills of mosquito larvae. J Insect Phys 10:1–15

The Genetics of Seasonal Polyphenism and the Evolution of "General Purpose Genotypes" in Butterflies

A.M. SHAPIRO[1]

Introduction

This paper is really a specialized appendix to Professor Scharloo's on "The Genetics of Adaptive Reactions." It deals with a particular set of such reactions - those of butterfly wing patterns to environmental factors - and asks whether those which seem adaptive are evolutionarily related to those which do not, and if so, how. Despite more than a century of interest in such phenomena, the answers are not yet in; we are only now able to do the carefully controlled experiments necessary to partition phenotypic variation into its environmental and genetic components and this work is still very much in progress. So this will be a very unsatisfying presentation - full of qualitative statements, long on speculation, short on hard data. If it serves as a provocation it will have done its duty.

It has been almost an article of Darwinian faith that butterfly wing patterns are adaptive; after all, mimicry - essentially a butterfly story - is a pillar of Darwinism. A glance through any butterfly book of the coffee table variety reveals an astonishing diversity of patterns. The fact is that we have only the remotest idea of the functional significance of any of them, as we were recently reminded by Silberglied et al. (1980). One reason is that bewildering diversity which defies rational classification; another is that it is almost impossible to relate a pattern to an ecological and behavioral context when observing a specimen set on a pin.

Attempts at a comprehensive classification of butterfly wing patterns break down because not all patterns can be reduced to homologous elements. Munroe (1982) reviewed such attempts. Wing patterns consist of a mixture of highly conserved elements and more or less radical ones; this makes them most comprehensible when approached developmentally. In the United States, Nijhout (1981) has been trying to revive this line of inquiry by looking at pattern ontogeny in the buckeye, *Precis coenia* (Nymphalidae). One of the first tasks in such work is to establish valid homologies. Munroe, surveying the confusion concerning valid and spurious homologies, concludes that "one of the strong lines of approach will certainly be the investigation of genetic differences ... capable of direct study in individual mutants, in sexes and polymorphs ..., in different populations ..., and in different species that are close enough in reproductive physiology to be hybridized." Such studies are only getting started.

1 Department of Zoology, University of California, Davis, CA 95616, USA

Population Biology and Evolution
Ed. by K. Wöhrmann and V. Loeschcke

Goldschmidt, back in 1938, enumerated eight categories of genetic effects on pattern. His classification, however noble in intent, is arbitrary and quickly fails to encompass the almost infinite complexity of the subject. The categories are not mutually exclusive; thus, the *"elymi"* aberrations in *Vanessa* involve regional melanization (Goldschmidt's category 4) and regional suppression of melanic pattern elements (category 7), with a hint of radical reorganization (category 8) in the most extreme examples - if we did not have the intermediates we would be hard pressed to explain where full-blown *elymi* comes from. The much maligned Californian Lepidopterist Gunder published prolifically on aberrations in the 1920's and 1930's, developing a classification employing such words as "albifusism" and "melanifusism." As we shall see, the physiological and developmental genetics of these situations probably foredoom any such system, but Gunder's attempt was a valiant one; he made the fatal mistake, however, of insisting on naming all the aberrations he collected lovingly, and at great cost, as the basis for his efforts.

We can see Goldschmidt's categories, more or less, not only at the intrapopulational level, but in comparisons among related taxa. Thus, in the tribe Scolitantidini of the Blues (Lycaenidae), the genus *Maculinea* shows a full spot pattern, as its name suggests, while such related genera as *Glaucopsyche* tend to lose it (category 7). But very rare aberrations can occur in *Glaucopsyche* which restore the full *Maculinea* pattern; one of these is illustrated in Shapiro (1983). Gunder (1927) called recurrent aberrations transition forms, because - thinking orthogenetically - he believed they expressed a drive in the indicated direction, a literal evolutionary *transition* in progress. Some other workers thought them atavistic (reversional), as in the *Glaucopsyche* case.

Seasonal and Geographic Patterns of Variation

A great many butterflies of both temperate and tropical climates have seasonal phenotypes which are often so different as to have been described as different species. As Munroe says, "the fundamental pattern and its homologies may be completely hidden or almost unrecognizably altered" in such situations (Goldschmidt's category 8). The most extreme cases of this sort occur in the Old World *Precis* (Nymphalidae) (cf. McLeod 1968).

I began some 18 years ago a systematic, empirical study of the environmental and physiological control of seasonal phenotypes - what I labeled *seasonal polyphenisms* in a 1976 review - in the Holarctic Pierini (Whites). I hasten to add that studies of pierine phenotypes go back to the last century and that the first evidence for photoperiodic influence was Ae (1957), working with *Colias eurytheme*, a Coliadine. This system was later expanded brilliantly by Watt and his associates, showing that seasonal phenotypes of this beast are thermoregulatorily adaptive (Watt 1968, 1969). (This, of course, is not equivalent to demonstrating that thermoregulation was the selective pressure giving rise to seasonal polyphenism - such a demonstration is probably formally impossible.) *Colias* melanizes its entire ventral hindwing in cold weather. So do some pierines, e.g. *Pieris rapae* and *P. brassicae*, but most melanize only their wing veins and there is no clear mechanism for the latter pattern to be thermoregulatorily efficient. Rather than reviewing the physics and physiology of butterfly thermoregulation here, I will simply note that thermoregulatory advantage is not yet proven for the seasonal phenotypes of any pierine, though strongly suspected; increased survivorship, or rather residence time, seems associated with them in one experiment, the only such ever reported (Shapiro 1976).

Whatever the adaptive significance of these patterns, there is compelling evidence that *they are adaptive.* This consists of the most striking global pattern of wing design ever recognized in butterflies - replete with parallelisms and convergences. Specifically, pierines have melanized wing veins in cold and cloudy climates and tend to lose this pattern in sunny and warm ones. The melanization is most pronounced on the ventral hindwing and this is coupled with the (unambiguously thermoregulatorily useful) dorsal bases of all wings. The highest elevation Andean pierines, genera *Hypsochila* and *Tatochila*, which occur commonly above 5,000 m, have reflective sheens beyond the dorsal-basal melanization, which may enhance its efficiency further.

The melanic tendency is repeated both within and between lineages along both latitudinal and altitudinal gradients *and on a seasonal basis within populations*, where the alternate phenotypes are under environmental control. In general, cold season phenotypes of multivoltine mid latitude taxa resemble the sole phenotypes of arctic and alpine univoltine relatives. We see close convergences in two Holarctic pierine lineages: the *Pieris napi* species group, forming part of the subgenus (or splitters' genus) *Artogeia*, and the *P. callidice* species group (*Synchloe* or *Pontieuchloia*). Most remarkably, we see the same things in the Andean-Neantarctic genus *Tatochila*, of which much more shortly; suffice to say that the well-known phenotypic gradient in the *napi* group, with yellow-melanized females in the Alps *(bryoniae)* and high latitudes (*adalwinda, hulda,* etc.) is repeated in the *T. sterodice* species group.

To understand how these gradients and convergences arise, we must follow Goldschmidt and Munroe into the thicket of wing pattern ontogeny. Several attempts have been made, as noted previously, to derive butterfly wing patterns from a *bauplan*. The Schwanwitsch (1924) system works fairly well for the Nymphaloid families, but not for the Lycaenoids, Papilionids, or Pierids. I have been led to an experiment of my own in pattern simulation, based on the sequence of pigment deposition in late pupal pierine wings. All the pattern elements found in temperate Pierini can be derived from two basic systems, shown in Figs. 1b,c and 2b,c. One is a pattern of black lines following the venation, including a (probably secondary) thickening at the end of each discal cell; the other is a pair of parallel black lines transverse to these. The two are superimposed in Figs. 1d and 2d, forming a remarkably good match to the dorsal pattern of the darkest morph of the female *Tatochila stigmadice* from northern Argentina (Fig. 11). Let us note some interesting features of this system. First, the vein-line pattern (Fig. 1b) is more complex than the actual, normal pierine venation (Fig. 1a). The additional elements appear to follow embryonic vein traces which are aborted early in wing ontogeny, but not before pigment precursor or some determining factor has been laid down in them. These same elements are very rarely found pigmented in *Pieris napi* as well (Shapiro, unpublished). The ontogeny of the venation is a "recapitulatory" or pseudorecapitulatory feature which has received little attention from Lepidopteran phylogeneticists. The extent to which

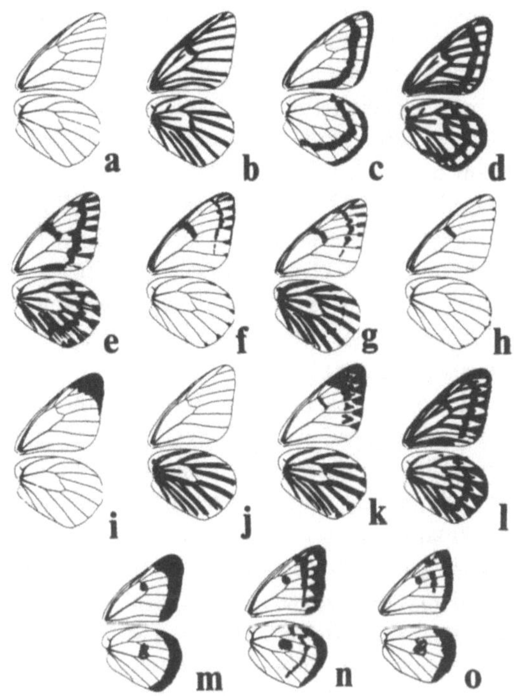

Fig. 1. *a* Pierine venation *(Tatochila vanvolxemii). b* vein-line pattern. *c* transverse-line pattern. *d* vein- and transverse-line patterns superimposed to generate "complete" Pierine pattern. *e* female *Tatochila vanvolxemii*, dorsal. *f* male *T. vanvolxemii*, dorsal. *g* male *T. vanvolxemii*, winter phenotype, ventral. *h* same, summer phenotype, ventral. *i* male *Theochila maenacte maenacte*, dorsal, winter phenotype. *j* same, ventral. *k* same, female, dorsal. *l* dark morph of female *Tatochila stigmadice* with complete pattern (cf. *d*). *m* summer phenotype of male *Colias eurytheme*, dorsal. *n* same, female. *o* same, aberrant spring male showing "feminized" forewing pattern

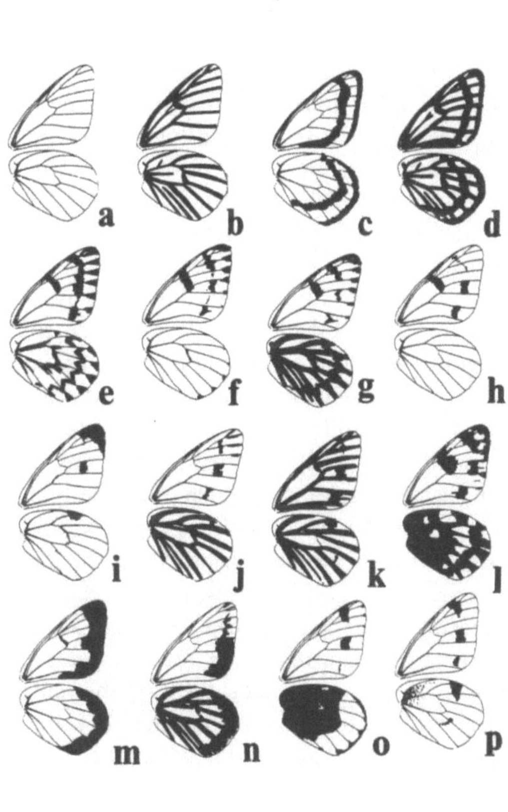

Fig. 2. *a–d* as in Fig. 1. *e* female *Pieris occidentalis*, dorsal. *f* same, male, dorsal; cf. Fig. 1e,f. *g* male *P. occidentalis*, winter phenotype, ventral. *h* same, summer phenotype, ventral; compare Figures 1g,h. *i Pieris rapae*, summer male, dorsal. *j Pieris napi*, winter female, ventral. *k Pieris bryoniae*, female, dorsal. *l Pontia daplidice*, winter male, ventral. *m Pieris napi* ab. *funebris*, female, dorsal. *n* same, ventral. *o Pieris krueperi*, winter phenotype, ventral. *p* same, summer phenotype, ventral

primitive elements are conserved in the wing pattern, though lost from the venation, varies among taxa. By sequential dissections it may be feasible to learn whether these differences are determined by the timing of vein trace abortion, precursor deposition, or both. Second, the two parallel transverse lines suggest derivation from a more primitive solid border, subsequently invaded and divided by the ground color. No extant pierine has such a pattern. There is a mutant of European *P. napi*, described by Lorković (1971), which *simulates* it (Fig. 2m,n). Here the entire wing distad of the inner transverse line is melanized on both surfaces, but the line itself is not involved. This is clearly shown in the figures, which are of a female; the two postmedian spots (in cells M_3 and Cu_2) are remnants of the inner line and are exterior to the melanized area, fusing with it on their outer sides. This is something new, not a reversional mutant.

The solid black border is, however, the normal state of affairs in many Coliadini. In the widespread genus *Colias*, males commonly have a solid black border, but in females it is interrupted or invaded by the ground color (Fig. 1m,n). No *Colias* has the male pattern in both sexes, but scattered species have the female pattern in both; this situation has arisen repeatedly, even in the presumably monophyletic and isolated Andean group of species. The sexual dimorphism occurs in the related genus (or subgenus) *Zerene*, basically Neotropical with two Nearctic outliers, and probably antedates its differentiation. There is a shibboleth in butterfly evolution to the effect that male patterns are more conservative than female as a result of the need for mate recognition – a hasty generalization from female-limited mimicry. In *Colias* the female pattern may well be the primitive character state. This is supported by modifications of the male pattern in the cold season phenotypes of multivoltine species; in both sexes the border recedes toward the margin and may almost disappear, but in the male the apical portion tends to separate into a truly apical and a slightly hooked subapical portion, as if it had two foci. Figure 1o is a wild-collected male aberration of *C. eurytheme* from California, figured in color in Shapiro (1983). It is a spring individual in which the hindwing border is normal, but the forewing has a well-developed double-line female pattern. In agriculturally associated *Colias* from the Urubamba Valley of the Department of Cusco, Peru, dry season males have small spots of ground color in the border near the apex, while wet season males do not (Shapiro, unpublished data). Overall, the solid border hypothesis for the origin of marginal patterns is as weak in *Colias* as in the Pierini.

Third, it should be emphasized that the elements represented in Figs. 1b,c and 2b,c are not arbitrary, but correspond to an ontogenetic sequence. They are clearly under different genetic controls, but neither is a simple single locus Mendelian system. For example, the discocellular spot, despite its connection with the vein lines, has been lost in the *Pieris rapae, napi,* and *brassicae* species groups (Lorkovic's aberration restores it feebly in *napi* females). In the *Tatochila sterodice* species group, the vein lines are suppressed in males, but the cell spot is polymorphic, especially in *T. vanvolxemii*, which has reduced spots like *Theochila maenacte* and *Ascia monuste* and also enlarged ones like *Ascia josephina. Rapae* and *brassicae* have lost their vein lines altogether, but *napi* retains them on the ventral hindwing except in summer forms of many multivoltine populations and they are expressed dorsally in spring females of many populations and routinely in the females of univoltine high-altitude and -latitude populations (Fig. 2j,k).

The relation between the vein line pattern and overall ventral hindwing melanization of the *Colias* type is dynamic and perhaps evolutionarily easy, as shown by the Andean Pierini. *Colias*-type melanization has arisen twice or perhaps three times in this group. All the species of *Piercolias* and *Pierphulia* have *Colias*-type patterns. The most extraordinary and illuminating example is an undescribed species of *Hypsochila* from Peru and southern Ecuador, of which a short series exists in the National Museum of Natural History in Lima. In this animal the vein line pattern, which is characteristic of *Hypsochila*, is only barely detectable because the interspaces are filled almost completely with dark scales, producing a close approximation to the *Piercolias* pattern. Astonishingly, this species has also fused its dorsal apical forewing pattern into a *rapae*-type spot.

We may briefly examine some other variations on the theme of pierine pattern before moving on to our consideration of the genetics of seasonal polyphenism. The *rapae* pattern (Fig. 2i) has the apical area filled in between the transverse lines, which may often be recognized especially in material from dry climates; the inner line is represented, as in *napi*, by a single postmedian spot in the male - lost in most populations of *P. brassicae*, but not in Rhodes, for example - and two in the female. In the male of the Neotropical *Theochila maenacte maenacte*, the same apical marking has evolved and there are no remnants of the transverse lines and, frequently, of the vein lines either. Ventrally the vein lines are conserved in cold weather males (Fig. 1j) and more or less consistently in the southern Brazilian subspecies *itatiayae*. The origin of the dorsal forewing pattern is explicit when the female (Fig. 1k) is examined: both transverse lines are present, and nearly, but not quite confluent in the apical region. Again, the male seems innovative relative to a conservative female and there is interhemispheric convergence.

In the subgenus (or splitters' genus) *Pontia*, the vein lines expand along with the transverse lines so as to enclose spots of ground color within the cells. In summer forms and in some desert or semidesert taxa the pattern contracts to isolated blotches of dark scales, with the vein lines dropping out. An intermediate condition can be seen in the populations of the *Pieris callidice* group from Turkey, Iran, etc., in which the submarginal chevrons (inner transverse line) are expanded in conjunction with an almost normal vein line pattern. In *Pieris krueperi* (Fig. 2o,p) the cold weather phenotype has essentially the whole wing basad of the inner transverse line melanized, but this pigment disappears in the summer phenotype - leaving only indications of the line itself, including the apical spot characteristic of *rapae, napi*, etc., still connected to its phylogenetic umbilicus. According to prevailing physiological wisdom, cold weather *krueperi* should be very good at thermoregulation; no one has taken their temperatures yet.

Convergence in Seasonal Phenotypes

This brief survey is, I think, fairly convincing that large phenotypic changes can be achieved rather easily in the Pierini and can be reconstructed by the comparative method. Strikingly, phenotypic differences among generations of polyphenic species are similar to those among species, as already discussed. They are much more easily studied and provide the best opportunity to understand how such differences evolve and are controlled developmentally and physiologically. Figures 1e–h should be compared with Figs. 2e–h to demonstrate the extent of convergence in seasonal polyphenisms between ecological equivalents. In each case Fig. e represents the dorsal female pattern, f the dorsal male, g the ventral cold season male, and h the ventral warm season male. Figure 1e–h are *Tatochila* from Argentina, Fig. 2e–h *Pieris* (*callidice* group) from North America. There are many differences between seasonal phenotypes within each taxon, but the most thoroughly quantitated to date is the ventral hindwing pattern, which includes both the vein line and transverse line components previously discussed. (The inner transverse line is represented in both *Pieris* and *Tatochila* by a series of chevron-shaped markings. In *Pieris* and in the odd monotypic *Reliquia* from northeastern Colombia these point toward the wing base, while in *Tatochila* and its close relatives they point outward. Has the chevron pattern evolved twice? Probably.) The discussion here will focus on the ventral hindwing pattern and primarily on the vein lines, which might conceivably function in thermoregulation, while the chevrons certainly do not.

In keeping with the systematic axiom that it is easier to lose something than to gain it, as well as the results of latent polyphenism studies, I infer that the reduced patterns of summer brood of temperate multivoltine Pierini are normally derivative relative to the more or less complete patterns of the cold season broods. Figure 3 shows how the pattern is suppressed in two Andean and two Holarctic taxa: the end result is the same, but the ways to it, developmentally speaking, are diverse, reflecting

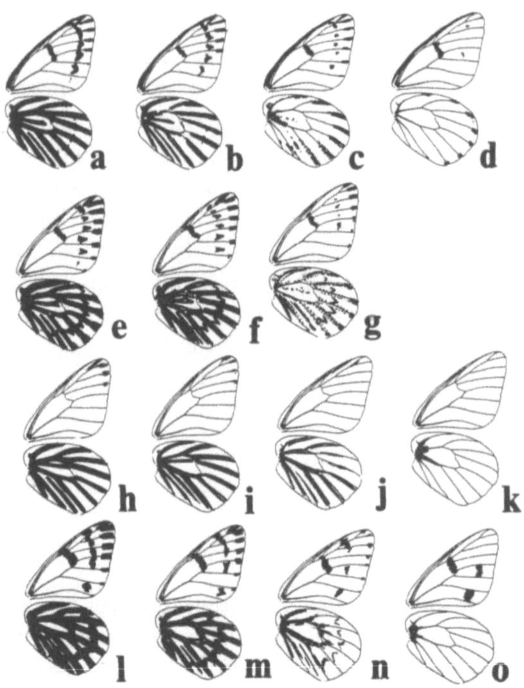

Fig. 3. Stages in lightening of ventral hindwing patterns of male Pierini. *a–d Tatochila vanvolxemii. e–g T. mercedis. h–k Pieris napi microstriata. l–o Pieris occidentalis*

again convergent evolution with a strong implication of functional utility. Figure 3a–d are *Tatochila vanvolxemii* from the Argentine grasslands and desert. In this species only the males are polyphenic, the females retaining the full pattern in all seasons; summer males may be totally immaculate except for the forewing discal spot, and even that may be reduced. (The overall effect is the reinvention of *Ascia monuste*.) Lightening is accomplished from the hindwing base outwards; the last thing to disappear is the series of dots at the vein tips (which may represent the outer transverse line, not the vein lines).

Figure 3e–g are *Tatochila mercedis* from Mediterranean Chile. In this beast both sexes are polyphenic and the vein lines disappear by becoming so pale overall that they hardly contrast, though the entire pattern is still discernible. (The chevrons are normally unaffected.) In *Pieris napi sens. lat.* (Fig. 3h–k are a California population) the vein lines retreat from the margin toward the base; the basal portion is the last to go (again, making the pigment potentially thermoregulatorily useful), and the males always retain more than their corresponding females. Pure white, immaculate ventral hindwings have evolved repeatedly in this complex in widely separated places with sunny, warm summers. Interestingly, the univoltine sibling species *P. virginiensis* from eastern North America produces a fully immaculate nondiapause summer generation when reared on continuous light at high temperature. This phenotype is indistinguishable from the summer phenotype of sympatric *napi* populations. Figure 3l–o represent *Pieris occidentalis*, a member of the *callidice* group. As in the *napi* group, its vein line pattern retreats basad. Males as light as in Fig. 3o are rare (the veins are still pigmented in dull yellow), but in the sibling species *P. protodice*, which replaces *occidentalis* at lower elevations in habitats similar to those of *Tatochila vanvolxemii*, summer males are completely immaculate on the ventral hindwing.

Experimental Genetics

The genetic control of the ability to respond phenotypically to photoperiod and temperature is approachable via hybridization studies, in which polyphenic and monophenic (demonstrably refractory) entities are crossed. There are not many systems in which the requisite genetic compatibility exists for quantitative analyses to be done. The *Tatochila sterodice* group is one. This group consists of three taxonomic species: *T. sterodice* itself, with four subspecies; *T. vanvolxemii;* and *T. mercedis*. The subspecies of *sterodice*, from north to south, are: *arctodice*, in the *páramo* and *puna* in Colombia through Ecuador to northern Peru, monophenic and strongly sexually dimorphic; *macrodice*, in the *puna* and *altiplano* from central Peru to northern Chile and northwestern Argentina, including all of highland Bolivia, monophenic and less dimorphic, usually above 3,500 m; nominate *sterodice*, from the central Argentine Andes to Fuegia and on the coast of southern Patagonia in towns, monophenic and moderately dimorphic; and *fueguensis*, poorly differentiated from *sterodice* and integrating northwand with it, from Fuegia, monophenic with strong sexual dimorphism. *Sterodice* contacts both *mercedis* and *vanvolxemii* in northwestern Patagonia (Neuquén and Río Negro). All three (and the others as well, except *fueguensis*, which we have not had in culture) hybridize readily in the lab with virtually no ill effects; indeed, I have done almost all the possible crosses to the F_2, F_3, and back-crosses. There is some spotty hybrid breakdown in the F_2 and F_3, but it is unpredictable and no more significant, I suspect, than the problems encountered with sterility and recessive lethals in fresh, pure cultures of *T. mercedis*. All these animals seem to be strong outbreeders. Wherever they meet in nature, they tend to form hybrid swarms. The principal contact zone is the narrow strip of mesic habitat along the eastern Andean foothills, between the rainy *Nothofagus* forests of the Andes and the dry, bleak steppe. A secondary zone (not involving *sterodice*) is farther north, in the canyon of the Rio Mendoza from Punta de Vacas down to Mendoza itself. Having manufactured lab hybrids from crosses of allopatric populations, my colleague Michael Collins and I have been able to use these as reference groups in order to analyze the composition of the hybrid swarms, using discriminant-function analysis; this work, still in progress, is another story. What is important is the compatibility of the entities which differ among themselves in their capacity to react phenotypically to rearing environment.

When *T. s. sterodice* is crossed with allopatric *T. vanvolxemii*, both reciprocal matings give surprisingly variable F_1s in which the vein line pattern is expressed to a greater or lesser degree even under continuous light at high temperature, a regime which gives immaculate male *vanvolxemii*. The F_2 is more variable still, as expected, but the variance is not symmetrical, there being an excess of darker animals. Shapiro (1984) presents the results of applying Lande's (1981) reformulation of the Castle-Wright "number of effective factors" approach to these data. In a nutshell, the number of "factors" is one - however, the data themselves make clear that modifiers are segregating in the background, and we must conclude that the factor-equality and perhaps also the simple additivity assumptions of the method are being violated. Where does the high F_1 variance come from? I crossed *vanvolxemii* with Ecuadorian *arctodice*, which is also monophenic. There are no modifier problems with *this* cross: for hindwing pattern, *arctodice* monophenism behaves as a simple Mendelian dominant, with absolutely no variation among rearing regimes or individuals in the F_1. We have reared, but not yet analyzed the F_2, F_3, and backcrosses. We have not yet crossed *arctodice* with *sterodice*. That is a high priority cross; there are enough pattern differences between these monophenic entities to allow us to do genetics, even though both have a full set of vein lines.

We are presently rearing a series of broods deriving from crosses of pure Chilean *mercedis* with allopatric Argentine *vanvolxemii* and with *vanvolxemii* from the Mendoza Canyon, where we think some introgression has occurred. The two sets of broods are slightly different, as expected, but they agree in the important respects - high phenotypic variance in the F_1 males (again), ranging from immaculate to fully patterned; and invariant females, indicating complete dominance by the *vanvolxemii* system which makes polyphenism sex-limited. (Due to this phenomenon, Mendoza populations would be classified routinely as *vanvolxemii*.)

In the meantime, we have some very complex hybrids involving four entities, whose patrimonies are too complex to discuss here, but which have been very instructive. In one we have decoupled ventral hindwing variance from environment, so that the full range of male phenotypes is found in all rearing regimes. In another, female plasticity has been restored by backcrossing a part-*mercedis* hybrid to that species. The data and formal genetic analyses for all these broods will eventually appear elsewhere. Overall, we have data for 21 pattern characters in a wide variety of crosses in three rearing regimes. Even at this "in progress" stage, some inferences can be drawn with a high degree of confidence:

1. Suppression of the vein line pattern in summer broods of *Tatochila* has been achieved in fairly simple ways, genetically speaking. The differences in the F_1 phenotypes of *arctodice* × *vanvolxemii vs sterodice* × *vanvolxemii* must reflect genetic differences between the monophenic stocks - specifically the existence of loci in *sterodice* which are capable of modifying the ventral pattern in hybrids, but are segregating invisibly in the pure population.
2. The phenotypic differences among the taxa of the *sterodice* group are not under supergene control and, in fact, are largely unlinked. The phenotypes are readily decomposed into numerous segregating traits. Ventral hindwing pattern is weakly coupled developmentally to the extent of apical forewing dorsal pattern. Transverse line derived pattern elements segregate independently of vein line derived ones in both mono- and polyphenic animals.
3. Dominance relationships in the control of plasticity and other characters are seemingly independent of the direction of evolution in the group, as inferred from biogeography. (The *sterodice* group is inferred to have come down the Andes from the north, evolving seasonal polyphenism twice as it colonized temperate lowland habitats, and losing a variety of high altitude specializations as it dedifferentiated into a secondarily generalized pierine type convergent with the Holarctic lineages.)

It is still too early to posit precise pathways for the evolution of polyphenism, such as gene duplication followed by coupling to a hormonal switch mechanism.

Another Case: The *Vanessa "Elymi"* Series

The Nymphalidae are a huge family, some of whose tropical members have spectacular seasonal polyphenisms. In temperate latitudes, however, there are few polyphenic Nymphalids (the European *Araschnia levana* is both striking and exceptionally well-studied, however). Among the Nymphalini we find a few polyphenic *Polygonia* spp., but the closely related genera *Nymphalis sens. lat.* and *Vanessa sens. lat.* are extremely conservative. Many species have vast geographic ranges, ranging on cosmopolitanism, but occur in a great variety of climates and seasons with essentially no variation whatsoever; monophenic Nymphalines often co-occur with exuberantly polyphenic pierines. This is not surprising if pierine phenotypes are thermoregulatorily adaptive, as those butterflies are obligate heliotherms, while Nymphalines are not.

Except for the Indo-Pacific taxa, *Vanessa* are monotonously similar in appearance. The same is true in *Nymphalis*; only a casual examination is necessary to recognize the pattern homologies of the two most aberrant species, *N. io* and *N. antiopa*, with their "tortoiseshell" congeners. No taxon in either genus is truly seasonally polyphenic. On the other hand, the patterns of all the taxa thus far tested are extraordinarily plastic when subjected to either heat or cold shock early in the pupal period; the responses, though not predictable on an individual basis, are quite repeatable in a statistical sense with large lots. The aberrant phenotypes are closely similar and inescapably homologous in congeners, even on different continents. These are presumably *nonadaptive* phenotypic responses to environments which are ecologically meaningless. Their study has a very venerable history, especially in Germany, where they contributed mightily to Goldschmidt's synthesis of evolutionary, developmental, and physiological perspectives. Shapiro (1976) reviews some of this history. Figures 4 and 5 show some of the phenotypes.

Nymphaline patterns are sufficiently complex that the sort of analysis done previously with pierines becomes a much more challenging task. One must do it, however, if the control of temperature-shock aberrations is to be understood. The *elymi* series of aberrations in *Vanessa* (despite the existence of many arbitrarily named forms) represents a continuum in which different pattern components vary in accord with Pictet's (1913) "law of melanization and albinization of parts." There is a good circumstantial case for the T-shock phenotypes being phenocopies of genetically determined examples collected from time to time in the wild. It is difficult to breed wild male *Vanessa*, aberrant or otherwise, but I have twice established cultures from *elymi* females of *V. annabella* taken wild at Suisun City, Solano County, California – an apparent "hot spot" for them. In both cases the father was unknown, the F_1 was

Fig. 4. Dorsal surfaces of some laboratory-reared *Vanessa cardui*. *Left* and *middle rows* contain cold-shocked specimens showing the transition from an extreme *elymi* phenotype to nearly normal. *Upper right* specimen is a normal control. *Lower right* specimen is a full *elymi* which appeared spontaneously in a reared brood without cold shock; all the others are from a different brood and are sibs

Fig. 5. Ventral surfaces of the specimens shown in Fig. 4

normal, and some moderately aberrant individuals occurred in the F_2 under normal rearing regimes. However, susceptibility to T shock was greatly enhanced in the F_1 and especially the F_2 - Figs. 6 and 7 suggest the difficulty of quantitating the response. The genetics of this situation are unlikely to be simple and straightforward. This is unfortunate, as is the fact that Nymphalines are far from ideal lab animals. The non-adaptive shock phenotypes of the Nymphalines may offer a model for the evolution of adaptive, pierine-style polyphenisms; if a predictable pattern change is advantageous in a particular environment and can get coupled to a good environmental predictor, the system is off and running. Our first try to do this failed due to logistical reasons (Shapiro 1976).

Some of the old writers on the Nymphaline aberrations, such as Standfuss, firmly believed they were Lamarckian in character. In the light of recent - and astonishing - developments in "gene automodulation" and heritable, pharmacologically induced changes in physiological traits (e.g., Campbell 1981), it is not inconceivable that something quasi-Lamarckian *was* going on in their experiments. They, unlike most of Kammerer's, *should* be repeated. Whatever the truth here, Gunder's notion of "transition forms" is not applicable; there is no evidence that internal factors are systematically turning all the Vanessas into *elymi*. Nor does a cladistic analysis of pattern suggest that *elymi* is a primitive, ancestral, Vanessine pattern, and the aberrations atavisms. The system is fascinating. It is certainly not elegant.

We are thus not yet in a position to answer the question posed at the beginning of this paper - whether there is an evolutionary relationship between nonadaptive (Nymphaline) and adaptive (pierine) responses to environment. We are, however, making progress in elucidating the genetic architecture of phenotypic plasticity, the

Fig. 6. A brood of 17 sibs of *Vanessa annabella*, cold-shocked at the same pupal age, showing the extent of variability in expression of the *elymi* tendency. The 18th specimen *(lower right)* is an unshocked control. Note apparent homology with Fig. 4

origin of the "general purpose genotypes" which enable a butterfly to be one kind of specialist in the summer and another in the winter without carrying a genetic load; and we may at last put an end to the denigration of variation which is "merely" under environmental control. That alone would be an achievement.

Addendum. After this paper was presented, Bowden (1983) published a discussion of Lorković's form *funebris* of *Pieris napi*, including a figure of a similar wild specimen of uncertain taxon from the Alai Mountains of western Central Asia (Avinoff 1910). Bowden discusses the possibility that this form is a reversion to a primitive pattern, illustrating Kautz's (1955) reconstruction of the Ur-*napi*, but does not reach any definite conclusion. Kautz's figure, which is based solely on inference from the Palaearctic *napi*-group, is crude but similar to Figs. 1 c, 2 c, and 1 d, 2 d of the present paper, which are based on a more general overview of the Pierini. Kautz was of the opinion that the Ur-*napi* was yellow and that increasing whiteness represented an orthogenetic tendency in the lineage. The same tendency would today be interpreted in terms of climatic adaptation *(supra)*.

Acknowledgements. The work reported here has been supported at various times by grants from the National Science Foundation (USA), the National Geographic Society, and at U.C.D., the

Fig. 7. Ventral surfaces of specimens in Fig. 6. Cf. Fig. 5

Department of Zoology and the Institute of Ecology; to all, and to my ongoing collaborator Michael M. Collins and my stalwart field assistant Adrienne R. Shapiro, many thanks. Gerardo Lamas M. kindly provided access to material at the National Museum of Natural History, Lima, Peru. The photographs are all by Samuel W. Woo.

References

Ae SA (1957) Effects of photoperiod on *Colias eurytheme*. Lepid News 11:207–214

Avinoff A (1910) New forms of Rhopalocera in the Fergan region. Horae Soc Ent Ross 39: 247–248, Pl. XIV

Bowden SR (1983) A palaemorph of *Artogeia*? – f. *funebris* Lorković (Lep., Pieridae). Proc Trans Br Ent Nat Hist Soc 16:76–80

Campbell JH (1981) Profane genes, physiological evolution, and the inheritance of experiential information. Symp Address, Am Soc Nat, Soc Stud Evol

Goldschmidt R (1938) Physiological genetics. McGraw-Hill, New York, pp 361

Gunder JD (1927) Transition forms (Lepid., Rhopalocera). Entomol News 38:263–271 (with plates v–x)

Kautz H (1955) Die Rückschlags- und Zukunftformen von *Pieris napi* L. Z. Wien Ent Ges 40: 151–156
Lande R (1981) The minimum number of Mendelian factors contributing to quantitative variation between and within populations. Genetics 99:541–553
Lorković Z (1971) *Pieris napi* (L.) morfa *funebris*, osebujna nova rekombinacija krizanja. Acta Entomol Jugosl 7:5–9
McLeod L (1968) Controlled environment experiments with *Precis octavia* Cram. (Nymphalidae). J Res Lepid 7:1–18
Munroe E (1982) Wing patterns in the Lepidoptera. Bull Entomol Soc Can 14(4):103–112
Nijhout HF (1981) The color patterns of butterflies and moths. Sci Am 245:140–151
Pictet A (1913) Recherches experimentales sur les méchanismes du mélanisme et de l'albinisme chez les Lepidoptères. Mem Soc Phys His Nat Genève 37:111–278
Schwanwitsch B (1924) On the ground plan of the wing-pattern in nymphalids and certain other families of Rhopalocerous Lepidoptera. Proc Zool Soc London 1924:509–528
Shapiro AM (1976) Seasonal polyphenism. Evolut Biol 9:259–333
Shapiro AM (1983) Notes on aberrations: an aberration of *Glaucopsyche lygdamus* with a complete Scolitantidine dorsal pattern. An apparent "intersexual" *Colias eurytheme*. J Res Lepid 20:240–244
Shapiro AM (1984) Studies in the evolution of seasonal polyphenism. In: Vane-Wright R, Ackery P (eds) The biology of butterflies. Academic Press, London New York (in press)
Silberglied RE, Aiello A, Lamas G (1980) Neotropical butterflies of the genus *Anartia*: systematics, life histories, and general biology. Psyche (Cambridge, Mass) 86:219–260
Watt WB (1968) Adaptive significance of pigment polymorphisms in *Colias* butterflies. I. Variation of melanin in relation to thermoregulation. Evolution 22:437–458
Watt WB (1969) Adaptive significance of pigment polymorphisms in *Colias* butterflies. II. Thermoregulation and photoperiodically controlled melanin variation in *C. eurytheme*. Proc Natl Acad Sci USA 63:767–774

How Much of the Phenotype is Necessary to Understand Evolution at the Level of the Gene?

S.C. STEARNS[1]

Introduction

Parts of ecology have been transformed in the last 20 years by an influx of evolutionary thinking to form a new field, evolutionary ecology. Within this new field, the well-developed specialties now include life-history evolution, sociobiology, and sex-allocation theory (Stearns 1982b). At a symposium on the relationship between population genetics and ecology, it is appropriate to ask what role population genetics has played in the transformation of ecology. In fact, three theoretical frameworks have been involved: game theory (e.g., Maynard Smith 1982), optimization theory (e.g., Oster and Wilson 1978), and modifications of parts of population genetics (e.g., Hamilton 1964, Charnov 1982). The contributions of population genetics are seen most clearly in the kin-selection models of sociobiology, where the focus remains selection acting on genes, and in the Shaw-Mohler equations at the core of sex-allocation theory, where recent developments are clearly based on classical contributions going back to Fisher (Charnov 1982). Because population geneticists do not regard kin-selection as one of the main themes of their field, this means that population genetics *sensu strictu* has not had much influence on evolutionary ecology (De Jong, pers. commun.).

Within sociobiology, Hamilton's insight concerning kin-selection will no doubt stand as an enduring contribution. However, a number of studies indicate that alternative explanations based on individual advantage remain plausible – for example, Alexander's (1974) hypothesis of parental manipulation as an explanation of the evolution of castes in the social insects, and Packer and Pusey's (1982) gametheoretical explanation of cooperative mating behavior in male lions. Both are strictly phenotypic explanations. The issue can be put a number of ways. Must one think of changes in gene frequency when dealing with the evolution of ecologically important traits at the level of the phenotype? Is it often – or ever – practical or productive to do so? What are the units of selection and what definition of fitness should be applied to them? Sociobiology has not yet settled these issues, for game theory can generate explanations in terms of individual advantage for most types of behavior that are explained with kin-selection (cf. Axelrod and Hamilton 1981).

Within sex-allocation theory, the contribution of population genetics would seem clearer, because the central Shaw-Mohler theorem comes directly out of population

1 Institute of Zoology, University of Basel, Rheinsprung 9, 4051 Basel, Switzerland

Population Biology and Evolution
Ed. by K. Wöhrmann and V. Loeschcke

genetics. However, sex-allocation theory can be just as fully developed with optimization techniques, as can be seen by comparing the alternative formulations of the central insight. In the language of population genetics, the Shaw-Mohler theorem states that selection favors a mutant gene that alters various life-history parameters if the percent gain in fitness through one sex exceeds the percent loss through the other sex. In the language of optimization, it states that selection will maximize the product of fitness gained through male function and fitness gained through female function (Charnov 1982). Since both approaches lead to the same predictions, the empirical content of the two frameworks is identical, and so far as sex-allocation theory is concerned, the choice of population genetics or optimality theory is a matter of taste.

Within life-history evolution, the role of population genetics is even less clear. Charlesworth (1980) has tried to establish the connection between the two fields, but most of his analysis dealt with the complications encountered when formulating traditional problems in population genetics while dealing with overlapping generations, age structure, and separate mortality and fecundity schedules for the two sexes. He used optimal control theory to get life-history predictions. Thus, population genetics is *consistent* with all the observations of evolutionary ecology, but it is not *necessary* for any of the major predictions, each of which can be couched in terms of whole organism alternatives, and it is *sufficient* for only some of the major predictions, such as the Shaw-Mohler theorem. In brief, as an evolutionary ecologist, I am not forced by my data to use population genetics in my work.

That does not mean that population genetics is not relevant. In fact, all the reasons that led to the emphasis on population genetics in the modern synthesis are still good reasons. In an outcrossing population, only the genes record the history of selection. Only the genes contain the information required for an egg to develop into an adult. Only the Mendelian model explains most of the facts of inheritance. Evolutionary ecology simply emphasizes that in addition, important mechanisms operate at the level of whole organisms and for equally good reasons. Only whole organisms survive and reproduce. Only whole organisms interact with the environment. The whole thrust of evolutionary ecology has been to discover the principles of whole organism design that maintain function in the face of ecological problems. Whereas population genetics emphasizes gene frequency change and de-emphasizes the phenotype, evolutionary ecology emphasizes the phenotype and de-emphasizes the gene. Because the simplifications of one are the complexities of the other, they would appear to be natural partners. At least one population geneticist agrees: "Evolutionary ecology and population genetics are indeed natural partners. The first tends to focus on the causation of fitness w from phenotype and environment; the second tends to focus on the numerical processes in the population, given certain fitness values w."

Despite impressive progress in both fields, technical and conceptual problems have prevented a synthesis and at this point at difficult question remains unsolved and little discussed: what are the general and minimal properties of the phenotype necessary to understand genetic change (Stearns 1983a)? This paper discusses two candidates for such phenotypic properties, developmental plasticity and population dynamics.

Developmental Plasticity for Age at Maturity

This example, taken from Stearns and Crandall (1983), makes several points. It shows that the developmental plasticity of an important life-history trait - age-at-maturity - can itself be considered an adaptation about which successful quantitative and qualitative predictions can be made. It develops an image of the phenotype as a set of surfaces along which the organism responds to stress, considerably altering the impact of selection on the genotype. Thus, it refines and extends the classical concept of the genotype-phenotype relation - the norm of reaction - on which it builds.

How should an organism encountering an unavoidable stress that results in slower growth alter its age-at-maturity to maximize fitness? Stearns and Crandall assumed that maturity may be delayed either because it results in a gain in fecundity or because it results in reduced juvenile mortality, or both. We also assumed that the primary impact of environmental stress is on growth rate, that changes in growth rate impose unavoidable changes in size at given ages, and that these changes in size have particular consequences for fecundity and for the juvenile mortality rates of the offspring produced. In addition, in some cases we assumed that changes in growth rate can serve as reliable *cues* to the direct impact of environmental change on juvenile and adult mortality, unmediated by size.

We defined functions for death rates and birth rates in terms of age and size, then assumed that r defines fitness by assuming stable age distributions and applying the Euler-Lotka relation. The Euler-Lotka relationship defines fitness (r) implicitly in terms of age-specific birth and death rates; from it, one can find the age-at-maturity that maximizes *I* for any given growth rate. We examined the sensitivity of the predicted optimal age-at-maturity to changes in growth rate, in the rate of gain of fecundity with size, in juvenile mortality rate, and in adult mortality rate.

The Predictions

The model makes two different types of qualitative predictions. The first type of prediction has to do with the relative *position* of the trajectory in age-size space and results from sensitivity analyses. Consider one example of a sensitivity analysis (Fig. 1). As the juvenile mortality rate is increased, the predicted trajectories of optimal age-at-maturity and size-at-maturity move to the right. This implies that an even increase in juvenile mortality across the entire range of environments inhabited by these organisms will select for a delay in maturity at all growth rates. Similar analyses indicate that if two populations are brought into the laboratory, then the one with the higher adult mortality should have a trajectory that lies below and to the left and the one with the higher rate of gain of fecundity with size should also lie below and to the left. Thus, the position of the trajectory depends on rate of gain of fecundity with size, adult mortality, and juvenile mortality.

Before discussing the second type of qualitative prediction, I want to emphasize that there are two types of information contained in Fig. 1. First, the positions of the trajectories for different levels of juvenile mortality represent predictions about evolutionary divergence with a genetic basis. Second, the shape of any single trajectory

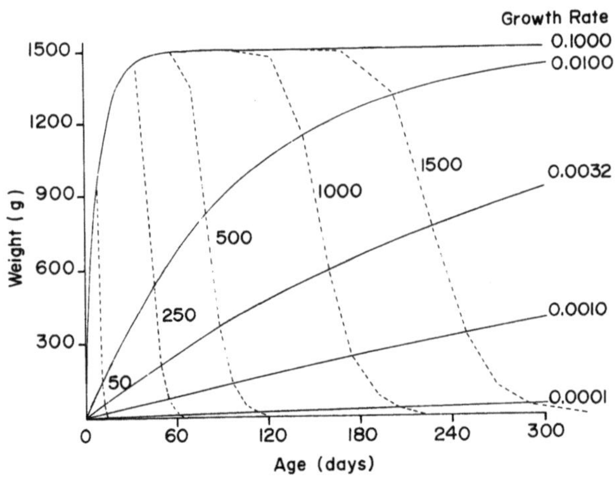

Fig. 1. An example of the type of sensitivity analysis carried out. Plastic trajectories are depicted with ***broken lines***, growth curves with ***solid lines***. Juvenile mortality increases in proportion with the numbers 50–1,500 that label the curves. The growth rate parameter, k, in the Von Bertalanffy growth equation is varied from 0.0001 to 0.1 to generate the growth curves. For moderate rates of fecundity gain with size, increases in juvenile mortality cause large delays in maturity

predicts the plastic response of one population with a given demographic history to some environmental stress that slows growth. Every point along such a trajectory represents the age and size at maturity of a single organism that maximizes fitness for the given growth rate. Thus, the analysis neatly separates the effects of nature and nurture.

The second prediction has to do with the *shape* of the trajectory followed by age and size at maturity as stress is applied. If increased stress only affects growth, with no systematic effect on mortality rates, the trajectory should be a mirror-image sigmoid curve (Fig. 2). In other words, as growth rates decrease, the organisms should delay maturity and mature at smaller sizes. If increased stress not only slows growth, but also increases adult mortality (i.e., stress not only slows growth rate, but is also a *cue* to other environmental changes), then the trajectory should be J-shaped. In other words, as growth rates decrease, the organisms should mature earlier and at smaller sizes. If increased stress not only slows growth, but also increases juvenile mortality, then the trajectory should either be L-shaped or the upper half of a paraboloid. In other words, as growth rates decrease, the organisms should delay maturity, but whether they mature later at larger sizes, or later at smaller sizes, depends on how rapidly juvenile mortality increases as growth rates decrease. Thus, the model suggests that there are at least four qualitatively different types of trajectories that should be expected in populations that have had different evolutionary histories of particular demographic conditions.

Tests

The quantitative predictions consist of a single point, rather than a trajectory, for each population, and are obtained by finding the age at maturity that results in maximal r for a population with the observed juvenile and adult mortality, growth rate, and rate of fecundity gain with size. They predicted age at maturity for 19 populations of fish. Direct estimates of juvenile mortality are hard to come by and in most cases they used a rough estimate derived by assuming that on average two offspring

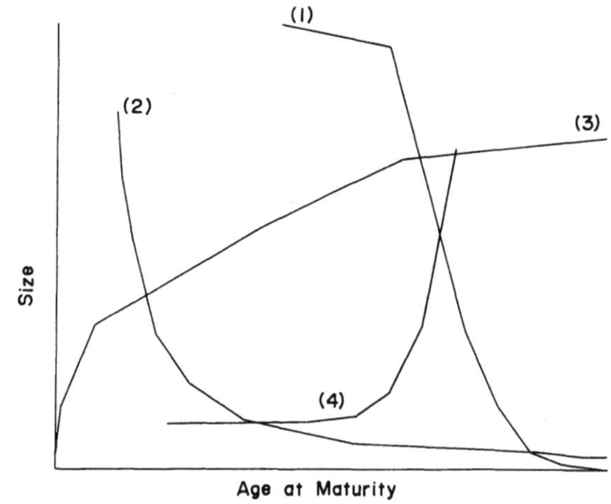

Fig. 2. The four qualitatively different types of age-size maturation trajectories predicted by the model. Every point on each line represents the optimal age and size at which to mature under the given assumptions. You should imagine that the plane is covered by a family of growth curves such that growth decreases from the upper left to the lower right. *1* Growth rates not correlated with juvenile or adult mortality. *2* Juvenile mortality increases slowly as growth rates decrease. *3* Juvenile mortality increases rapidly as growth rates decrease. *4* Adult mortality increases rapidly as growth rates decrease. In (*1–3*) maturity is delayed as growth rate decreases; only in (*4*) is maturity advanced when growth rate decreases. In (*1*) and (*2*), later maturity occurs at smaller sizes; in (*3*) and (*4*), later maturity occurs at larger sizes

survived to maturity and that population size was approximately constant. In the 19 populations analyzed, the correlation of prediction with observation was strongly positive ($r = 0.91, p < 0.01$).

The second case demonstrates that trajectories do exist and have shapes that correspond to one of the four predicted. McKenzie et al. (1983) raised male platyfish (*Xiphophorus maculatus*) of two genotypes under an extreme range of conditions. One genotype matures later and at a larger size than the other genotype. They plotted the ages and sizes at maturity for both genotypes in age-size space and discovered that the data could be described by two L-shaped, nonoverlapping envelopes, one centered at 84 days and the other at 140 days (Fig. 3). The shape of the envelopes indicates that juvenile mortality increases as growth rates decrease in this species. The average distance between the trajectories cannot be accounted for by differences in instantaneous juvenile mortality rates or in rate of gain of fecundity with size, leaving differences in adult mortality as a potential explanation. If the earlier maturing males have significantly higher adult mortality, which would be consistent with their smaller size, the observed pattern would agree with the predictions of the model.

The optimization model can explain the shape of the two envelopes and perhaps the distance between them, but it cannot explain why there are *two*. A frequency-dependent model using an ESS approach of the sort taken by Gross and Charnov (1982) to explain male dimorphism in bluegill sunfish will probably be necessary to explain the presence of the two genotypes.

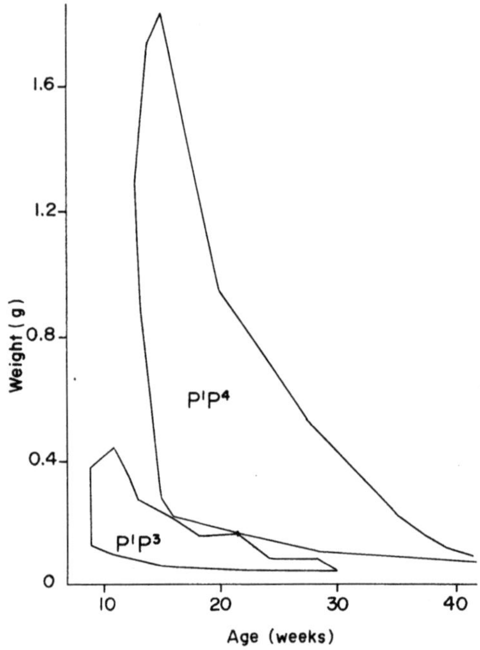

Fig. 3. The two envelopes enclose the ages and sizes at maturity for two genotypes of male platyfish, *Xiphophorus maculatus.* They were raised under a wide range of growth conditions. The shapes of the envelopes are consistent with portions of trajectories *1* and *2* (see Fig. 2). (Adapted from McKenzie et al. 1983)

In addition to these two cases, data on the plasticity of age at maturity in Hawaiian mosquitofish, human females, laboratory rats, Norwegian moose, and Atlantic eels are all consistent with the qualitative predictions of this model (Stearns and Crandall 1983, Saether pers. commun., Helfman pers. commun.).

Conclusions

This model makes a number of points within the context of evolutionary ecology, but it also suggests an experiment that links genotype and phenotype. Within evolutionary ecology, it suggests a resolution of the disagreement over whether organisms mature at a fixed size or at a fixed age. Over some portions of the age-size maturation trajectory, organisms appear to mature at a fixed size; over other portions, organisms appear to mature at a fixed age. Only by considering maturation events occurring across a wide range of growth rates can the shape of the whole trajectory be perceived. Instead of the single general pattern suggested by Alm (1959), it predicts at least four types of trajectories that correspond to particular demographic conditions.

Moreover, it unifies genetic and environmental explanations of variation in age and size at maturity. In this model, the environment imposes changes in growth rates and serves as a cue to short-term (single generation) shifts in juvenile and adult mortality rates. These short-term changes determine the position of the maturation event along a single trajectory. In addition, populations have encountered certain average demographic conditions over evolutionary time and particular trajectories of age- and size-at-maturity, with characteristic shapes and position, have evolved in each population and represent the genetically fixed and adaptive component of the organisms' response.

The model suggests that an experiment be done to test one of its critical assumptions - that there is heritable variability in plastic trajectories. This could be tested by performing a full-sib/half-sib cross on an organism that is a facultative parthenogen, cloning the progeny of the cross, and rearing the cloned sibs in a range of environments. If this procedure indicated that there was a significant amount of additive genetic variability for an age-size maturation trajectory, then a selection experiment could be carried out to test the assertion that different demographic conditions lead to plastic trajectories with different shapes.

What is an Organism and What is a Trait?

Two concepts often accepted as familiar and well-defined - organism and trait - may be subtler than they appear. The view of an organism that one develops after working with the concept of plastic trajectories is quite different from the one habitually used. In this framework, an organism is a set of potentials expressed as n-dimensional trajectories. The trajectories precisely describe compensations that maintain fitness at a high level in the face of the inevitable compromises forced upon the organism by environmental stress and the nature of the chemicals out of which it is built. Thus, an organism may not present to selection the furry or scaley object that one holds in one's hand, but rather its complete set of potential responses to environmental conditions. This is not news to anyone familiar with the concept of norms of reaction, but the precision with which plastic trajectories can be predicted imparts new force to these ideas.

The definition of a trait may also be problematic. In standard life-history theory, age and size at maturity are considered to be separable traits, but if they are only components of a plastic trajectory that is itself the primary object of selection - the trait - then their original status as separable traits is cast into doubt. This one case of changing the definition of what constitutes a trait raises the general problem of specifying the conditions under which one decides that one is dealing with a separable component of the organism. In nature, the object under selection is the entire unfolding of the life-history of the organism, not any specific structure or part of the life-history. The partitioning of the organism into a set of traits is an act carried out for the convenience of our analysis and there is no guarantee that it represents biological reality. I cannot specify yet what difference this observation makes to how one does population genetics or life-history theory. It reminds us that the organism is a sophisticated unit of coherent design whose parts are intimately tied together by function and by genetic, developmental, and physiological mechanisms. If one carries out an analysis that assumes that the organism can be broken into a set of independent traits, each of which is under selection, then one is making many assumptions that may not hold.

An Evaluation of Fitness Measures for Whole Organisms

This section explores the consequences of another phenotypic property, population dynamics, that has direct consequences for population genetics because it leads to a reexamination of the definition of fitness. Recent discussions of the fitness concept (e.g., Dawkins 1982) usually refer to the choice of a unit of selection - gene, organism, group, or species. If one approaches the question as a population geneticist, then there is rich literature in which the concept of fitness is clearly defined, either as relative rate of effective reproduction per generation (e.g., Haldane 1949), as the rate of increase of an allele in an age-structured population (e.g., Fisher 1930), or as some similar measure designed to include the effects of genes passed through relatives (Hamilton 1964). These are all short-term measures of relative abundance. In this context, the method of judging the utility of fitness measures has been to assume the Mendelian mechanism, then explore the capacity of various measures to predict the rate of gene frequency change (cf. Charlesworth 1980). An implicit assumption is that changes in gene frequencies are the stuff of evolution and that fitness measures are properly defined primarily by reference to such changes.

Within evolutionary ecology, both because it is a younger field and because it asks different questions, the problem of defining fitness remains an open question. As in population genetics, it is used in models to represent the action of natural selection. However, the question asked is not, how do gene frequencies change, but what are the equilibrium properties of the phenotype. Thus, it is natural to consider definitions of fitness that reflect the long-term consequences of population dynamics and to entertain the possibility that short-term success is not necessarily a good predictor of long-term success.

Therefore, I adopt here the alternative view set out in the introduction. While only genes record the history of selection, only organisms survive and reproduce. Mechanisms other than recombination have consequences that must be taken into account in describing selection and defining fitness; they include the interactions that constitute the demography and ecology of whole organisms. These interactions, determining fitness at the level of the whole organism, are a necessary component of whatever global definition of fitness one might want to make. Whatever other effects a gene might have, it must at least modify the phenotype in such a way that whole organism fitness is increased if the gene is to increase and persist in the population. By dealing only with whole organisms and ignoring the complications of recombination, kin-selection, and group selection, one can fix ideas clearly and reexamine traditional fitness measures.

This approach was taken by Crandall, Stearns, and Dudman (1983). Their procedure was, in formal outline, similar to the one used by Charlesworth (1980). First, we defined a criterion against which to judge the performance of various fitness measures. This criterion is the property of being the last phenotype to disappear from the population (in discrete, microscopic models) or of being the phenotype that drives all other phenotypes out of the population (in continuous, macroscopic models). All these models are stochastic, rather than deterministic. The property one seeks is a form of evolutionary stability (cf. Maynard Smith 1982), and the question asked is, what measure of fitness will predict the evolutionarily stable strategy (ESS)?

In the process of finding the ESS, one must explore the consequences of various life-histories for population dynamics and thus satisfy, at least in part, the requirement that models of life-history evolution be consistent with plausible population dynamics. By carrying out an explicit evaluation of *long-term* winners, this method deals explicitly with the notion of fitness that short-term definitions are assumed to approximate. In addition, this method makes predictions about the evolution of life-histories without any use of the Lotka-Euler equation, thus avoiding the assumption of a stable age distribution.

The Model

The model deals with organisms that mature at age one (arbitrary units) and that reproduce repeatedly thereafter at time intervals one unit apart, suffering a cost of reproduction in the form of increased adult mortality. These organisms send out progeny that compete for spaces in which only one survives to maturity. In the finite models, there are N such spaces. We examined the behavior of six intrinsic measures of fitness, including those traditionally used in life-history work, r and K. Here *intrinsic* refers to measures defined when the organisms live with no other types present. Using both analytical methods based on Markov models and computer simulation tournaments in which many types competed for representation in a finite environment, we evaluated the capacity of the intrinsic measures to predict the long-term winners. At this point, I ask your indulgence in allowing me simply to introduce the measures without going into the details of the model from which they naturally arise (cf. Crandall et al. 1983). The measures were:

Malthusian growth rate:	r	$= b - d$
Carrying capacity:	K	$= \max[0, (1 - d/b)/(1 + d)]$
Productivity:	π	$= b/d - 1$
Persistence:	σ^{-1}	$= (b + d - d^2)^{-1/2}$
Stability:	Σ^{-1}	$= [d(2 - d - (b - d))/(1 + b)]^{-1/2}$.

The sixth measure, log-time-to-extinction, has no exact formula for arbitrary N, but has been computed exactly in some cases. The first two measures are familiar; only the fact that one can get an exact expression for K in terms of birth and death rates is new. Note that it is standard practice in population genetics to define K by counting the zygotes produced in each generation, to avoid complications of age-structure in a density-dependent situation. We defined K as the total number of organisms present, but since all organisms were adults and all adults had a fixed fecundity, this was equivalent to counting zygotes, differing only by a scaling factor. The third measure, productivity, is very nearly births divided by deaths, suggesting that very low birth rates may be associated with high fitness if they imply even lower death rates. Persistence measures fluctuations in density at low numbers and is large when fluctuations are small. Stability measures fluctuations in density at carrying capacity and is large when fluctuations are small.

Principal Results

Recall that in these models, everything depends on birth rate alone. If the organisms increase their birth rate, they send more offspring out into the environment each season, but they are also more likely to die before the next season. If they decrease their birthrate, then their death rates also fall and organisms with very low birth rates are likely to persist for a long time.

The first result is that neither r nor K predicts the winners of computer tournaments, in either the finite, combinatorial setting or in the infinite, continuous setting. In the continuous models, the winner is always the type with highest productivity (b/d; Fig. 4). In the finite models, the outcome cannot be described as the success of a single optimal type. Instead, there is a certain probability of winning associated with each birth rate. Types with birth rates that imply high productivity and high carrying capacity also have relatively high probabilities of winning, but types that have very low birth rates also have a moderately high probability of winning (Fig. 5).

Thus, in the finite setting, the capacity to persist at low population numbers by having very high adult survival rates becomes important. The implication for life-history evolution is that iteroparity is favored in unpredictable environments, but that some small percentage of the time one might find semelparous organisms, because even semelparous types will win some small proportion of the competitions.

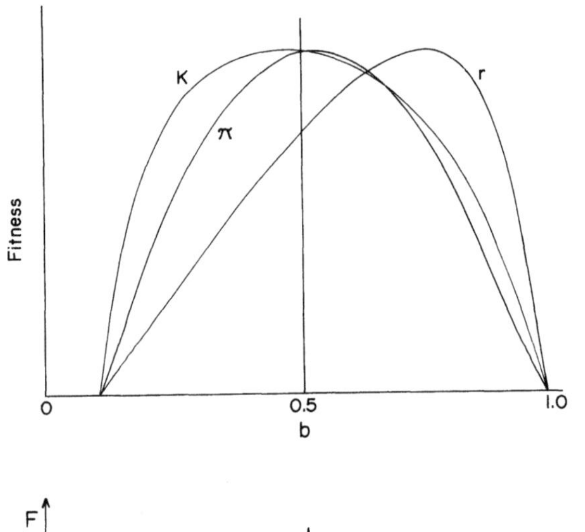

Fig. 4. Plot of growth rate (r), carrying capacity (K), and productivity π as intrinsic fitness measures in the COR model. The winning type in multiway competition in a continuous environment has birth rate (b) = 0.5; thus, productivity is the best predictor of winners. The y-axis is scaled in arbitrary units as a percentage of the maximum value of each measure

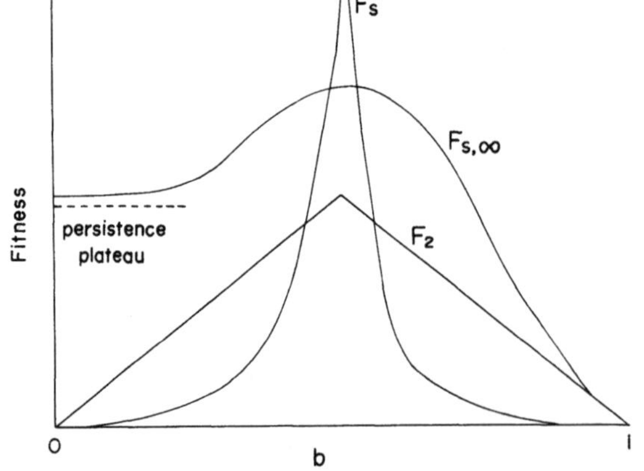

Fig. 5. General behavior of unified fitness measure for an optimizable model. Macroscopic unified fitness becomes cusp-like for greater numbers of types competing, peaking at the optimal b (which maximizes productivity). The microscopic behavior is completely different, because in the large-N limit there are contributions from random fertility parameters of low value, which correspond to high intrinsic persistence

This result confirms earlier work on the forces that select for iteroparity and it does so using the mechanical consequences of population dynamics, rather than static optimization procedures, to arrive at the conclusion. It also does not assume a stable age distribution, an assumption that must be made to carry out almost all work on optimality models of life-histories.

The second result is more mathematical in nature. The predictions of the finite and continuous approaches do not converge in the limit of large N (number of spaces for organisms to live). This is unusual in mathematics, but it can be explained in these models as the result of qualitative differences in what happens at low population densities. In the continuous models, partial organisms are possible and if the population gets down to one organism, it can decrease to 0.5 or 0.01 organism and still recover. In the finite models, only whole numbers of organisms are possible and the transition from one organism to zero organisms is accomplished in a single step that spells extinction. Thus, in the finite models, survival depends on having life-history characteristics that imply very little fluctuation in numbers at low population densities and in these models, the way to achieve small fluctuations is to have low birth rates and very good adult survival rates.

It might seem that the failure of convergence is of interest only to mathematicians, but it has general significance for work in population biology. Real organisms are born and die as whole units; thus, the assumption of finite mathematics correspond much more closely to the real world than do those of continuous mathematics. If the two approaches yield different results, then one might be inclined to prefer the prediction of the finite models on the grounds that their assumptions are more plausible. Moreover, the two approaches lead to quite different views of the world. The combinatorial, finite approach leads one to expect probability distributions of types of life-histories in given environments (Fig. 6). The continuous approach leads one to expect a single optimal life-history. That is a profound difference in expectations.

If the finite approach is correct, then types that have life-histories with low persistence measures of fitness should go extinct more frequently than types with high persistence measures. In these models, persistence was high when birth rates were low and the organisms had very long adult reproductive lives. If that assumption holds in nature, we should expect to see a pattern in which the main stems of phylogenetic lineages are primarily iteroparous, with semelparity occurring out at the tips of the phylogenetic tree. Similarly, we should expect low reproductive effort as the ancestral condition, and high reproductive effort as the derived condition.

The argument here is precisely the one given for the distribution of sexual and asexual types across lineages by Maynard Smith (1978). If one type goes extinct more frequently than the other, it should typically be found as the derived condition. In both cases, the type with the higher extinction rate – the asexual type, the semelparous type – also has associated with it a higher abundance measure of fitness (for example, the Malthusian parameter r). Thus, the observations on the distribution of asexuality in nature – thoroughly documented by Bell (1982) – can be taken as a confirmation of this model, because both iteroparity and sexuality can be viewed as adaptations that increase persistence measures of fitness. In addition, the major cases of semelparity – Pacific salmon, bamboos – are found as derived conditions in iteroparous lineages – trout, grasses.

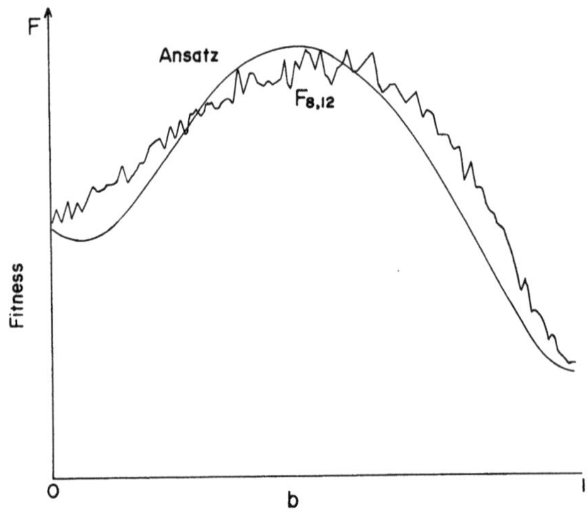

Fig. 6. Typical attempt at approximating unified fitness $F_{8,12}$ with an ansatz linear combination of intrinsic (one-type) measures. Here, the approximation is a linear combination of persistence and log-time-to-extinction, but $F_{8,12}$ was computed with Monte Carlo techniques. $F_{8,12}$ indicates the outcome of 8 types competing for 12 places. Note both the persistence plateau at low birth rates and the positive intercept for very high birth rates. This implies that some types with very low birth rates will occasionally win the competition and that somewhat less frequently types with very high birth rates – semelparous types – will also win the competition

However, just as the bdelloid rotifers stand out as an ancient asexual lineage, so do cephalopods stand out as an ancient semelparous lineage. In both those cases, the organisms probably have evolved other adaptations that help them overcome the disadvantage – with regard to persistence, at any rate – implied by those major features of their life-histories.

This argument may explain the otherwise puzzling existence of semelparity in large trees occupying the canopy of rain forests. Even in an environment thought to select strongly for iteroparity (Stearns and Crandall 1981), one expects to find a small proportion of semelparous types that may achieve high density for a relatively short period of time because of their higher abundance component of fitness. Foster has discovered semelparity in *Tachigalia*, a fast-growing canopy species up to 50 m tall found in Central and South America (Batten 1983), groups of which flower and die approximately every 4 years. I suspect that they have evolved relatively recently and will go extinct before their iteroparous relatives in the same family. I would also expect to find other semelparous species at low frequency on the lists of species in environments thought to select for iteroparity.

Comparative evidence of this sort is suggestive, but a strong test would involve competition trials among clones with different birth and death rates. Since the prediction concerns the shape of a probability distribution (Fig. 6) of outcomes, many replicates would be needed. The critical piece of evidence would be the presence or absence of the plateau in fitness at low birth rates. If it is really there, the combinatorial models are appropriate. If it is not, then the continuous models are appropriate and productivity would be suggested as the measure of fitness.

Discussion

Everyone who has done much work with theory in biology has encountered the criticism that the results could not possibly hold because the models were too simple to capture a complex reality. That is, of course, not a valid objection. If simple models make successful predictions and integrate previous knowledge in a novel way, then there is little reason to make them more complex and the models of population genetics have the great virtue of being relatively simple. Nevertheless, I have suggested that complex properties of the phenotype - developmental plasticity and stochastic population dynamics - can contribute to our understanding of the kinds of genetic change that occur in nature. Certainly population geneticists deserve some substantial reasons for accepting this recommendation, which makes life much more complex.

The first reason is the justification for studying life-history evolution itself: life-history traits are the phenotypic components of fitness. Whatever definition of fitness one uses must involve life-history traits, for only through reproduction and survival can organisms increase in abundance and persist and only through the differential survival and reproduction of the organisms that carry them can genes change systematically in frequency. The long-term consequences of differences in life-histories are realized in population dynamics. Thus, the fitness of genes is in fact defined by the population dynamics of the organisms that carry them.

The second reason has to do with developmental plasticity. Wright (1931) argued that developmental plasticity is "of the greatest significance as a factor of evolution in damping the effects of selection," and Stearns (1982a) suggested that developmental plasticity can uncouple phenotype from genotype, slowing the rate of evolutionary change. To the degree that plasticity is effective in maintaining fitness at the highest available level within the physiological constraints on the organism, it will greatly decrease the strength of selection on the plastic trait. However, I was wrong in suggesting that it could uncouple genotype from phenotype, for recent work on the evolution of life-history traits in mosquitofish in Hawaii since 1905 suggests that more plastic traits have also undergone more divergence among populations (Stearns 1983a,b). There are at least two explanations. Selection may have been sufficiently stronger on the plastic trait to overcome the damping effects to which Wright referred, or the plastic trait - in this case, age-at-maturity - may not have been the trait under selection. If the trait under selection is not a component of a plastic trajectory, but the trajectory itself, then the components could undergo genetic change despite their plasticity so long as there was heritable variability for the trajectory. In other words, Wrigth's comment retains its strength only if the plastic trait the biologist measures is also the component of the organism on which natural selection has its impact.

There are two additional reasons to be concerned with plasticity. Scharloo (pers. commun.) has pointed out that developmental plasticity extends the range of habitats that can be occupied by any given genotype, thus increasing the range and diversity of selection pressures encountered by it. Ashby (1956) suggested that it is plasticity in some traits that compensate for rigidity - canalization - in others. Caswell (1983) has recently explored some of the consequences of that idea. Both ideas suggest subtle effects in which genes affecting one trait have impact on other traits not directly through development or physiology, but by changing the manner in which the whole organism interacts with its environment and thus modifying the selection pressures on the traits rather than the traits themselves.

There is a deeper reason for suggesting that one must study phenotypic properties to understand genetic change. The organism is a coherent unit of complex design, with many parts tied together by physiology, by developmental origin, and ultimately by gene products. For example, the use of hormones for integration and regulation has resulted in pervasive pleiotropy for most whole organism traits. This integration is reflected in the evolution of delayed maturity, repeated reproduction, viviparity, and parental care, where organisms encounter tradeoffs with consequences that are just as precise and inevitable at that level as recombination is at the genetic level.

We study population genetics and evolutionary ecology to understand the general features of evolution. Population genetics and evolutionary ecology are natural partners that both need clarification of the connections between changes in gene frequences and changes in phenotypic traits. However, the complexity of development and physiology makes clear that we have only illuminated two small sites on which selection most clearly acts – organism and gene. We have yet to define precisely which intermediary mechanisms – developmental, endocrinological – must be understood to connect the two fields.

Acknowledgements. I thank the Deutsche Forschungsgemeinschaft for support, Richard Crandall for mathematical collaboration, and Gerdien de Jong and my students at Reed College for their probing questions.

References

Alexander RD (1974) The evolution of social behavior. Annu Rev Ecol Syst 5:325–383

Alm G (1959) Connection between maturity, size, and age in fishes. Rep Inst Freshwater Res, Drottningholm 40:1–145

Ashby WR (1956) An introduction to cybernetics. Chapman & Hall, London, p 295

Axelrod R, Hamilton WD (1981) The evolution of cooperation. Science 211:1390–1396

Batten M (1983) The rush is on to study jungles. Int Wildlife 13:16–19

Bell G (1982) The masterpiece of nature: the evolution and genetics of sexuality. Univ Cal Press, Berkeley

Caswell H (1983) Phenotypic plasticity in life-history traits: demographic effects and evolutionary consequences. Am Zool 23:35–46

Charlesworth B (1980) Evolution in age-structured populations. Univ Press, Cambridge, p 300

Charnov E (1982) The theory of sex allocation. Univ Press, Princeton, p 355

Crandall RE, Stearns SC, Dudman JR (1983) An evaluation of fitness measures for whole organisms: a combinatorial approach. (submitted)

Dawkins R (1982) The extended phenotype. Freeman & Co, San Francisco, p 307

Fisher RA (1930) The genetical theory of natural selection. Univ Press, Oxford

Gross MR, Charnov EL (1982) Alternative male life histories in bluegill sunfish. Proc Natl Acad Sci USA 77:6937–6940

Haldane JBS (1949) Parental and fraternal correlates of fitness. Ann Eugen 14:288–292

Hamilton WJ (1964) The genetical evolution of social behavior. I and II. J Theor Biol 7:1–16; 17–32

Maynard Smith J (1978) The evolution of sex. Univ Press, Cambridge, p 222

Maynard Smith J (1982) Evolution and the theory of games. Univ Press, Cambridge, p 224

McKenzie W, Crews D, Kallman KD, Policansky D, Sohn JJ (1983) Age, weight, and the genetics of sexual maturation in the platyfish *Xiphophorus maculatus*. Copeia 1983:770–773

Oster G, Wilson EO (1978) Caste and ecology in the social insects. Univ Press, Princeton, p 352

Packer C, Pusey AE (1982) Cooperation and competition within coalitions of male lions: kin selection or game theory? Nature (London) 296:740–742

Stearns SC (1982a) The role of development in the evolution of life histories. In: Bonner JT (ed) Evolution and development. Springer, Berlin Heidelberg New York, pp 237–258

Stearns SC (1982b) The emergence of evolutionary and community ecology as experimental sciences. Perspect Biol Med 25:621–648

Stearns SC (1983a) Introduction to the symposium: The interface of life-history evolution, whole-organism ontogeny, and quantitative genetics. Am Zool 23:3–4

Stearns SC (1983b) The evolution of life-history traits in mosquitofish since their introduction to Hawaii in 1905: Rates of evolution, heritabilities, and developmental plasticity. Am Zool 23: 65–76

Stearns SC, Crandall RE (1981) Bet-hedging and persistence as adaptation of colonizers. In: Scudder GGE, Reveal JL (eds) Evolution today. Carnegie Inst, pp 371–383

Stearns SC, Crandall RE (1983) Plasticity for age and size at sexual maturity: A life-history response to unavoidable stress. In: Wootton RJ (ed) Fish reproduction. Academic Press, London New York, pp 13–33

Wright S (1931) Evolution in Mendelian populations. Genetics 16:97–159

Quantitative Characters and Selection in Natural Populations

Genetic Changes in the Morphological Differentiation of Darwin's Ground Finches

T.D. PRICE[1], P.R. GRANT[1], and P.T. BOAG[2]

Speciation involves both the differentiation of populations and the establishment of reproductive isolation. One current debate in evolutionary biology stems from differences in opinion about the extent to which the two processes are coupled, especially when population differentiation is largely a result of natural selection. Reproductive isolation can occur before, concurrently with, or after, any visible divergence. The establishment of reproductive isolation often involves some form of sexual selection or assortative mating. Indeed, sexual selection may often be the driving force in both morphological divergence and reproductive isolation, thereby coupling the two processes (Lande 1981).

Darwin's finches constitute a classical example of adaptive radiation (Lack 1947, Grant 1981a). Thirteen species evolved on the Galápagos Islands from a single ancestral species (Lack 1947, Grant 1981a, 1984a,b), differing most strikingly in beak size and shape. Differences between species in beak size are positively correlated with differences in diet (Abbott et al. 1977, Smith et al. 1978, Grant and Grant 1982, Schluter 1982a). Furthermore, combinations of coexisting species are nonrandom with respect to beak size: species occurring commonly together differ markedly in beak size, and hence in diet, while those occurring rarely together or not at all differ much less (Grant 1981a, Grant and Schluter 1984). These results, and several others (see Abbott et al. 1977, Grant and Abbott 1980, Grant 1981a, 1984a, Schluter and Grant 1982, 1984, Boag and Grant 1984a), suggest that the evolution of species and their coexistence has been determined by processes of adaptation to food supply in allopatry and subsequent competition for food, in sympatry. In addition to these inferences about the role of natural selection, there is one observed and documented case of natural selection on heritable variation in beak size (Boag and Grant 1981).

In this paper we use measurements of genetic and ecological parameters obtained during a 6-year study of two small populations of Darwin's ground finches *(Geospiza)* to characterize and quantify the processes of morphological evolution in this genus. We then use observations on sexual selection to suggest a connection between morphological evolution and reproductive isolation, hence speciation.

1 Division of Biological Sciences, University of Michigan, Ann Arbor, MI 48109-1048, USA
2 Department of Biology, Queen's University, Kingston, Ontario, K7L 3N6, Canada

Population Biology and Evolution
Ed. by K. Wöhrmann and V. Loeschcke

Modes of Speciation

We first present a description of the species and an assessment of the dominant mode of speciation, based on the classification of Templeton (1981). There are six species of *Geospiza*, but here we will consider just four (Fig. 1). The other two species (*G. conirostris* and *G. difficilis*) are largely intermediate to the species we consider; in the morphological space depicted in Fig. 1, *G. conirostris* lies above and *G. difficilis* lies below the *G. fortis* - *G. scandens* positions on a trajectory from *G. fuliginosa* to *G. magnirostris*. The same principles almost certainly apply to all six species.

The differences between species resides in beak size and shape and in body size (Lack 1947). Characteristics of their behavior (Orr 1945, Lack 1947, Ratcliffe 1981), plumage (Lack 1947),polymorphic enzymes (Yang and Patton 1981), and chromosomes (Jo 1976) are all very similar. Differences in electrophoretically detectable enzyme polymorphisms are very small, suggesting recent divergence from a common ancestor (Yang and Patton 1981).

"Divergence" through adaptation (Templeton 1981) appears to have been the mode of speciation, resulting from natural selection on beak and body characters. Barrowclough (1983) proposed "transilience" as an important mechanism of speciation in the finches. Transilience connotes genetic instability, such as might accompany the founding of a new population by a few individuals. It is described as occurring "despite selection" (Templeton 1981) and is generally expected to be very rare in nature, requiring special population structure (notably founder effects) and special genetic systems (Templeton 1980) not known to exist in *Geospiza* species.

Among divergence modes of speciation the overwhelming evidence points to divergence in allopatry (Lack 1947, Grant 1981a). Ford et al. (1973) and Grant and Grant (1979) raised the possibility of sympatric (habitat) speciation, based mainly on the discovery of size or shape biomodality in morphological frequency distributions of certain species. In the second of these studies a detailed ecological investigation has shown that the required stringent conditions are only partly and temporarily met (Grant and Grant 1983). On some large islands species show altitudinal variation (S.J. Millington and D. Schluter, pers. commun.) and latitudinal variation (Lack 1947)

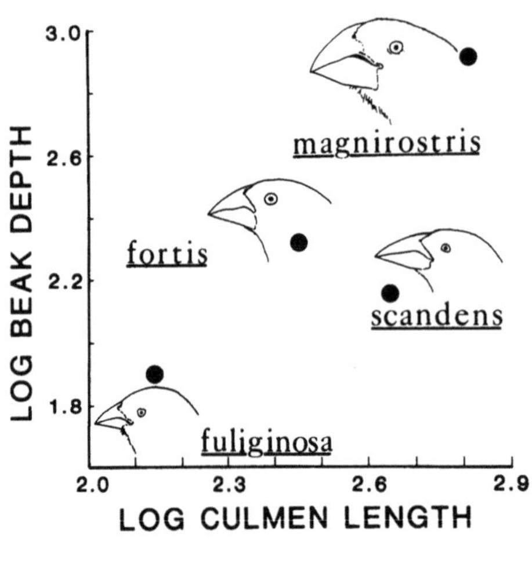

Fig. 1. The four species of *Geospiza*: original measurements were in mm

in morphology, but there is no evidence that the variation is strong enough to have given rise to parapatric speciation on these islands. Extensive clinal differentiation is prevented by migration in the breeding season (gene flow) and in the nonbreeding season, with the result that phenotypes experience similar selective environments (Grant and Boag 1980, Schluter 1982b).

Detailed Observations on Natural Populations

Models dealing with the process of morphological differentiation under natural selection incorporate four parameters that can be measured in nature (e.g., Lande 1980): genetic and phenotypic covariance structure, effective population size, selection and migration. We present estimates of each from the populations of Darwin's finches on I. Daphne Major, Galápagos.

Since 1975 we have been studying populations of the medium ground finch *(G. fortis)* and the cactus finch *(G. scandens)* on Isla Daphne Major (Grant et al. 1975, Boag and Grant 1981, 1984a,b). No other finch species breeds on the island. Finches have been captured in mist nets, measured, uniquely color-ringed and released. Seven measurements have been taken, as described in Grant et al. (1975) and Boag (1983). For simplicity we restrict our analysis to four measurements: weight (in grams), beak length, beak depth, and beak width (all in mm).

Population sizes have been estimated by regularly searching the island for all color-ringed birds. Both species breed in response to rain, which falls in varying quantities between December and May of each year (Grant and Boag 1980). Nest histories have been obtained for many pairs and all nestlings have been color-ringed. Seed abundance has been assessed twice yearly by counting seeds on plants and in the soil in 501 m^2 quadrats (Boag and Grant 1981, 1984b).

1. Variances and Covariances Between Characters. In Table 1 we present phenotypic variances and correlations for the four species. In Table 2 we give three independent estimates of heritabilities and genetic correlations.

G. fortis displays high genetic variances and genetic correlations. The adults in 1981 were from several age cohorts and had been measured over a period of 6 years by several observers. Thus, the estimates of heritability obtained in this year, which are lower than those from the pooled 1976/1978 data, are probably close to a lower bound. However, none of the heritability or correlation estimates are significantly different between the two periods.

Significant heritabilities were not detected in *G. scandens* in 1976 or 1978 (Boag 1983). In 1981 beak characters were found to be significantly heritable (Table 2), although not significantly different from the 1976/1978 measures. No heritability for weight was found. Measuring error contributes relatively largely to the low phenotypic variances in this species (Table 1). Repeatability problems are particularly apparent with the character weight: in 1981, adults, which had been measured over a period of 6 years, were nearly twice as variable as the juvenile cohort, measured over a period of 2 weeks.

Table 1. Descriptive statistics for the four *Geospiza* species discussed in the text

	Means and Variances[a]							
	G. fuliginosa[c]		*G. fortis*[b,d]		*G. scandens*[b,d]		*G. magnirostris*[c]	
	Mean	Var	Mean	Var	Mean	Var	Mean	Var
Weight, g	11.8	1.95	17.4	1.18	20.7	0.63	29.8	0.76
Beak Length, mm	8.2	0.23	10.9	0.56	14.6	0.19	15.3	0.30
Beak Depth, mm	6.6	0.33	9.7	0.80	9.2	0.25	17.0	0.25
Beak Width, mm	6.5	0.32	8.8	0.58	8.6	0.22	14.3	0.31

	Phenotypic Correlations[e]							
	G. fuliginosa				*G. magnirostris*			
	1	2	3	4	1	2	3	4
1. Weight		0.34	0.52	0.75		0.66	0.63	0.71
2. Beak Length	0.74		0.56	0.57	0.52		0.70	0.60
3. Beak Depth	0.75	0.82		0.73	0.47	0.28		0.79
4. Beak Width	0.78	0.82	0.96		0.42	0.33	0.80	
	G. fortis				*G. scandens*			

[a] Means are given in untransformed units. Phenotypic variances (multiplied by 100) and correlations are based on log (base e) transformations

[b] *G. fortis* and *G. scandens* were individuals breeding on Daphne: equal numbers of males and females were included in the sample

[c] *G. magnirostris* and *G. fuliginosa* were immigrants to the island and were mostly unsexed, therefore it was not possible to make the same sex-correction

[d] In *G. fortis* and *G. scandens* the effect of pooling the sexes is to lower variances by approximately 20% for weight and 10% for beak characters and to lower correlations amongst beak characters by 3% and by 7% for beak characters with weight

[e] For the correlations, significance at the 0.05 level is achieved when $r = 0.31$. Sample sizes are 40 for each species

The inconsistent and often low estimates of genetic variance make it impossible to estimate genetic correlations accurately. In *G. scandens,* unlike in *G. fortis,* we found reciprocal parent-offspring cross-covariances of nonhomologous characters to vary substantially (see Van Noordwijk, this volume). Nevertheless it is clear that *G. scandens* has lower genetic variances than *G. fortis. G. scandens* may have lower genetic correlations as well, a possibility supported by low phenotypic correlations (Table 1), if environmental effects are similar in both species.

2. Population Sizes. Estimated population sizes for *G. fortis* and *G. scandens* on I. Daphne are shown in Fig. 2. Population sizes were high at the beginning of the study, but crashed during a drought in 1977 and did not recover until 1983. In early 1979 there were just 21 breeding pairs of *G. scandens* and 28 breeding pairs of *G. fortis* on the island. Droughts are likely to occur fairly frequently (Grant and Boag 1980, Grant 1984c), suggesting that effective population sizes in both species may be low, from a few tens to a few hundred. There is also a real possibility of extinction; indeed

Table 2. Genetic correlations and heritabilities (along the diagonal), based on the arithmetic mean of midparent-offspring crosscovariances of $\log_e$-transformed data[a] (Falconer 1981)

	G. fortis (1976, 1978)[b]				*G. fortis* (1981)[c]				*G. scandens* (1981)[c,d]			
	1	2	3	4	1	2	3	4	1	2	3	4
1. Weight	0.89				0.55				0.02[1]			
2. Beak Length	0.95	0.71			0.80	0.57			–	0.58*		
3. Beak Depth	0.87	0.90	0.82		0.88	0.72	0.74		–	0.28[1]	0.80*	
4. Beak Width	0.94	0.89	0.93	0.92	0.87	0.75	1.00	0.83	–	0.53*	0.55*	0.56*

a All correlations and heritabilities are significant at the 0.01 level except those indicated by * ($p < 0.05$) and [1] (not significant)

b *G. fortis* data, pooled over 1976 and 1978, are taken from Boag (1983): N = 39 families, 82 offspring. No significant *G. scandens* heritabilities were found in these years, making the calculations of genetic correlations meaningless (Boag 1983)

c The 1981 data are based on N = 44 families, 74 offspring for *G. fortis* and N = 29 families, 40 offspring for *G. scandens*

d Lack of significant heritability for weight in *G. scandens* precludes the calculation of genetic correlations between this character and the others

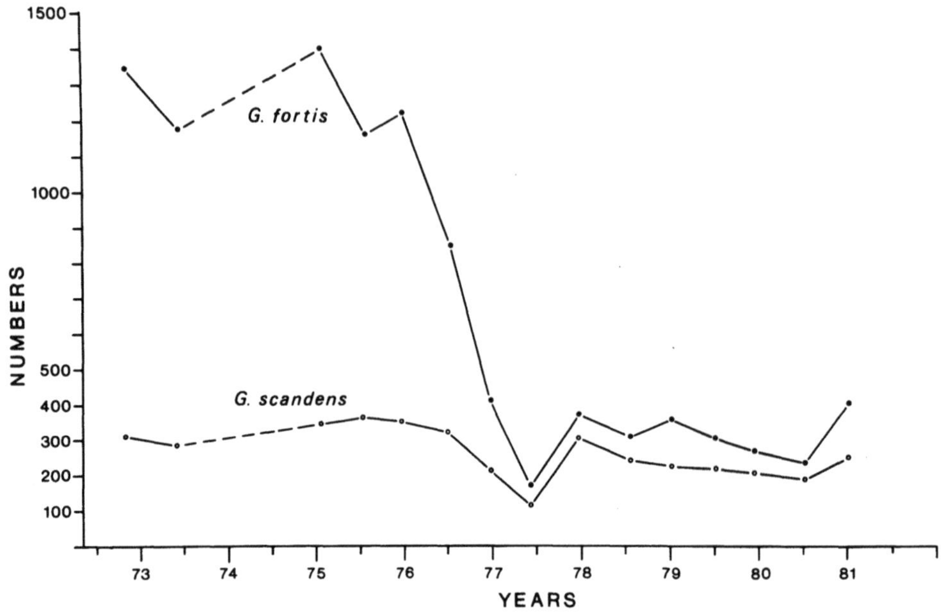

Fig. 2. Estimated population sizes on I. Daphne from 1973 to 1981. The population was not studied in 1974

G. scandens was known to be very rare on I. Daphne earlier this century (Grant 1983b).

3. Immigration and Hybridization. G. magnirostris, G. fortis, and *G. fuliginosa* immigrate regularly from I. Santa Cruz (8 km distant). Immigrant *G. fortis* and *G. fuliginosa* females have been recorded mating with male *G. fortis* on Daphne. The *G. fortis* and *G. scandens* populations also hybridize (Table 3).

Table 3. The number of nests with eggs which had immigrant and/or hybrid parents in four different breeding seasons[a]

	1976	1978	1980	1981
G. fuliginosa	5	3	1	4
G. scandens	1	0	2	0
G. fortis[b] from Santa Cruz	0	0	0	1
G. fortis from Daphne	257	89	101	114

[a] The male was always *G. fortis*, the female is indicated in the Table

[b] The number of these pairings was probably underestimated because female *G. fortis* from Santa Cruz and Daphne cannot always be distinguished by morphological traits

Offspring from all types of hybrid pairs have fledged, but potential breeding success of these offspring is unknown because most have died in the dry years of 1977 and 1980. Two male offspring from *G. fortis* × *G. fuliginosa* pairs in 1978 held territories in 1981, although neither acquired a mate. It appears likely that the hybrids could successfully breed and, in fact, an individual with intermediate morphology, and therefore suspected of being a hybrid, bred in 1976.

The effect of hybridization and immigration is strikingly illustrated among the male *G. fortis* hatched in 1978 (Fig. 3). The three smallest males are from the *G. fuliginosa* × *G. fortis* matings. Two other males fall well outside the general distribution of size and shape (Fig. 3). The parents of both these males were known and fell within the size/shape distributions for their sex, although the large male (male **a**, Fig. 3) was descended from relatively large parents. It is possible that male **b** with the relatively long beak was a first generation hybrid, resulting from a "stolen" copulation by a male *G. scandens*. Male **a**, which resembles a *G. fortis* from I. Santa Cruz, probably represents a second or later generation genetic segregation. In support of this interpretation, male **b** is a significant outlier with respect to shape (using Grubbs's test for the ratio of beak length to beak depth, $p < 0.05$; Dunn and Clark 1974), while male **a** is not a significant outlier with respect to beak depth or length (Grubb's test, $p > 0.1$). Both males sired offspring in 1981 which were morphologically similar to their fathers.

It is clear that hybridization and further breeding of hybrid offspring could produce a diversity of beak and body sizes and shapes in this population.

4. Selection. Our current view of selection in the *G. fortis* population is presented in Fig. 4. Natural selection favoring large size in adults appears to occur frequently, having been observed in 1977 accompanying 85% mortality (Table 4) and again in 1980 accompanying 23% mortality, although the latter case only applies to the cohort hatched in 1978 (Price et al. unpublished). Selection for large size is at least partly

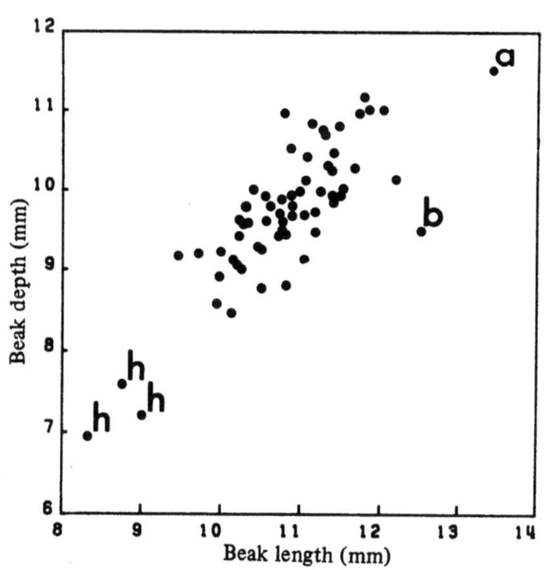

Fig. 3. Scatter plot of beak depth against beak length for all *G. fortis* males known to have hatched on I. Daphne in 1978. Three *G. fortis* – *G. fuliginosa* hybrid offspring are marked *h* and two other unusual males are shown as *a* and *b*

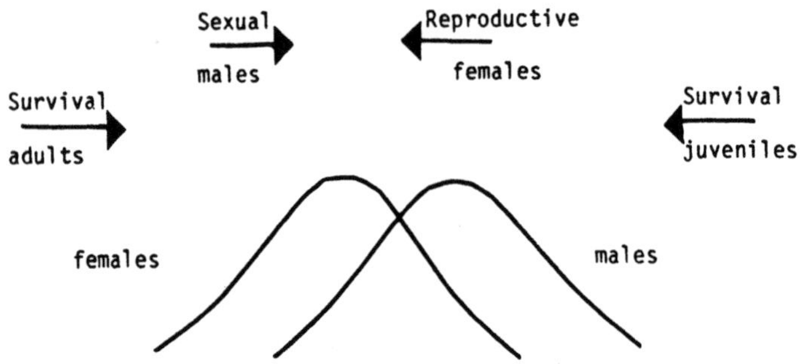

Fig. 4. Forces of selection identified by arrows in the *G. fortis* population and the subsection of the population on which they act. Male and female size-frequency distributions are diagrammed separately. Males average 5% larger than females for most characters

Table 4. The standardized selection gradient on the first principal component, estimated from the covariance matrix of $\log_e$-transformed morphological variables following the methods of Lande and Arnold (1983), during episodes of natural selection in 1977 and sexual selection in 1979 and 1981[a]

		1977 Survival Selection			1979 Sexual Selection			1981 Sexual Selection	
	N	Dir.	Stab.	N	Dir.	Stab.	N	Dir.	Stab.
G. fortis									
Males	198(59)	0.44**	0.10	82(28)	0.34*	0.23	105(46)	0.33**	0.22
Females	66(18)	0.39*	0.02	–	–	–	–	–	–
G. scandens									
Males	74(35)	0.14	0.49[1]	64(20)	0.36[1]	0.30	61(25)	0.28[1]	0.08
Females	34(17)	0.12	0.49*	–	–	–	–	–	–

[a] The first principal component typically accounted for about 80% of the variance in *G. fortis* and 60% in *G. scandens*. In both species it was strongly correlated positively with all the characters and can be considered a measure of overall size. The principal component scores were transformed to zero mean and unit variance prior to each comparison. The 1979 sexual selection compares unmated with mated males. The 1981 sexual selection compares males acquiring mates in that year with those that did not (Price 1984b). Sample sizes refer to the total and the number surviving or the number acquiring mates are in parentheses.
Dir. = Directional; Stab. = Stabilizing; [1] $p < 0.1$; * $p < 0.05$; ** $p < 0.01$

the result of the ability of large finches to crack large seeds (Boag and Grant 1981, Grant 1981b) which are relatively common after droughts (Fig. 5). Mortality among very old adults and/or those that have successfully survived a previous drought may not be selective: drought survivors from 1977 suffered 16% mortality in 1980 and mortality was not differential with respect to size (unpublished observations).

Natural selection may favor small size in juveniles, but this is much less well established. It has been inferred from the discovery that juveniles when fully grown

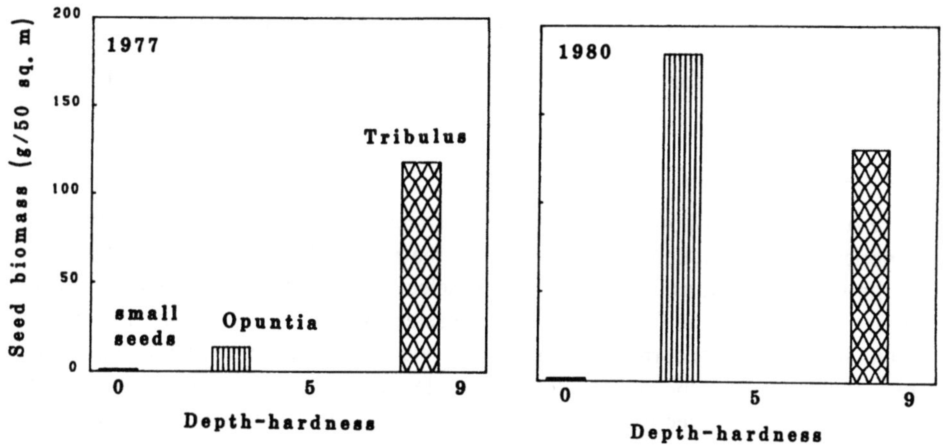

Fig. 5. Histograms of seed biomass in 50 randomly chosen 1 m^2 quadrats in December 1977 and December 1980. The depth-hardness index on the X axis is the square root of the product of seed depth (in mm) and hardness (in kgf required to crack a seed). (Abbott et al. 1977)

are often substantially smaller than the adults from which they were descended. In addition, weak selection favoring small size from age 2-4 months up to age 10 to 12 months has been directly observed (Price and Grant 1984). Small birds may be at an advantage when young because their energy requirements are lower than are those of large birds (cf. Downhower 1976). All juveniles forage on small soft seeds because their beaks are too soft to crush large seeds during their first few months of life (Price and Grant 1984). It may take them over a year to achieve maximal efficiency on such seeds (Millington and Grant 1984).

Two selection pressures which are sex specific have also been identified (Fig. 4). Both arise from causes similar to those giving rise to overall selection. One is termed reproductive selection: some females breed when less than 1-year-old and these females are smaller than those that do not breed (Price 1984a). The other is sexual selection: among males that are 2 or more years old, those that are larger are preferred as mates (first year males have never been recorded breeding) (Table 4). Sexual selection is discussed later in connection with the origin of reproductive isolation.

Selection pressures in *G. scandens* may be weaker than in *G. fortis*, but it is not easy to compare them directly because of smaller sample sizes, lower variances, and lower repeatabilities of characters. Natural selection associated with adult mortality appears to be mainly stabilizing (Table 4). Juveniles are often smaller than the adult population from which they were descended (unpublished observations), but it is difficult to disentangle and allow for effects of incomplete growth. The conditions have not arisen for reproductive selection on females. Directional and possibly stabilizing sexual selection have been identified (Table 4). The coefficients in Table 4 show that selection over single life-history stages in both species can be quite intense.

Reconstructing the Process of Morphological Divergence

The general formula for multivariate evolution of the mean phenotype under selection (Lande 1979) is:

$$G^{-1} \Delta \bar{z} = \nabla \mathrm{Ln}\bar{w} \ , \qquad (1)$$

where $\Delta \bar{z}$ is the vector of mean differences between the characters of two populations or species; G is the genetic variance-covariance matrix which is assumed to be the same for the two populations; and $\nabla \mathrm{Ln}\bar{w}$ is the selection gradient, a vector whose entries can be considered as the forces of directional selection acting directly on each character and independent of correlated responses to selection on other measured characters. Thus, evolutionary responses to directional selection depend on both the intensities of selection acting directly on characters and the genetic variances and covariances between characters. We consider the genetic parameters first.

1. Genetic Variances and Covariances: The Role of Hybridization. Genetic variances and covariances can be affected by a number of factors including mutation, hybridization, selection, and drift. The generally high and always positive genetic correlations between characters presumably reflects both pleiotropic mutations and selection associated with allometric and other functional relationships among characters (e.g., Cheverud 1982). We cannot assess possible differences in selection regimes among species which could affect variances and covariances. However, the difference in genetic variances between *G. fortis* and *G. scandens* can be attributed to differences in frequencies of hybridization (Grant and Price 1981). Both the high genetic variances (Grant and Price 1981) and covariances in the *G. fortis* population are an expected result of hybridization with immigrating *G. fuliginosa*. Hybridization will lead to an increase in genetic variances, as a simple consequence of the populations having different mean values. It is also likely that different alleles will be segregating in the two populations. This is necessary if the hybridization is to facilitate evolution under selection beyond phenotypes observed both in the population receiving the immigrants and in the population from which the immigrants are derived.

We have no direct evidence of successful introgression of *Geospiza* genes into the *scandens* population. In the absence of hybridization, drift could appreciably lower genetic variances (Lande 1980) and this may explain the variable and often low estimates of *scandens* heritabilities.

2. Selection Involved in Species Transitions. If the genetic covariance matrix is known, the net forces of selection involved in species transitions can be reconstructed using Eq. (1) (Lande 1979). In this reconstruction it is important to bear in mind that selection will also be occurring on other unmeasured characters and that the genetic covariance matrix is assumed to have been determined without error and to have remained constant (see below).

The selection gradients for all species transitions are presented in Table 5 (columns 1-6). These are based on the first genetic covariance matrix in the Appendix. Trials with sample matrices showed that the results were fairly robust to fluctuations in genetic correlations, provided all characters are reasonably strongly and positively correlated genetically ($0.95 > r > 0.5$).

Table 5. Selection gradients (divided by 100) for six species transitions based on the *G. fortis* genetic variance-covariance matrix from 1976,and 1978 and for the selection on *G. fortis* in 1977 (Appendix)

	Species Transitions[a]						Observed Selection in 1977[b]		
	fuliginosa – *fortis*	*fuliginosa* – *magnirostris*	*fortis* – *magnirostris*	*fuliginosa* – *scandens*	*fortis* – *scandens*	*magnirostris* – *scandens*	*fortis* Males[c]	*fortis* Females	*fortis* Average
$\sqrt[3]{\text{Weight}}$	-2.02	0.01	2.03	-12.61	-10.58	-12.61	0.11*	0.04	0.08
Beak Length	0.93	-0.54	-1.47	11.16	10.23	11.70	-0.04	-0.01	-0.03
Beak Depth	0.32	1.46	1.14	- 3.92	- 4.24	- 5.38	0.09**	0.06	0.08
Beak Width	0.38	0.56	0.18	1.87	1.49	1.31	-0.05	-0.03	-0.04
Sum of absolute values of vector entries	3.65	2.57	4.82	29.56	26.54	31.00	0.29	0.13	0.23

a Log (base e) transformed data

b Sample sizes for the selection event in 1977 are 198 males and 66 females: 59 males and 18 females survived

c Significance for the male and female selection gradients

* $p < 0.05$; ** $p < 0.01$

The *G. fuliginosa* – *G. magnirostris* transition apparently requires less total selection, as measured here by the sum of absolute values of entries in the selection gradient, than is involved in the transition from either species to the intermediate sized *G. fortis*. However, the differences are small and the results are affected by the magnitude of the genetic variance, which, as discussed above and suggested by phenotypic measures (Table 1), may be unusually large in the Daphne *G. fortis* population.

A large amount of selection is required in transitions from any of these three species to *G. scandens*. A consequence of strong positive genetic correlations between beak length and other characters is that a large amount of antagonistic selection is required to produce the greatly elongated beak of *G. scandens*. Progress under antagonistic selection is facilitated by low genetic correlations between characters (Table 6), but is small when genetic variances are low. The *G. scandens* matrix exhibits both of these characteristics and their effects cancel out.

Table 6. Selection gradients involved in the *G. fortis* – *G. scandens* transition based on (1) *G. fortis* genetic covariance matrix of 1976 and 1978; (2) the *G. scandens* genetic covariance matrix of 1981 (see Appendix)[a]

	(1)	(2)
Beak Length	6.32	5.16
Beak Depth	– 2.64	–0.19
Beak Width	– 1.84	–3.31
Sum of Entries	10.80	8.66

[a] The *G. fortis* matrix for 1981 was not used because it is almost singular (Table 2)

An alternative way to view the difference in required selection intensities is to consider that transitions among *fuliginosa, fortis,* and *magnirostris* occur largely along the first principal component derived from the genetic covariance matrix, whereas transitions to *scandens* occur largely along the second principal component (Jayakar, pers. comm.). The first principal component typically accounts for 4–5 times more of the genetic variance than does the second (Boag 1983, Grant 1983b).

Evolution Under Selection and Drift

The potential progress of morphological evolution under the directional selection pressures described above can be estimated by using observations on present environmental conditions and the ecology of the finches. The overall inferred stabilizing selection on the *G. fortis* population on I. Daphne can be transformed into directional selection by relaxation of selection at one stage of the life cycle or intensification at another (Fig. 4). Given the inherent density dependence of mortality associated with starvation, the two processes may be coupled: for example, the strong directional selection associated with high adult mortality in 1977 resulted in very low densities of finches and this was followed by no selection on the 1978 cohort of juveniles (Price and Grant 1984).

The selection gradient (Lande and Arnold 1983) associated with selection for large size during the 1977 drought is presented in Table 5 (columns 7–9). The direction of selection was most similar to the selection gradient in the *G. fortis* - *G. magnirostris* transition. The vector correlation between the selection gradient estimated for this species transition and the selection gradient observed (averaged across the sexes) is r = 0.88. This fits with ecological observations: surviving *G. fortis* fed on seeds that *G. magnirostris* also consumed although the *G. magnirostris* were much more efficient at exploiting them (Grant 1981b). The intensity of selection, as measured by the sum of the absolute values of vector entries, was 5% of the total selection required in the transition from *G. fortis* to *G. magnirostris* (Table 5).

For continued evolution, changed environmental conditions must not be so severe that the population goes extinct. Histograms of seed abundance at the end of the drought years in 1977 and 1980 (Fig. 5) illustrate that populations may not be able to respond evolutionarily to severe changes in environmental conditions. There were basically three seed types on the island. The several types of small seeds are exploited most efficiently by *G. fuliginosa* (Schluter 1982b), but this species is unable to exploit the next largest seed, *Opuntia echios* (Schluter 1982b; unpublished observations). *G. fuliginosa* numbers declined to near zero in these 2 years. In general, gaps of this magnitude in the seed depth-hardness distribution are not bridged by population variation either in the finches or seeds. An exception to this is the pair of seed types *Tribulus cistoides* and *Opuntia echios* (Fig. 5). Finches able to handle *Opuntia* seeds (i.e., most *G. fortis* individuals) are able to extract seeds from a *Tribulus* fruit. Efficiency at dealing with *Tribulus* increases at least up to the size of *G. magnirostris* (Grant 1981b). However, *Tribulus* is a recent introduction to the Galápagos (Porter 1967) and played no part in the adaptive radiation.

One can imagine an "adaptive surface" of the environment, with a population held by stabilizing selection around one seed type (e.g., *G. fuliginosa* on small seeds), and a "valley" of low fitness between seed types. This is similar to a situation envisaged by Wright (e.g., 1982), who suggested that genetic drift may be important in moving a population mean across an adaptive valley to a new adaptive peak against the forces of selection, i.e., to a new regime of stabilizing selection. This process has been modelled for polygenic characters by Lande (1976). Genetic drift can counter the force of stabilizing selection, but this can only realistically occur when valleys are shallow and peaks are not greatly separated. Although the Daphne populations are certainly of the size where drift could be important, measurements of the intensity of selection (Tables 4 and 5) suggest that valleys are deep, i.e., selection can be very strong and transitions through drift are effectively impossible.

On larger islands in the archipelago seeds are more diverse and adaptive peaks may be closer (Abbott et al. 1977, Schluter and Grant 1984), but it is doubtful if population sizes are ever sufficiently reduced for drift to be important. Movement between peaks on one island are more easily effected by invasion of a partly differentiated species from another island. Thus, small islands, such as Daphne are potentially important in that their reduced diversity of seed types, selects for intermediate morphologies. When birds with these intermediate morphologies disperse to a large island they will not be adapted to a peak there, but selection on them and their descendents will complete the process of a peak shift. The strength of observed selection and the high observed genetic variances implies that periods of directional evolution of new optima may be very rapid (Boag and Grant 1981). We suggest that this mechanism of evolution accounts for the *G. fuliginosa* – *G. fortis* – *G. magnirostris* transitions.

Other mechanisms may also operate. For example, the long beak of *G. scandens* is used for foraging on *Opuntia* flowers and for opening *Opuntia* fruits (Grant and Grant 1981). Since the other three species can all exploit these resources to some extent, the long beak could have evolved as a response to continued directional selection for more efficient exploitation of resources, i.e., there was no valley to cross.

A third possibility for generating new morphological types involves predator-prey coevolution. The *G. magnirostris* on I. Floreana was particularly large and exploited the largest *Opuntia* seed on the archipelago before going extinct in the 19th century. This seed may have evolved in the Galápagos (Wiggins and Porter 1971) possibly in response to predation pressure (Grant and Grant 1982).

The Establishment of Reproductive Isolation

Experiments have shown that finches of both sexes distinguish between conspecifics and heterospecifics on the basis of body size and beak characteristics (Ratcliffe and Grant 1983a,b). Thus, the characters involved in ecological isolation are also involved in reproductive isolation. This suggests that morphological evolution and reproductive isolation may have been coupled during speciation events.

Consideration of observed sexual selection suggests how this could have been brought about. Sexual selection is likely to be generally detectable whenever the sex ratio is unequal, either because males remain unmated when they are in excess or because females mate polygamously when they are in excess. The former situation has been present on Daphne since the drought of 1977 and sexual selection has been detected (Table 4).

Sexual selection for large size could be related to the possible dominance of large males (i.e., male-male competition), but other causes of the sexual selection could also be involved (Price 1984b). In both species the pattern of sexual selection was similar to previous natural selection in that males of large size were favored (Table 4). One interpretation of this is that males with the highest survival probability over the period of natural selection were favored as mates (Price 1984b). Females may identify such high quality males by cues other than morphology. In the *G. fortis* population two such cues were identified: plumage variation and territory size. However, after these cues were statistically removed in a multiple regression analysis, evidence remained of substantial selection directly on body size (Price 1984b).

Although other unmeasured cues which were correlated with body size may have been present, it is possible that females chose males directly on the basis of either body size or their beak size/shape. Such female preferences may be genetic or transmitted through imprinting (Immelman 1975, Cooke et al. 1976). In either case, the preferences are expected to become correlated with morphology (e.g., Lande 1981), so that selection in both sexes will result in a correlated response in mate preferences. Thus, long continued evolution under directional selection will lead to automatic reproductive isolation. If mating preferences are genetically based, isolation may be asymmetric, i.e., females in both populations prefer one sort of male, but if they are based on imprinting there is no reason to expect asymmetry.

Discussion

In this paper we have extended Lack's (1947) description of the adaptive radiation in Darwin's finches by applying quantitative genetics techniques to the problem of estimating the changes occurring during speciation. To do this we have used results from a study of two populations. Similar results have been obtained in a second population study on the Galápagos (Grant and Grant 1983, Grant 1983b), encouraging us to believe that the results are general. Thus, hybridization is now believed to occur episodically on I. Genovesa between *G. conirostris* and two sympatric congeners, *G. magnirostris* and *G. difficilis* (Grant and Grant 1982, 1983). Estimates of genetic variances and covariances are high for this population (Grant 1983b). Selection has not been directly identified because of difficulties in following the fates of all individuals, although it has been inferred (Grant and Grant 1983).

The intensity of natural selection observed in the finch populations on I. Daphne implies that evolution of mean phenotypes could have occurred rapidly and that stabilizing selection is powerful in maintaining mean positions. Observations of sexual selection show that natural selection on morphology and speciation through the establishment of reproductive isolation may be tightly coupled. It is worth emphasizing that Darwin's finches are unusual in that the same characters are involved in both ecological and reproductive isolation between species. In many avian genera, for instance those in which plumage patterns serve to distinguish species, sexual selection may lead to reproductive isolation, with adaptive (naturally selected) differences being rather incidental.

On the other hand we are unable to ascribe an important role to genetic drift. Drift possibly reduces genetic variances. This will generally act to slow rather than enhance any morphological evolution in small populations not subject to hybridization.

Our results and interpretations are largely in accord with a theoretical analysis of morphological divergence in allopatry by Lande (1980). Based on measurements of selection in nature and estimated mutation rates in polygenic characters, Lande concluded that genetic variance could be maintained and substantial evolution could occur in small populations. Our results are consistent with the view (Grant and Price 1981) that despite more intense selection and smaller population sizes than Lande assumed in his models, high genetic variance can be maintained through hybridization.

A major purpose of genetic studies of natural populations is to understand the process of evolution (e.g., see papers in Karlin and Nevo 1976, Christiansen and Fenchel 1977). Application of new quantitative genetics techniques and theory has enabled us to draw nontrivial and fairly robust conclusions about the process of morphological evolution in a group of Darwin's finches. Newly developed and relatively untested quantitative genetic techniques should be applied to natural populations with caution, of course (Van Noordwijk, this volume). We hope our results and conclusions stimulate refinement and elaboration of those techniques, and provide encouragement for others to apply them to different organisms.

Acknowledgements. This article was prepared while the senior author was supported by Biomedical Review Council funds at the University of Chicago. Field research was supported by National Science Foundation grants DEB77-23377 and DEB79-21229. We thank R. Lande for showing us an unpublished manuscript and for extensive discussion, D. Schluter, P.E. Smouse, and A.J. van Noordwijk for excellent comments on the manuscript, and the Deutsche Forschungsgemeinschaft for enabling us to attend the symposium. Mao kindly typed the first draft.

Appendix. We present the $\bar{z}$ values and genetic variance-covariance matrices used in the calculation of the selection gradients

Mean ($\bar{z}$) Values

	G. fuliginosa	*G. fortis*	*G. magnirostris*	*G. scandens*
$\sqrt[3]{\text{Weight}}$	0.8240	0.9210	1.1242	1.0016
Beak Length	2.1647	2.3715	2.7218	2.6763
Beak Depth	1.9238	2.2373	2.8220	2.2164
Beak Width	1.9195	2.1573	2.6539	2.1564

Genetic Covariance Matrices

	1976/1978 *(G. fortis)*				1981 *(G. scandens)*			
	1.	2.	3.	4.	1.	2.	3.	4.
1. $\sqrt[3]{\text{Weight}}$	0.0852				–			
2. Beak Length	0.1501	0.2930			–	0.0829		
3. Beak Depth	0.1927	0.3697	0.5760		–	0.0247	0.0947	
4. Beak Width	0.1776	0.3120	0.4570	0.4193	–	0.0378	0.0418	0.0609

References

Abbott I, Abbott LK, Grant PR (1977) Comparative ecology of Galápagos ground finches (*Geospiza* Gould): evaluation of the importance of floristic diversity and interspecific competition. Ecol Monogr 47:151–184

Barrowclough GF (1983) Biochemical studies on microevolutionary processes. In: Brush AH, Clark GA (eds) Perspectives in ornithology. Univ Press, Cambridge

Boag PT (1981) Morphological variation in the Darwin's finches (Geospizinae) of Daphne Major island, Galápagos. Unpubl PhD Thes, McGill Univ, Montreal

Boag PT (1983) The heritability of external morphology in the Darwin's finches (Geospizinae) of Daphne Major Island, Galápagos. Evolution 37:877–894

Boag PT, Grant PR (1981) Intense natural selection in a population of Darwin's Finches (Geospizinae) in the Galápagos. Science 214:82–85

Boag PT, Grant PR (1984a) The classical case of character release' Darwin's finches *(Geospiza)* on Isla Daphne Major, Galápagos. Biol J Linn Soc 21 (in press)

Boag PT, Grant PR (1984b) Darwin's finches *(Geospiza)* on Isla Daphne Major, Galápagos: breeding and feeding ecology in a climatically variable environment. Ecol Monogr 54 (in press)

Cheverud JM (1982) Phenotypic, genetic and environmental integration in the cranium. Evolution 36:499–516

Christiansen FB, Fenchel TM (eds) (1977) Measuring selection in natural populations. Springer, Berlin Heidelberg New York

Cooke F, Finney GH, Rockwell RF (1976) Assortative mating in the snow geese *(Anser caerulescens)*. Behav Genet 6:127–140

Downhower JF (1976) Darwin's finches and the evolution of sexual dimorphism in body size. Nature (London) 263:558–563

Dunn OJ, Clark VA (1974) Applied statistics: analysis of variance and regression. Wiley, New York

Falconer DS (1981) Introduction to quantitative genetics, 2nd edn. Ronald, New York

Ford HA, Parkin DT, Ewing AW (1973) Divergence and evolution in Darwin's finches. Biol J Linn Soc 5:289–295

Grant BR, Grant PR (1979) Darwin's finches: population variation and sympatric speciation. Proc Natl Acad Sci USA 76:2359–2363

Grant BR, Grant PR (1981) Exploitation of *Opuntia* cactus by birds on the Galápagos. Oecologia 49:179–187

Grant BR, Grant PR (1982) Niche shifts and competition among Darwin's finches: *Geospiza conirostris* and congeners. Evolution 36:637–657

Grant BR, Grant PR (1983) Fission and fusion in a population of Darwin's finches: an example of the value of studying individuals in ecology. Oikos 41:530–547

Grant PR (1981a) Speciation and the adaptive radiation of Darwin's finches. Am Sci 69:653–663

Grant PR (1981b) The feeding of Darwin's finches on *Tribulus cistoides* (L.) seeds. Anim Behav 29:785–793

Grant PR (1984a) Recent research on the evolution of land birds on the Galápagos. Biol J Linn Soc 21 (in press)

Grant PR (1984b) The endemic landbirds. In: Perry RL (ed) The Galápagos. Key environment series, vol I. Pergamon Press, Oxford, pp 175–189

Grant PR (1983a) The role of interspecific competition in the adaptive radiation of Darwin's finches. In: Bowman RI, Berton M, Leviton AE (eds) Patterns of evolution in Galápagos organisms. AAAS Spec Publ, AAAS Pac Div, San Francisco, Cal, pp 187–199

Grant PR (1983b) The inheritance of size and shape in a population of Darwin's Finches. Proc R Soc London Ser B 220:219–236

Grant PR (1984c) Past and present climatic fluctuations of the Galápagos islands and their influence on Darwin's Finches. Ornithol Monogr (in press)

Grant PR, Abbott I (1980) Interspecific competition, island biogeography and null hypotheses. Evolution 34:332–341

Grant PR, Boag PT (1980) Rainfall on the Galápagos and the demography of Darwin's finches. Auk 97:227–244

Grant PR, Price TD (1981) Population variation in continuously varying traits as an ecological genetics problem. Am Zool 21:795–811

Grant PR, Schluter D (1984) Interspecific competition inferred from patterns of guild structure. In: Strong DR, Simberloff D, Abele LG (eds) Ecological communities: conceptual issues and the evidence. Univ Press, Princeton, NJ

Grant PR, Smith JNM, Grant BR, Abbott IJ, Abbott LK (1975) Finch numbers, owl predation and food supply on Isla Daphne Major, Galápagos. Oecologia 19:239–259

Immelmann (1975) Ecological significance of imprinting and early learning. Annu Revu Ecol Syst 6:15–37

Jo N (1976) Karyotypic analysis of Darwin's finches. Unpul MSc Thes, San Francisco State Univ, San Francisco

Karlin S, Nevo E (eds) (1976) Population genetics and ecology. Academic Press, London New York

Lack D (1947) Darwin's finches. Univ Press, Cambridge

Lande R (1976) Natural selection and random genetic drift in phenotypic evolution. Evolution 30:314–334

Lande R (1979) Quantitative genetic analysis of multivariate evolution, applied to brain: body size allometry. Evolution 33:402–416

Lande R (1980) Genetic variation and phenotypic evolution during allopatric speciation. Am Nat 116:463–479
Lande R (1981) Models of speciation by sexual selection on polygenic traits. Proc Natl Acad Sci USA 78:3721–3725
Lande R, Arnold SJ (1983) The measurement of selection on correlated characters. Evolution 37:1210–1226
Millington SJ, Grant PR (1984) The breeding ecology of the cactus finch, *Geospiza scandens* on Isla Daphne, Galápagos Ardea (in press)
Orr, RT (1945) A study of capative Galápagos finches of the genus *Geospiza*. Condor 47:177–201
Porter DM (1967) Another *Tribulus* adventive in the New World. Rhodora 69:455–456
Price TD (1984a) The evolution of sexual size dimorphism in Darwin's finches. Am Nat 123: 500–518
Price TD (1984b) Sexual selection on body size, plumage and territory variables in a population of Darwin's finches. Evolution 38 (in press)
Price TD, Grant PR (1984) Life history traits and natural selection for small body size in a population of Darwin's finches. Evolution 38 (in press)
Ratcliffe LM (1981) Species recognition in Darwin's Ground finches (*Geospiza*, Gould). Unpubl PhD Thes, Mc Gill Univ, Montreal
Ratcliffe LM, PR Grant (1983a) Species recognition in Darwin's finches. I. Discrimination by morphological cues. Anim Behav 31:1139–1153
Ratcliffe LM, PR Grant (1983b) Species recognition in Darwin's finches. II. Geographic variation in mate preference. Anim Behav 31:1154–1165
Schluter D (1982a) Distributions of Galápagos ground finches along an altitudinal gradient: the importance of food supply. Ecology 63:1504–1517
Schluter D (1982b) Seed and patch selection by Galápagos ground finches: relation to foraging efficiency and food supply. Ecology 63:1106–1120
Schluter D, Grant PR (1982) The distribution of *Geospiza difficilis* in relation to *G. fuliginosa* in the Galápagos Islands: tests of three hypotheses. Evolution 36:1213–1226
Schluter D, PR Grant (1984) Determinants of morphological patterns in communities of Darwin's Finches. Am Nat 123:175–196
Smith JNM, Grant PR, Grant BR, Abbott IJ, Abbott LK (1978) Seasonal variation in feeding habits of Darwin's ground finches. Ecology 59:1137–1150
Templeton AR (1980) Macroevolution (book review). Evolution 34:1224–1227
Templeton AR (1981) Mechanisms of speciation – a population genetic approach. Annu Rev Ecol Syst 12:23–48
Wiggins IL, Porter DM (1971) Flora of the Galápagos Islands. Univ Press, Stanford, Cal
Wright S (1982) Character change, speciation, and the higher taxa. Evolution 36:427–443
Yang SY, Patton JL (1981) Genic variability and differentiation in Galápagos finches. Auk 98: 230–242

Quantitative Genetics in Natural Populations of Birds Illustrated with Examples from the Great Tit, *Parus major*

A.J. VAN NOORDWIJK[1]

Introduction

There are many reasons to study quantitative genetics in natural populations, and birds have the great advantage that pedigrees can be determined easily in natural populations. Therefore, it is not surprising that nearly all quantitative genetic studies where the observations were made in the natural environment of the study species concern birds. This is important because the environment in which the observations have been made is very critical in quantitative genetic studies. At the same time, natural environments and natural breeding systems pose several methodological problems and in general the conditions are not optimal for efficient estimation of parameters.

It is therefore good to examine the reasons for studying quantitative genetics in natural populations more closely. Depending on the aim of the study different trade-offs between practical and theoretical considerations suggest themselves. Some of the initial questions might be:

1. One wishes to study natural selection and genetic variation simultaneously. In other words, one wishes to study the interactions between the two elements of the microevolutionary process.
2. In evolutionary ecology one wishes to give estimates for the potential rates of change of the traits under investigation.
3. One wishes to study the effects of variable environments on the ontogeny of traits and the possible ways in which this process is affected by genetic variation.
4. One wishes to test certain theories about the process of evolution, for example that long periods of stasis are intermingled with relatively short periods in which rapid evolution occurs. This may be explained either from lack of the right sort of genetic variation being present, in which view the leaps occur immediately after the occurrence of mutations with a large effect on, e.g. growth or allometric rates, or it may be explained from discontinuities in the selective forces operating on the traits.

In the latter case one expects that additive genetic variation is present for many traits, but that the mean value is at some sort of local optimum and that therefore no systematic changes take place. If it is generally true that there is a large potential to react to changes in environmental conditions in a microevolutionary way, it could

1 Department of Population and Evolutionary Biology, University, 3504 CH Utrecht, and Institute for Ecological Research, 6666 GA Heteren, The Netherlands

Population Biology and Evolution
Ed. by K. Wöhrmann and V. Loeschcke

have great consequences for the effects of human interference with the environmental conditions for all species on earth.

The General Philosophy

The methods of quantitative genetics were developed for use in animal and plant breeding. In many respects the application of these methods relies heavily on the elimination of potential complications through the experimental design. With data from natural populations, factors such as environmental heterogeneity or genotype-environment correlations give rise to problems. The best way to avoid these difficulties is to return to the basic ideas underlying the theory of quantitative genetics. In its most elementary form one could say that there are two main causes for a resemblance between relatives; namely, relatives share genes and/or environments. This makes it possible to make statements about the role of genetic variation by first measuring the resemblance between relatives and then explicitly testing the hypothesis that the resemblance is not due to sharing environments. There are several ways in which this can be done.

Using the available knowledge about the trait involved and about the species studied, it is possible to devise several controls (Table 1). In general, these controls consist of calculating a resemblance between individuals that share environmental conditions in the same way as relatives, but being unrelated they do not share genes. The bird breeding in the next nestbox in the same year can be a suitable control for the spatial and the year-to-year variation in environmental conditions. The same method is used in the calculation of several repeatabilities for reproductive traits; one may calculate the repeatability of pairs, of females with different males, of males with different females,

Table 1. Explicit tests of familial resemblance: genetic or environmental?

1. Comparing the repeatability of:	Pairs Females with different males Males with different females Territories with different birds
2. Calculating "control resemblances":	Regression of offspring on next-door neighbors Regression of sons on fathers (for female traits)
3. Calculation of discriminative resemblances: such as:	regression on $\frac{\text{Maternal}}{\text{Paternal}}$ grandparent correlation between $\frac{\text{Maternal}}{\text{Paternal}}$ half-sibs
4. Checking the consistency of different types of family relationships such as:	Comparing parent-offspring and sib-analysis Comparing analysis of half-sibs and full-sibs, etc.
5. Manipulation:	Cross-fostering Controlling environmental variation

or of nest boxes with different occupants. In the study of the Great Tit *(Parus major)* it was shown that males have no effect on the clutch size or on the date of laying of their partner by calculating the relevant repeatabilities (Van Noordwijk et al. 1981a,b). It was then possible to attribute the traits to the male and use the father-son regressions as a control for the mother-daughter regressions. This supposes that the female partner of the son is not related to the mother, and hence, that a resemblance is caused by nongenetic factors. In the single case, where a positive and significant father-son regression was found, this could be attributed to the fact that the females involved were related or even the same individuals in several mother-son pairs (Van Noordwijk et al. 1981a).

In situations where a small effect of sharing environments is found, the difference between the resemblance of relatives and the resemblance in the controls can be used as a heritability estimate (Van Noordwijk et al. 1981a,b). The use of such a difference results in a larger standard error of the estimate and is likely to give a somewhat conservative estimate of the heritability if there are shifts in the population mean due to shifts in the genotypic mean value of the population. Nevertheless, the estimate based on such a difference is more realistic and closer to the original meaning of heritability than an estimate that also contains elements of resemblance caused by sharing of environments by relatives.

In this way it was possible to demonstrate that genetic variation is present in Dutch Great Tit populations for clutch size, egg size, body size, bill dimensions, and date of first egg laying. All of these are traits of ecological importance to the individual and are likely to be subjected to natural selection (Van Noordwijk et al. 1980, 1981a,b,c, 1983). Very similar results have been obtained in a British population (Jones 1973, Garnett 1981) and a Finnish population (Ojanen et al. 1979). The demonstration that genetic variation is present is a first step toward an integrated study of the genetic and the ecological factors having an effect on the phenotypes. It opens up the road for a second type of study where the two sorts of relevant processes are intertwined and where knowledge from both disciplines is necessary.

In this contribution I would like to illustrate some problems and some possibilities encountered in this second type of study where the questions are no longer either genetic or ecological. I shall do so with four examples, all taken from the Great Tit studies and each of which illustrates a point that I would like to make in setting out the perspectives for integrated research. The first two cases are more or less complete, the last two cases are (at present) open-ended. The four cases are: (1) The heritability of dispersal, as an illustration of the difficulties in identifying the proper null hypothesis and as an illustration of resemblance of relatives that seems to be entirely caused by the sharing of environments. (2) The within-clutch variation in egg size, as an illustration of what problems are encountered in trying to investigate the environmental variation in a case where about 70% of the phenotypic variation is genetic. (3) The genetic correlation between egg size and body size, as an illustration of interactions between variables that are too weak to be picked up directly, but can nevertheless frustrate the calculation of quantitative genetic parameters, which are very sensitive to this kind of interaction. (4) The heritability of nestling body size, where there are several complications and discrepancies between what is seemingly the case and what is theoretically possible. Finally, I will try to indicate the common aspects of these four cases. At the same time I will point out the direction in which further developments are expected.

Case 1: The Heritability of Dispersal

When the coordinates of the nest boxes where a bird is born and where it breeds are known, it is possible to calculate the distance moved. If similar data are available for a number of parent-offspring pairs, it is possible to calculate the resemblance in the distance moved by parents and offspring. Greenwood et al. (1979) performed such an analysis and found four positive values for the regressions of either daughters or sons on either fathers or mothers. Their interpretation that this resemblance had a genetic basis was criticized by Van Noordwijk (1984) who argued that the resemblance was likely to be a property of the nest boxes in which the parents bred rather than a property of the birds involved.

This can be demonstrated by a simple Monte Carlo simulation in which birds move randomly from the box where they were born to any other box in the study area. The resulting distances moved by "parents" and "offspring" show patterns of correlation that are very similar to the real data. To understand the cause of these patterns, it is important to remember the source of the data. It is possible to describe the full set of possible observations, namely, the distances between every pair of nest boxes. Given that a bird is born in a particular box, all possible observations for that bird belong to a subset, namely, all the distances from that particular nest box to other nest boxes. These subsets can be very different between nest boxes, which is to some extent dependent on the geography of the study area. If one starts at the other end of the movement and describes the subset of possible observations on birds moving *to* that box, the result is similar in that different boxes have different distributions of possibly observed distances. In the case of random movement, the subsets of going *to* and coming *from* a box are identical. If movements are not entirely random, e.g., because shorter distances are more likely, the two distributions for each nest box are still similar. Because the offspring are born in the box where the parents breed, a similarity in random distances moved *to* and moved *from* a box will result in similarities between parents and offspring in their distances moved between site of birth and site of breeding.

Case 2: The Within-Clutch Variation in Egg Size

The size of eggs is a property of the female by which they are laid. Given that a large proportion of the phenotypic variation in egg size is genetic (Jones 1973, Van Noordwijk et al. 1981c, Ojanen et al. 1979), it would seem that the within-clutch variation provides a good opportunity to elucidate the effect of environmental factors on egg size. By regarding the variation within the clutch, the genetic component is excluded and therefore the possibility is excluded that differences in mean egg size between two periods with a different temperature could be the result of genetically different females laying during these periods.

Despite this increased sensitivity, the analysis introduced several problems that were not anticipated. The clearest picture of the effect of temperature is obtained by simply plotting the mean size of all eggs laid on a particular day against the temperature variable (the mean of the temperatures of the preceding 3 days). It is evident from Fig. 1 that the results for the 2 years are as different as the temperatures that were observed in those years. This presentation of the data makes no allowance for the possibility that different females were laying at different temperatures. Our first attempt to eliminate this effect, following Jones (1973), was based on the size of the egg relative to the mean egg size in the clutch. This relative size was then plotted against the absolute temperature. The results were very inconsistent, which is due to the effect of temperature on egg size, and thereby on the mean egg size in the clutch. In other words; if eggs are large when temperature has been high in the preceding days, an egg laid at 8 °C will be relatively large in a clutch laid at lower temperatures, while the same egg will be relatively small in a clutch laid at higher temperatures. This problem could, in principle, be overcome if only clutches that were started on the same day were used, but this reduces sample sizes to unacceptable levels. There is another way to overcome the problem, namely, by expressing the size of the egg relative to the size of the egg laid on the previous day. The difference in egg size can be related to the difference in temperature. There are, however, problems with scaling. There are strong correlations between mean and variance for egg sizes within clutches, in that clutches with low means have high variances (Van Noordwijk et al. 1981c), while there is an opposite trend in the variance with the mean size of all eggs laid on a single day. To avoid such problems with scaling, the proportion of females laying a larger and the proportion of females laying a smaller egg than they laid on the previous day, was plotted against the difference in temperature (Fig. 2). The two proportions are complementary apart from a third category where the difference in egg volume is less than 0.03 cm^3, which is the precision of measurement. It is evident that for differences in temperature of more than 2 °C all females change their egg volumes in the same direction. The temperature variable used in this analysis is the mean temperature of the 3 days preceding the day of laying. This effect of temperature on egg

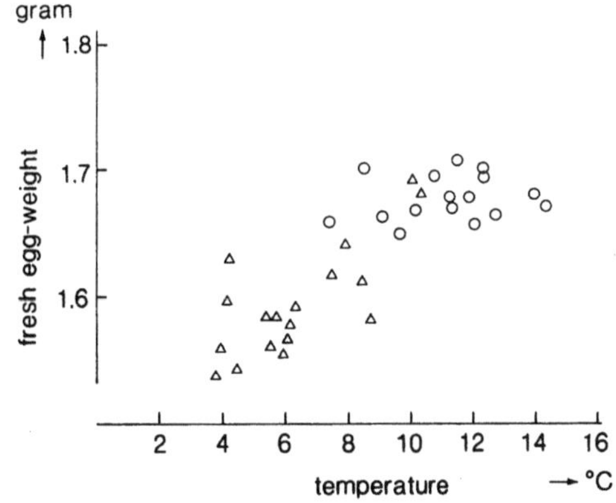

Fig. 1. The mean size of all eggs laid on a particular day, as a function of the average ambient temperatures in the preceding 3 days. ○ = data for 1978, $r = 0.16$ (N = 16, NS), △ = data for 1979, $r = 0.74$ (N = 18, $p < 0.01$)

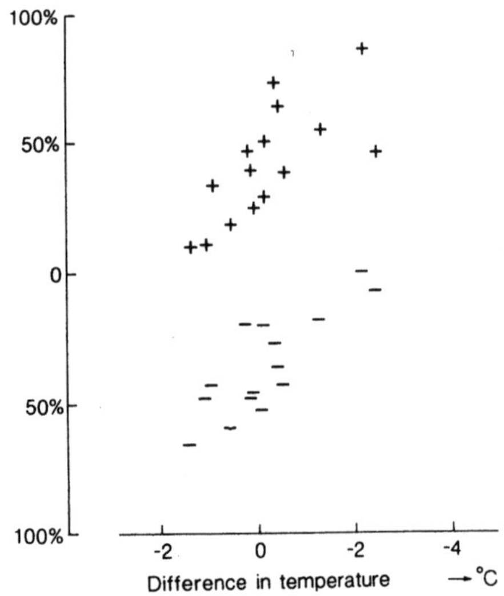

Fig. 2. The proportion of females laying eggs that are larger (+) or smaller (–) than the eggs laid by the same female on the previous day as a function of the difference in the temperature variable (see text)

size is only found at temperatures below 10 °C (see Fig. 1). This leads to an explanation of patterns in the repeatability for egg size and probably in the heritability as well. Although the differences are not statistically significant, the heritability estimates based on egg sizes in both first and second clutches are higher than those based on first clutches only. When temperatures below 10 °C occur during the laying period of first clutches, the repeatability comparing egg size in first and in second clutches is appreciably lower than in years with higher temperatures during the laying period of first clutches. Furthermore, the fact that the effect of temperature on egg size is restricted to low temperatures also explains why the repeatability estimates comparing egg size in second clutches between years is higher than the repeatabilities comparing egg sizes in first clutches, since temperatures are higher than 10 °C in the laying period of second clutches. It is thus possible to explain a part of the variability in egg parameters using quantitative genetic theory from variation in environmental parameters that have an effect on phenotypic variation. This may provide a good basis for further study of interactions and reaction norms in these traits. It is one of the axioms of quantitative genetics that relative estimates, such as heritabilities and repeatabilities, are only valid for the population and the environment in which they have been measured, but to my knowledge this case is one of the first where variation in these parameters can be interpreted and are understood well enough to enable prediction.

Case 3: The Genetic Correlation Between Egg Size and Body Size

Inferences about the genetic nature of variation in quantitative traits are based on the results of a statistical analysis. It is generally not known which genes are involved. It is, therefore, also unknown whether two quantitative traits are affected by the same genes. It is possible, however, to estimate the degree to which two quantitative traits are affected by the same set of genes. This is expressed in the genetic correlation. By analogy to the division of the total phenotypic variance into a genetic and an environmental component, the phenotypic correlation between two traits can also be split into a component attributable to genetic effects and a component that represents the similarity in the effect of environmental factors on the two traits. Genetic correlations are important in predicting the modification of traits as a result of selection. If selection acts on variation in one trait, the genetically correlated traits are expected to change along with the primary trait. In artificial selection, this property of the genetic correlation can be used to estimate them. In an analysis of data on natural selection, the fact that changes in the distribution of a trait may have been caused by selection on a correlated trait makes it necessary to provide evidence that the trait under investigation is also the target of selection. Thus, genetic correlations are not only important from the point of view of coevolution, but also in any study of the effects of natural selection acting on quantitative traits.

Unfortunately, the problems encountered in the calculation of genetic correlations on data from natural populations may well be greater than the problems which can be solved with knowledge of the values. There are two ways to estimate genetic correlations; through the correlated response to selection, i.e., the change in a trait as a result of selection on a genetically correlated trait and through the cross-covariance, i.e., the covariance between one trait in one individual and the other trait in a relative. Using the first technique, it is common practice to analyze the two different correlations resulting from selecting for each of the traits. Using the second technique, it is less common to analyze the correlations resulting from the two cross-covariances separately. Instead either the geometric or the arithmetic mean of the two is considered. The formula for the standard error of a genetic correlation applies to either mean of the two estimates (Reeve 1955). However, the assumptions underlying the concept of genetic correlation allow some discrepancy between the two estimates. Furthermore, a separate analysis of the two estimates based on the two cross-covariances is useful for checking possible contaminations of the estimates arising from similarities between relatives caused by common environments. In Table 2, pairs of estimates for genetic correlations between body size and egg size traits are given, one based on the first trait in the mother and the second trait in the daughter and the other estimate based on the reciprocal situation. There are great differences between the two corresponding estimates. For comparison the phenotypic correlations between the same traits are also given in Table 2. They show discrepancies that are expected for these sample sizes.

It appears that the correlation between egg size of the mother and body size of the daughter is higher than the other way round. This is to be expected if there is a direct (phenotypic) effect of the size of an egg on the size of the individual hatching and growing out of this egg. This effect can not be demonstrated directly in our data, the correlation between the size of an egg and the body size of the individual hatching from it is close to 1.0 at hatching and drops rapidly with increasing age to less than 0.1 before nestlings have reached their fledging weights. There is a phenotypic effect of body size of the female on the size of her eggs. This last correlation is less than 0.3 (Van Noordwijk et al. 1984). Therefore, a small correlation between the size of an individual and the size of the egg it hatched from may be expected from the two traits being heritable and the correlation between body size and the size of the eggs

Table 2. Phenotypic and genetic correlations between body size and egg size traits

	Phenotypic Correlations[a]			Genetic Correlations[b]	
	Mothers	Pop.	Daughters	Est. 1	Est. 2
a.w. – tars.[c]	0.52	0.41	0.42	0.09 (0.32)[c]	1.22 (.......)
a.w. – e.l.	0.18	0.25	0.19	–0.07 (0.32)	0.82 (0.11)
a.w. – e.v.	0.28	0.20	– 0.04	–0.01 (0.26)	0.09 (0.26)
tars. – e.v.	0.20	0.08	0.13	0.11 (0.26)	0.61 (0.16)
e.l. – e.b.	0.54	0.41	0.36	0.69 (0.13)	0.12 (0.24)
sample size	56–70	165–213	65–75		

[a] The phenotypic correlations are given separately for the group of mothers, the total population and the group of daughters

[b] The genetic correlations are based on the cross-covariances and are given separately for the first trait in the mother and the second trait in the daughter and vice versa. The values in parentheses are the result of (incorrectly) applying the formula for the standard error of the genetic correlation to the separate estimates

[c] Abbreviations: a.w. = adult weight, tars. = tarsus length, e.l. = egg length, e.b. = egg breadth, e.v. = egg volume

laid by that individual. This strengthens the conclusion that the effect that is likely to cause the discrepancy between the two estimates of the genetic correlation can not be demonstrated directly. Thus, small phenotypic effects can nevertheless influence the calculated genetic correlations in such a way that they become meaningless. Research into the variability and sensitivity of genetic correlations to slight deviations from the assumptions, implicit in data from natural populations, is badly needed. Until clear results of such sensitivity studies emerge, the use of genetic correlations derived from data on natural populations is likely to give misleading results.

Case 4: The Heritability of Nestling Body Size

From the practical point of view, body size and morphology in general are more easily studied than such breeding traits as clutch size, egg size, and timing of laying. This is simply a consequence of the fact that it is much easier to collect data; both sexes can be used and one is not limited to the survivors till breeding age. Further, some structures have already reached their full adult size in nestlings that can still be handled without danger of causing premature fledging. The measurements made on fully grown nestlings and measurements made on adults can be considered as the same trait (see below). Both tarsus length and body weight belong to this group of traits and both are commonly used as an index of body size. Because relatively large numbers of measurements can be collected, it is possible to compare heritability estimates made in single breeding seasons and to relate the differences between these estimates to differences in environmental conditions during nestling growth. Several detailed studies show very strong relations between nestling growth and environmental conditions, especially the availability of food (e.g., Bryant 1978, Tinbergen 1981). It was observed that in an extremely poor breeding season no resemblance between offspring and parents was present, while normally heritability estimates of about 0.5 are observed (Van Noordwijk 1982, Van Noordwijk et al. in prep.). However, even in a situation where a reasonably strong resemblance of nestling body size to parental body size was observed, the resemblance of individuals that survived for at least 3 months after fledging to their parents was stronger. Moreover, calculations using the slope of the regression and the observed phenotypic variance suggest that the genetic variance is the same for both nestlings and survivors, but that a considerable reduction has occurred in the environmental variance (Van Noordwijk 1982). These observations led to experiments from which some preliminary findings will be presented below, but they also lead to some theoretical considerations, because an explanation in terms of a higher heritability due to a lower environmental variance is too simplistic. It can be seen as an example of using definitions out of context and ignoring some of the underlying assumptions. These are usually met in the standard application of the method, which can lead to carelessness in checking whether they are met in every case.

In general, it is a truism that a lower heritability and a higher environmental variance go together if the additive genetic variance is constant (and if interactions are absent). There is, however, also a basic assumption that the distribution of the parental values and of the offspring values is the same. In calculating a heritability estimate from a regression of offspring on parents, the environmental variance used is that represented in the parents and not that in the offspring values. In this respect, the observation mentioned earlier that heritabilities based on recapture weights are higher than those based on nestling weights poses a problem. The nestlings that are later recaptured are larger (and have a lower phenotypic variance) than those that disappear. The group of parents used in both estimates is exactly the same and there is no evidence for differences in survival after fledging related to the parental body size. Thus, the difference in slope of regression must be (and is) due to an increase in the covariance between parents and offspring. This increase in covariance implies, however, that there is an interaction between the parental values and the absence of part of the environmental component in the offspring values included in the set of survivors. This contradicts the absence of selection relative to the parental phenotype.

Apart from the theoretical considerations, which are by no means resolved, the interaction between conditions during growth and genetic variation for body size is also open to experimentation. It is thought that the amount of food that is available for the nestlings is the critical factor. In principle it is possible to manipulate the amount of food in the habitat, but this is impractical. Under some testable assumptions the amount of food available per nestling can also be influenced by manipulating brood size. The main assumption involved is that parental feeding behavior does not fully compensate the changes in demand for food.

In 1982 experiments were carried out in the Liesbos study area involving manipulation of brood size simultaneously with cross-fostering. The experimental design is shown in Fig. 3. The Liesbos near Breda in the southern part of the Netherlands is a deciduous wood on rich soil and is considered to be prime habitat for Great Tits (for a description of the study area see Van Balen 1973). Furthermore, 1982 was a year with high densities of caterpillars throughout the Netherlands. This may explain why no significant effect of brood size on the mean nestling size was observed (Table 3).

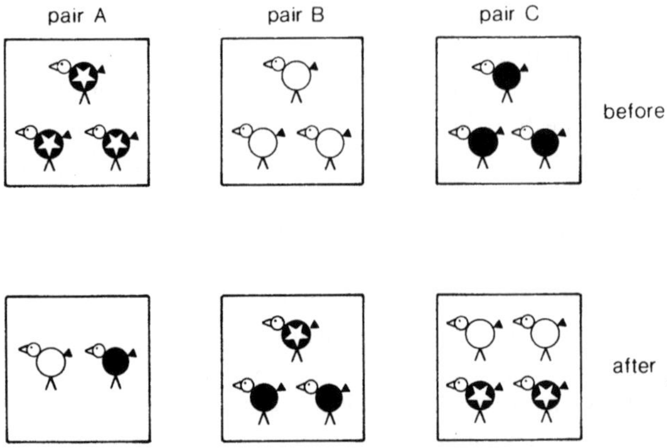

Fig. 3. The experimental design employed in Liesbos in 1982. Every symbol stands for one-third of the brood. The manipulation, performed shortly after hatching, results in all nestlings being brought up by foster parents and brood sizes of 2/3, 3/3, and 4/3 of the original size

Table 3. The mean values for the three experimental groups of 15-day old nestlings in the experiment in Liesbos 1982

Experimental group	Weight (g)		Tarsus (mm)	
	Female	Male	Female	Male
Enlarged (+)	17.7	18.2	18.8	19.1
Control (0)	18.1	18.6	18.8	19.1
Reduced (–)	18.1	18.7	18.8	19.1

It is possible that the parents were able to compensate for the increased demand for food in enlarged broods. It would thus seem that the conditions were not favorable to test the hypothesis. Yet, there is a clear trend in the mean of the variances within groups of true sibs brought up together in that the variances are smaller with smaller brood sizes and smaller with an increase in age (Table 4). It must be emphasized that these results are preliminary and based on small sample sizes, but the observations (Table 5) are in agreement with the prediction that higher heritabilities would be found in the broods reduced in size than in the enlarged broods. It must be remembered that none of the offspring were brought up by their biological parents, so that nongenetic sources of covariance were thought to be eliminated. In these results there

Table 4. Mean of the within real sib-group variances at different ages in the Liesbos (1982) experiment

	Weight			Tarsus		
	5	10	15	5	10	15
Enlarged (+)	18.3	19.4	10.4	1.2	0.7	0.3
Control (0)	13.2	11.4	10.0	1.3	0.4	0.3
Reduced (–)	10.2	11.4	5.8	1.0	0.3	0.2

Table 5. Heritability estimates for 15-day-old nestlings in the experiment in Liesbos (1982)

	Weight	Tarsus	Sample size (broods)
Enlarged (+)	0.56	0.35	10
Control (0)	1.16	−0.09	10
Reduced (−)	1.19	0.99	10
Total	0.87	0.39	30

is, however, also a resemblance to the foster parents and a positive correlation between the size of biological and foster parents. It is at present not possible to judge whether this is just bad luck and partly due to the small sample sizes or whether it has a different background. It is clear that more data are needed and the experiment will be repeated in a poorer habitat. These first results do look promising and it seems that, at least in principle, experiments involving classical approaches from evolutionary ecology combined with a genetic analysis are possible. They are a good way to investigate interactions between genetic and ecological processes.

The Similarities in All These Cases

In applying the methods of quantitative genetics to data from natural populations, one is outside the realm for which the methods were developed and in which the methods have been tested and shown to have practical usefulness. The similarity between the four cases that have been presented is therefore mainly on the methodological level.

In plant and animal breeding all methods have acquired a default null hypothesis, that is, in a typical application of these methods several potential problems can be ignored, because they are unlikely to be important in that situation or because they have been excluded through the experimental design. In all four cases some of these problems are in fact present. It is therefore necessary to check the underlying definitions and assumptions explicitly and very carefully. In many situations the interactions between slight deviations from the assumptions can not be seen immediately. This makes explicit modelling of the null hypothesis a necessity. In the first two cases such explicit models, a simulation model and an explicit conceptual model, have helped to clarify the problems addressed. The latter two cases are still incomplete in that problems have been identified, but no satisfactory solution has yet been found. Yet the identification of the problem is a first and essential step.

The Future of Genetical Ecology

I have used the term genetical ecology for the application of quantitative genetics to data of natural populations to set it apart from ecological genetics, which is nearly always concerned with major genes (Van Noordwijk 1980). In a nutshell one could say that ecological genetics deals with traits that are genetically simple, but ecologically complex, while the reverse is true in genetic ecology. In both, however, the aim is to understand the interactions between genetical and ecological processes in natural populations.

Genetical ecology is a rather new field. The first stage has been to demonstrate that significant amounts of genetic variation for traits of ecological importance (life-history traits) are present in natural populations under natural conditions. These findings are contrary to the expectation that significant amounts of additive variation will only be found for traits with a loose connection with fitness (Falconer 1981). This expectation is based on an unjustified extension of "Fishers Fundamental Theorem." This theorem deals with overall fitness in one specific life-history and in a specific ecological situation. It is assumed that environmental conditions are stable and suddenly change to another stable situation which remains constant for at least tens of generations. Such an ecological situation may be useful for genetic calculations, but it is not very realistic for most higher organisms. The realization that genetic modelling has so far covered only very limited ecological situations is relatively new (see Jayakar, De Jong, Loeschke, this symposium). The results from the first generation of studies in genetical ecology are among the sources to inspire this interest in trying to build ecologically more interesting situations into population genetic models. It is difficult to say when these models will become population models rather than population genetical models.

I think that genetical ecology can play a very important role in stimulating this conversion from separate genetical and ecological models to integrated population models. The reason is that genetical ecology is a suitable tool to elucidate important interactions between genetical and ecological processes. In three of the four cases presented above such interactions are important. Of course, the results from such studies are interesting for their own sake as well as for providing ideas for new laboratory experiments and new models. At the same time the interactions will nearly always necessitate some modelling to allow interpretation of the results, if only to construct null hypotheses. It is possible that in due time new standard applications of quantitative genetic methods will evolve in which it is then no longer necessary to check basic assumptions explicitly in every instance. Before that time, however, many more case studies have to be worked out to raise any hope of seeing possible generalizations.

Acknowledgements. I wish to thank Paul Keizer, who collected the data on intraclutch variation in egg size and Joep Schoemaker, who performed the experiment in Liesbos in 1982. I thank J.H. van Balen, P.R. Grant, T.D. Price, and J. Vijverberg for their critical remarks on an earlier version of this manuscript.

References

Balen Van JH (1973) A comparative study of the breeding ecology of the Great Tit *Parus major* in different habitats. Ardea 61:1–93

Bryant DM (1978) Environmental influences on the growth and survival of nestling House Martins *Delichon urbica*. Ibis 120:271–283

Falconer DS (1981) Introduction to quantitative genetics, 2nd edn. Longman, London

Garnett MC (1981) Body size, its heritability and influence on juvenile survival among Great Tits *Parus major*. Ibis 123:31–41

Greenwood PJ, Harvey PH, Perrins CM (1979) The role of dispersal in the Great Tit *(Parus major)*: The causes, consequences, and heritability of natal dispersal. J Anim Ecol 48:123–142

Jones PJ (1973) Some aspects of the feeding ecology of the Great Tit *Parus major*. D Philos Thes, Univ Oxford

Noordwijk Van AJ (1980) On the genetical ecology of the Great Tit *(Parus major, L.)*. Thes, Univ Utrecht

Noordwijk Van AJ (1982) Variation in body weight of the Great Tit, heritability and condition. Verh K Ned Akad Wet Afd Natuurkd 2e Reeks 79, Prog Rep IOO:9–12

Noordwijk Van AJ (1984) Problems in the analysis of dispersal and a critique on its "heritability" in the Great Tit. J Anim Ecol (in press)

Noordwijk Van AJ, Balen Van JH, Scharloo W (1980) Heritability of ecologically important traits in the Great Tit. Ardea 68:193–203

Noordwijk Van AJ, Balen Van JH, Scharloo W (1981a) Genetic and environmental variation in clutch size of the Great Tit. Neth J Zool 31:342–372

Noordwijk Van AJ, Balen Van JH, Scharloo W (1981b) Genetic variation in the timing of reproduction in the Great Tit. Oecologia (Berlin) 49:158–166

Noordwijk Van AJ, Keizer LCP, Balen Van JH, Scharloo W (1981c) Genetic variation in egg dimensions in natural populations of the Great Tit. Genetica 55:221–232

Noordwijk Van AJ, Klerks PLM (1983) Heritability of bill dimensions in the Great Tit. Verh K Ned Akad Wet Afd./Natuurkd 2e Reeks 81, Prog Rep IOO:7–12

Noordwijk Van AJ, Balen Van JH, Scharloo W (1984) Heritability of body size in a natural population of the Great Tit and its relation to age and environmental conditions during growth (submitted)

Ojanen M, Orell M, Vaisanen RA (1979) Role of heredity in egg size variation in the Great Tit *Parus major* and the Pied Flycatcher *Ficedula hypoleuca*. Ornis Scand 10:22–28

Reeve ECR (1955) The variance of the genetic correlation coefficient. Biometrics 11:357–374

Tinbergen JM (1981) Foraging decisions in Starlings *Sturnus vulgaris L.* Ardea 69:1–67

Theoretical Aspects of Density Regulation and Life Histories

The Delayed Effect on Adult Fertility of Immature Crowding: Population Dynamics

T. PROUT[1]

It is well known that in many species of insects and other organisms that the environmental conditions of immatures - larvae, nymphs - have permanent effects which persist into the adult stage of the life history (Anderson and Cummins 1979, Azam and Anderson 1969, Bates 1949, Brown and Chippendale 1973, Caligari 1980, Citus 1935, Dunlap-Pianka et al. 1979, Englemann 1970, Harshman 1983, Howard 1978, Istock 1966, Lloyd and White 1976, Marks 1982, Marlatt 1907, Otto 1974, Pritchard 1978, Ryan 1941, Southwood 1966, Steinwascher 1978, Sweeney and Vannote 1978, Uhlenhuth 1919, Vannote 1978, Ward and Cummins 1979, Wilbur 1977, Wilbur and Collins 1973). Relevant to population dynamics, the adult fertility component of growth can be affected by limiting resources of immatures. For example, it is abundantly documented in the *Drosophila* literature that the degree of larval crowding not only affects larval survival, but also affects the size and fertility of the resulting adult females (see below, for citations). Crowded larvae result in females with fewer ovarioles (Nöthel, pers. comm.).

The aspects of this phenomenon, which are of concern here, are the consequences for theoretical and experimental studies of population dynamics.

On the theoretical side, letting n denote egg density and N, adults, then the direct effect of immature crowding on survival can be represented as follows:

$$N = S(n)n \,, \tag{1}$$

where $0 \leqslant S(n) \leqslant 1$, and is a decreasing function of n. The delayed effect of immature crowding on adult females can also be represented as a decreasing function of n, namely F(n) = eggs/female. For organisms with discrete generations these formulations result in the following egg-to-egg recursion function:

$$n_{t+1} = F(n_t)\,\frac{1}{2}\,S(n_t)n_t \,. \tag{2}$$

Even using simple functions for $F(n_t)$ and $S(n_t)$ can result in a recursion function which is mathematically awkward (except for exponential functions); for instance, if F(n) and S(n) are both linear, then the recursion would be as follows:

$$n_{t+1} = (\hat{F} - fn_t)\,(\hat{S} - sn_t)n_t \,. \tag{3}$$

---

1 Department of Genetics, University of California, Davis, CA 95616, USA

Population Biology and Evolution
Ed. by K. Wöhrmann and V. Loeschcke

($\hat{F}$ and $\hat{S}$ are maximum fertility and survival constants, respectively, and f and s are sensitivity parameters.) So, the recursions is a third order polynomial. For strictly "metaphorical" theory, the complications produced by Eq. (1) might well be dismissed as a fine point, especially when the conclusions of a theoretical treatment are robust.

On the other hand, in experimental work designed to deduce population dynamical models, this delayed effect, embodied in F(n), cannot be ignored. One of the several complications that recursion (1) presents is how to census a natural or experimental population. This arises from the fact that it is difficult or even impossible to convert Eq. (1) into an adult-to-adult recursion, $N_{t+1} = N_t f(N_t)$. For example, Eq. (1) represents the number of adults, N, produced by eggs n. There are ample data showing an increasing n causing N to pass through a maximum as graphically demonstrated by Fig. 1.

The consequence of this situation is that the adult recursion $N_{t+1} = N_t f(N_t)$ is not defined. This is because $f(N_t)$ must contain a "parameter" accounting for fertility. of N_t, [F(n)] depends, on the number of eggs, n, which produced N_t adults. In the case of Fig. 1 the number of eggs that produced N_t is ambiguous. Thus, this delayed effect can result in serious problems in the dynamical interpretation of adult census data in populations or experiments designed to elucidate population dynamics.

The *Drosophila* experimental literature seems to be dichotomized in such a way as to just avoid the formulation of population dynamics in a form represented by Eq. (1): On the one hand, there is extensive literature reporting studies of the effects of larval density and quality of larval nutrition on the size and fertility of the resulting adults. Caligari (1980) and Marks (1982) represent two recent studies. There is the somewhat earlier extensive work of Barker (Barker and Podger 1970 and earlier work cited therein). Still earlier is such work as Robertson (1957) and Robertson and Sang (1944) and possibly the first demonstration of this effects was shown by Pearl (1932). However, in none of this work were the data used for incorporation into a formal representation of population dynamics.

On the other branch of the dichotomy are a number of more recent experiments (Thomas et al. 1980, for example) where population dynamical models are constructed from experimental data on adult numbers, N. Specifically, there is a common design which uses adult-to-adult one generation recursion experiments. Different densities of adults, N_t, are counted out into vials or bottles and compared with the number of adult offspring produced, N_{t+1} and a recursion function is produced by these data. The problem with such an experiment is that the fertility of the initial adults, N_t, is determined by the cultures from which they were taken. The result is that a number of published experiments have been so designed that all of the adults

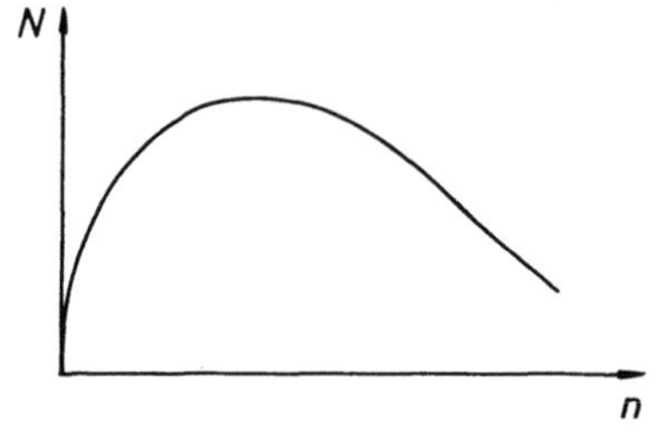

Fig. 1. Adult density (N), as a function of egg density (n)

introduced at different initial densities will have the *same* fertility, whereas in the running population whose dynamics are being sought, fertility will vary with density in not necessarily simple ways.

This same problem applies to similar experiments using adults which are used to construct the dynamics of a population of two competing species or genotypes. In this case adult mixtures of adults $N_{t,1}$, $N_{t,2}$ are introduced and compared with $N_{t+1,1}$, $N_{t+1,2}$ progeny adults and a set of different mixtures and densities is used for constructing the population dynamical model. However, for the same reasons given above, many of these experiments do not account for the larval crowding effect nor, in this case, competition effects among larvae on adult fertility, $F_i(n_i n_j)$. An example of this kind of experiment can be found in Ayala (1971).

Caligari (1980) and Barker and Podger (1970), on the other branch of the dichotomy, have clearly shown that larval density of a given type not only affects the fertility of the adults of that type, but also the fertility of the competing type.

In addition to *Drosophila* work, Hassell et al. (1976) deduced population dynamics from the life history data of some 26 species of insects. Although all cases evidently experience density compensation, these authors assumed fertility to be a constant. A perusal of the literature cited at the beginning of this article will suggest that the delayed effects on fertility of limiting resources directly affecting survival is of very common occurrence in a variety of insect species and even extends to some anurans as well.

These considerations lead to two general conclusions: First, it seems apparent that experimental and observational data should be obtained in such a way as to be able to detect this delay effect and, if it is important, it should be incorporated into any theory which represents the dynamics of the population. Secondly, it would be useful and rewarding to investigate the theoretical properties of systems which have the form of recursion Eq. (1) and also to study formal properties of age structured populations which contain this delayed effect. Some work on this latter problem has been done by Frauenthal (1975) who analyzed a human demographic phenomenon wherein the birth rate of female cohorts is inversely related to their size. However, a correlation of this kind could only occur when adult numbers N [= S(n)n] is a strictly increasing function of n, and, so does not exhibit a maximum as represented by Fig. 1.

More details and additional information on this subject will appear elsewhere.

References

Anderson NH, Cummins KW (1979) Influences of diet on the life histories of aquatic insects. J Fish Res Bol Can 36:335–342

Ayala FJ (1971) Competition between species: Frequency dependence. Science 171:820–824

Azam KM, Anderson NH (1969) Life history and habits of *Sialis rotunda* and *Sialis californica* in western Oregon. Ann Ent Soc Am 62:549–558

Barker JSF, Podger RN (1970) Interspecific competition between *D. melanogaster* and *D. simulans*. Effects of larval density and short term adult starvation on fecundity, egg hatchability and adult viability. Ecology 51:855–864

Bates M (1949) The natural history of mosquitoes. MacMillan Co, New York

Brown JJ, Chippendale GM (1973) Nature and fate of nutrient reserves of the periodical (17-year) cicada. J Insect Physiol 19:607–614

Caligari PDS (1980) Competitive interactions in *D. melanogaster*. I. Monocultures. Heredity 45: 219–231

Citus M (1935) The effect of temperature on the development of *Rana pipiens*. Physiol Zool 8: 290–310

Dunlap-Pianka H, Boggs CL, Gilbert LE (1979) Ovarian dynamics in Heliconius butterflies: Programmed senescence vs eternal youth. Science 197:487–490

Englemann F (1970) The physiology of insect reproduction. Pergamon Press, Oxford New York

Frauenthal J (1975) A dynamic model for human population growth. Theor Popul Biol 8:64–73

Harshman L (1983) Ecological genetics of *Alsophila punctaria*. SUNY, Stony Brook, PhD Thes

Hassell MP, Lawton JH, May RM (1976) Patterns of dynamical behavior in single-species populations. J Anim Ecol 45:471–486

Howard RD (1978) The influence of male-defended oviposition sites on early embryo mortality in bullfrogs. Ecology 59:789–798

Istock CA (1966) The evolution of complex life cycle phenomena: an ecological perspective. Evolution 21:592–605

Lloyd M, White J (1976) Sympatry of periodical cicada broods and the hypothetical four-year acceleration. Evolution 30:768–801

Marks RW (1982) Genetic variability for response to density of three components of fitness. Genetics 101:301–316

Marlatt CL (1907) The periodical cicada. USDA, Bur Entomol Bull 71:1–181

Otto C (1974) Growth and energetics in a larval population of *Potamophylax cingulatus* (Steph.) Tricoptera in a south Swedish stream. J Anim Ecol 43:339–361

Pearl R (1932) The influence of density of population upon egg production in *Drosophila melanogaster*. J Exp Zool 63:57–84

Pritchard G (1978) Study of dynamics of populations of aquatic insects: the problem of variability in life history exemplified by *Tipula sacra* Alexander (Diptera, Tipulidae) Verb Int Verlin Lininal 20:2634–2640

Robertson FW (1957) Studies of quantitative inheritance. XI. Genetic and environmental correlation between body size and egg production. J Genet 55:428–443

Robertson FW, Sang JH (1944) Ecological determinants of population growth. I. Fecundity of adult flies. Proc R Soc London Ser B 132:258–277

Ryan FJ (1941) Temperature changes and subsequent rate of development. J Exp Zool 88:25 to 54

Southwood TRE (1966) Ecological methods. Methuen, London

Steinwascher K (1978) The effect of coprophagy on the growth of *Rana catesbeiana* tadpoles. Copeia 1978:130–134

Sweeney BW, Vannote RL (1978) Size variation and the distribution of hemimetabolous aquatic insects: two thermal equilibrium hypotheses. Science 200:444–446

Thomas WR, Pomerantz MJ, Gilpin M (1980) Chaos, asymmetric growth and group selection for dynamical stability. Ecology 61:1312–1320

Uhlenhuth E (1919) Relation between metamorphosis and other developmental phenomena in amphibia. J Gen Physiol 1:525–544

Vannote RL (1978) A geometric model describing a quasi-equilibrium of energy flow in populations of stream insects. PNAS 75:381–384

Ward GM, Cummins KW (1979) Effects of food quality on growth of a stream detritivore, *Paratendlipes albimanus* (Meigen) (Diptera: Chironomidae). Ecology 58:206–209

Wilbur HM (1977) Interactions of food level and population density in *Rana sylvatica*. Ecology 58:206–209

Wilbur HM, Collins JP (1973) Ecological aspects of amphibian metamorphosis. Science 182:1305 to 1314

Selection and Numbers in Models of Life Histories

G. DE JONG[1]

Introduction

In integrating the models of population genetics and population dynamics, the main avenue opened is for models of the evolution of the ecological parameters determining the dynamics of a population. The first effort in this direction is perhaps that of MacArthur (1962). Coming from ecology, he based himself upon the familiar logistic equation. His supposition is that the fitness of population genetics and the r of the logistic equation refer to the same quantity. MacArthur proposes another population genetics, where fitness would be K. The substance of MacArthur's article is repeated in Chapter 7 of MacArthur and Wilson's *The Theory of Island Biogeography* (1967). The model of MacArthur and Wilson compares genes to species, though it can be doubted whether this a legitimate approach. Moreover, it is not at all clear that the verbal description of r- and K-selection in Chapter 7 is consistent with the bit of model presented.

At about the same time, Wallace (1970) introduced in his book *Genetic Load* a population dynamics model to explain the absence of genetic load in a population where polymorphism was maintained by heterozygote overdominance. This model was mainly presented graphically; it links several older concepts, to wit overdominance, density dependent selection, and some ideas on the exchangeability of causes of death. The ideas of Wallace seem to have attracted far less notice than the ideas of MacArthur and Wilson. Yet the ideas of Wallace were more biological than the ideas of MacArthur and Wilson.

Models of density-dependent selection followed soon (Anderson 1971, Roughgarden 1971, Charlesworth 1971, Clarke 1972). The most important theorem to be proposed was, however, first stated by Roughgarden (1976). In a model with random mating of gametes and only purely density dependent fitnesses [$W_{ij}(N_t) > 0$; $W'_{ij} < 0$], where numbers are censused at the start of density dependence, the numbers at the census point are described by

$$\begin{aligned} &\text{a) } N_{t+1} = N_t \overline{W}_t(N_t, p) \\ &\text{b) } p_{t+1} = p_t \overline{w}_1(N_t, p) / \overline{w}_t(N_t, p) \\ &\text{c) } W_{ij}(N_t) = w_{ij}(N_t) \\ &\text{d) Hardy Weinberg equilibrium in zygotes.} \end{aligned} \tag{1}$$

1 Department of Population and Evolutionary Biology, University of Utrecht, 3508 TB Utrecht, The Netherlands

Population Biology and Evolution
Ed. by K. Wöhrmann and V. Loeschcke

"By placing the problem of the evolution of ecological parameters in this particular population genetics context we can justify the following principle:

Principle 1. Natural selection causes the parameters of a single species population dynamic model eventually to assume those values which produce the largest equilibrium population size." (Roughgarden 1976). Interest is focused upon the parameters of the fitness function $W_{ij}(N_t)$. The principle is valid if there is only density dependent selection present during the life cycle, if there is only one density dependent stage during the life cycle as far as the model is concerned, if there is random mating of gametes. That is, the principle is valid if we suppose a simple life history. Changing assumptions about the life history leads us to discover cases in which the principle is no longer generally valid. Density independent selection between becoming independent individuals and the start of the nonselective density dependent stage leads to frequency dependent selection and therefore to cases where equilibrium population size is not maximized (Prout 1980). In fact, this applies to cases with density dependent selection followed by a separate nonselective density dependent stage, too. If there are several age classes independently subject to density dependent selection, the equilibrium number in no one age class is maximized (Léon and Charlesworth 1978, Charlesworth 1981). If not gametes, but individuals mate at random, equilibrium number is not necessarily maximized (De Jong 1982).

Therefore, it is not immediately clear in how far Roughgarden's principle is general and in how far the principle is restricted by the implicit assumption of some restricted life history. Some part of the role the assumptions play have already been mentioned. We will now examine the role of yet other implicit assumptions. Implicit in the model of Roughgarden (1976) and in all other models mentioned, is the assumption that individuals are independent and interact homogeneously during the density dependent stage - that is, that an individual experiences the presence of all other individuals equally. Now the situation might be that individuals find themselves grouped during the density dependent stage, such that any one individual experiences only the presence of the other individuals in the same group. The interaction is no longer homogeneous even if all individuals are independent. Such would be the case if density dependence is effective during a juvenile stage where the juveniles found themselves restricted by a patchily distributed food supply. On the other hand, individuals are not independent of each other in sib competition. Nonhomogeneous interactions between juveniles competing for food occurs too in sib competition if the sizes of all sib groups are not the same. The question is whether these two life history phenomena, sib competition and nonhomogeneous interaction between juveniles, lead to other conclusions regarding the maximization of equilibrium population size. And if so, whether it is sib competition or nonhomogeneous interaction of juveniles or both together that is of importance.

Models of the evolution of ecological parameters determining the dynamics of a population are often focused upon the evolution of equilibrium number. Another feature of the dynamics of a population is stability. There are a few models of evolution towards numerical stability (Heckel and Roughgarden 1980, Turelli and Petry 1980). These models deal with a fluctuating environment, as in their models selection in a stable environment does not act directly on parameters that determine local stability. In these models, it is possible to define quantities that determine local stability in a constant environment, but come under control of individual selection in a fluctuating environment. The problem is twofold. First, how to get local stability for a constant environment under the control of individual selection as opposed to group selection. Group selection is supposed necessary to get stability in a constant environment by Thomas et al. (1980). Second, whether the quantity indicating local stability can decrease under individual selection. A low value of the local stability quantity indicates that the number in the population converges toward the equilibrium number. The problem is therefore whether individual selection in a constant environment can bring about values of the local stability quantity such that the population always converges to the equilibrium number.

Here we will start with explicit life histories and examine the effect of natural selection on the parameters determining the dynamics of a population. Fitness is made up from fitness components in a way specific to the life history. In directional selection we end up with the type with higher fitness; therefore, we know what the value of the fitness components is at the end of selection. The components involved in fitness are involved too in equilibrium number and in the quantity indicating local stability. Therefore, we can examine whether equilibrium number increases and whether local stability comes to indicate convergence towards the equilibrium number under natural selection.

One Food Supply: Genetics and Dynamics

The simplest life history involves all individuals homogeneously competing for one food supply (Fig. 1a). Mating is random at the level of the gametes. Roughgarden (1979; Chap. 17) sketches this type of life history as a background to his simultaneous models of population genetics and population dynamics. He takes fish to fix people's minds. Generations are discrete; all zygotes start life at the same time. Juveniles are randomized over the one food supply and all juveniles compete homo-

Fig. 1 a–c. Distribution of competitors over food. **a** One total food supply; all juveniles compete homogeneously with all other juveniles. **b** The total food supply is subdivided in as many equal shares as there are nests; juveniles only compete with sibs. **c** The total food supply is subdivided in environmental patches of equal size; juveniles compete with sibs from the same clutch and with juveniles from a few other clutches

geneously with all other juveniles. Adult mating is at random at the level of gametes. All individuals are independent from zygote on. Absolute fitnesses have a form derived from the logistic equation, with r and K dependent upon genotype. The implicit assumption is that the product of female viability and number of eggs per female equals the product of male viability and male mating advantage for each genotype. Such a life history can be described by model 1.

We set out to analyze simultaneous models of population genetics and population dynamics in terms of life history components. Therefore, we will write the model for this type of life history in terms of viability and fecundity. Notation is summarized in Table 1. The number of zygotes is denoted by n. Viability covers the period from

Table 1.

n	Number of zygotes
n^*	Equilibrium number of zygotes
N	Number of adults
N^*	Equilibrium number of adults
K_{ij}	Equilibrium number per genotype
M	Total amount of juvenile food available
v_{ij}	Viability = probability of survival per genotype
$v(n/M)$	Viability as a function of density of zygotes n/M
$N = n \cdot v(n/M)$	Number of adults as a function of density of juveniles: survival curve
X	Local number of (nonsib) juveniles
$2\,F_{ij}$	Female fecundity per genotype
F^*	Optimum fecundity per individual
W_{ij}	Absolute fitness per genotype
w_{ij}	Relative fitness per genotype
x_{ij}	Genotype frequency
$p, q = 1 - p$	Gene frequency of gene A_1, A_2
h	Number of food patches
$\bar{W}$	Average absolute fitness: $x_{11}W_{11} + x_{12}W_{12} + x_{22}W_{22}$
$\bar{w}$	Average relative fitness: $x_{11}w_{11} + x_{12}w_{12} + x_{22}w_{22}$
$\bar{w}_i$	Average fitness of gene i
$\overline{vF}$	Average product of viability and fecundity: $x_{11}v_{11}F_{11} + x_{12}v_{12}F_{12} + x_{22}v_{22}F_{22}$

zygote to adult. Viability is the probability of survival and therefore always between 0 and 1. We will suppose that viability decreases monotonously with density n; this is very plausible, but not in fact a necessary feature of biological probabilities of survival. Viability increases with the amount of food available to all the juveniles, M. Viability is held to be a function of the form v(n/M). Viability is identical for males and females. The number of adults is therefore $N = n \cdot v(n/M)$; the number of adults as a function of the number of zygotes is called the survival curve. The sex ratio equals 1. Female fecundity is denoted by 2 F and is constant. Male mating advantage is equal to female fecundity. The number of zygotes in the next generation is therefore $n_{t+1} = n_t \cdot v(n/M) \cdot F$, if only one genotype is present. In the case of one locus with two alleles, A_1 and A_2, the frequencies of the three genotypes in zygotes are x_{11}, x_{12}, and x_{22}. Both viability and female fecundity (and male mating ability) can depend upon the genotype: $v_{ij}(n/M)$, and F_{ij}. The heterozygote has both viability and fecundity identical to one of the homozygotes. We are only interested in directional selection.

In this life history with random mating of gametes and homogeneous interaction between all juveniles during the density dependent stage, fitness of a genotype is unequivocally defined, as male mating advantage is assumed equal to female fecundity. The simultaneous recurrence equations for population genetics and population dynamics can be written directly in the numbers of zygotes of each genotype in the next generation as

$$\begin{aligned} n_{t+1}\, x_{11,t+1} &= n_t \cdot \quad (x_{11}v_{11}F_{11} + \tfrac{1}{2}x_{12}v_{12}F_{12})^2 / \overline{vF} \\ n_{t+1}\, x_{12,t+1} &= n_t \cdot 2\, (x_{11}v_{11}F_{11} + \tfrac{1}{2}x_{12}v_{12}F_{12}) \cdot \\ &\qquad (x_{22}v_{22}F_{22} + \tfrac{1}{2}x_{12}v_{12}F_{12}) / \overline{vF} \\ n_{t+1}\, x_{22,t+1} &= n_t \cdot \quad (x_{22}v_{22}F_{22} + \tfrac{1}{2}x_{12}v_{12}F_{12})^2 / \overline{vF} \\ p_t &= (n_t x_{11} + \tfrac{1}{2} n_t x_{12})/n_t \end{aligned} \tag{2}$$

Absolute fitness W and relative fitness w both are equal to $v_{ij}(n/M) \cdot F_{ij}$. From the numbers of the three genotypes in the next generation and the relation between genotype frequency and gene frequency model (1) can be found. Since at the end of selection the product vF equals 1 at the equilibrium number K_{ij} and since the viabilities are monotonously decreasing, higher equilibrium number of zygotes is the consequence of natural selection.

The principle of maximization of equilibrium number is graphically shown for the case of convergence in number (Fig. 2a). If we start with a population of A_1A_1 homozygotes, at their equilibrium number K_{11}, gene A_2 can invade. Over the whole range of numbers between K_{11} and K_{22}, gene A_2 is selected for. The homozygote A_2A_2 gets fixed at an equilibrium number of adults of K_{22}. What we use are five properties of the system:

$W_{ij}(N_t) \;=\; w_{ij}(N_t)$;
$W_{ij}(N_t)$ not frequency dependent ;
$W'_{ij} \;<\; 0$;
$W_{ij}(K_{ij}) \;=\; 1$, at equilibrium number K_{ij} ;
$w_{22} > w_{11}$ over the interval K_{11}, K_{22} .

Together this means that $K_{22} > K_{11}$. We maximize the equilibrium number of zygotes.

In both parts of Fig. 2, v_{11} is greater than v_{22} and F_{11} is less than F_{22}; not because such is necessarily the case, but to show that the absolute or relative fitness at very low numbers plays no role in the outcome of natural selection. This means that the r of the logistic equation plays no role in this sort of model. Selection can bring about both higher and lower fecundity, stability, or instability, but it does maximize equilibrium number under this model (1).

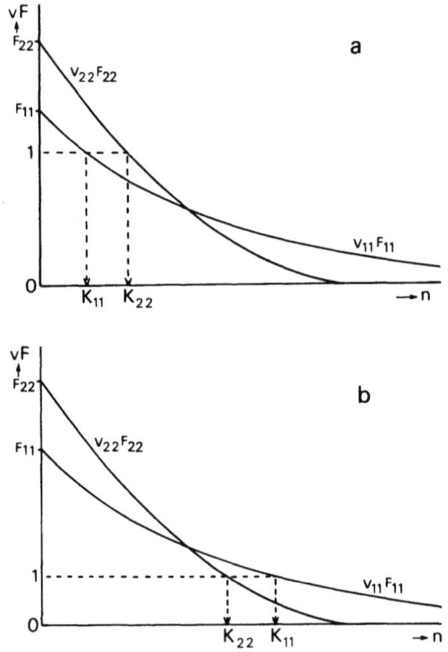

Fig. 2a,b. Maximization of equilibrium number under natural selection. Carrying capacity of a genotype is found at $W_{ij} = 1$. Absolute and relative fitnesses of a genotype are identical. **a** At K_{11} gene A_2 can invade; selection over the numerical range $K_{11}-K_{22}$ leads to fixation of A_2A_2. **b** At K_{22} gene A_1 can invade; selection over the numerical range $K_{22}-K_{11}$ leads to fixation of A_1A_1. The fitness relations at very low n are not of interest for the outcome of selection at the range of numbers including the carrying capacities. Drawn is $v_{11} > v_{22}$, $F_{11} < F_{22}$; this for demonstration, not as this is necessarily so

All individuals are fully independent from the age of zygote on. The appropriate census point is at the age of zygotes. It is possible to count at other ages, but not without giving some twist to the formulation of population genetics and population dynamics. Counting at other ages introduces frequency dependence somewhere in the formulation. Counting at the earliest age of independent individuals introduces the simplest formulation of the simultaneous equations. Therefore, this is the appropriate census point.

What happens to the numbers of adults? We cannot say without further information on the viability. Depending on the exact viability function, there are even for monotonously decreasing viability functions three possibilities for the numbers of adults: they might indefinitely increase, increase to an asymptotic value, or decrease again after an initial increase. The first possibility, indefinite increase of adults with increasing number of zygotes (Prout 1980), means that maximization of zygotes is accompanied by maximization of adults. The second possibility is that the number of adults rises to an asymptote with increasing number of zygotes. The third survival curve has a maximum at an intermediate number of zygotes. This is the survival curve most attention will be given to here. For one thing because this type of survival curve actually often occurs in nature. Secondly, because a simple probability model of competition for food leads to this survival curve. Thirdly, because this is the survival curve that generates instability of number at high fecundity. The evolution of numerical stability is only a problem with this type of survival curve, the other two survival curves lead always to stable numbers.

The survival curve is closely related to the local stability indicator b:

$$b = - \left.\frac{\partial n_{t+1}}{\partial n_t}\right|_{n^*} = - F \left.\frac{\partial n \cdot v(n)}{\partial n}\right|_{n^*} .$$

Monotone convergence to equilibrium number occurs with $-1 < b \leqslant 0$, a stable equilibrium number is reached for $0 < b \leqslant +1$ by decreasing fluctuations, and instability occurs at $b > +1$. The first two survival curves lead to convergence and numerical stability at all fecundities. Numerical instability is only possible with a survival curve of the third shape. With this survival curve, b and numerical instability increase with increasing fecundity F. Natural selection increases fecundity F in the life history represented by model (2) at equal viability between the genotypes. There is no way to stop natural selection at the value of F that gives an equilibrium number n^* such that the local stability indicator $b = 0$. That is at an equilibrium number n^* such that the maximum number of adults is reached. The maximum in the survival curve plays no role in the description of the life history.

The problem then is to find a life history that makes use of the maximum of the survival curve. It clearly involves keeping fecundity at an intermediate level. Of course, we have long known such a life history. Lack's explanation of the evolution of clutch size provides the clues for a model that does just that.

Clutches in Nests: Genetics and Dynamics

Let us suppose animals that care for a clutch of young in a nest. The parents provide the food for their own young (Fig. 1b). The number of adults is again N. There are $\frac{1}{2}$ N pairs and therefore $\frac{1}{2}$ N clutches to be fed from the food supply of size M. Each pair gets a share $M/\frac{1}{2}N$ from the common juvenile food supply to feed its young. For a pair with 2 F juveniles in their nest, the viability of those juveniles will be $v[2F/(M/\frac{1}{2}N)] = v(FN/M)$. If there is only one genotype in the population, the dynamics will be $N_{t+1} = N_t F v(NF/M)$ and this is equivalent to $n_{t+1} = n_t v(n/M)F$. For a single monomorphic species the dynamics are the same whether all juveniles are in one group themselves competing for the food supply or in clutches competing for the share of the food supply their parents bring. It all changes however when there are genotypic differences in fecundity.

With genotypic differences in female fecundity, $2F_{ij}$, the viability of a clutch becomes $v[2F_{ij}/(M/\frac{1}{2}N)] = v(F_{ij}N/M)$. There are $\frac{1}{2}N_t$ pairs. The simultaneous recurrence equations for the genetics and dynamics of the population become:

$$
\begin{aligned}
N_{t+1}\, x_{11,t+1} &= N_t \cdot \quad p \cdot (x_{11}W_{11} + \tfrac{1}{2} x_{12}W_{12}) \\
N_{t+1}\, x_{12,t+1} &= N_t \cdot [\, p \cdot (x_{22}W_{22} + \tfrac{1}{2} x_{12}W_{12}) + \\
&\qquad + q \cdot (x_{11}W_{11} + \tfrac{1}{2} x_{12}W_{12})] \qquad (4) \\
N_{t+1}\, x_{22,t+1} &= N_t \cdot \quad q \cdot (x_{22}W_{22} + \tfrac{1}{2} x_{12}W_{12}) \\
p_t &= (N_t x_{11} + \tfrac{1}{2} N_t x_{12})/N_t
\end{aligned}
$$

with x_{ij} the frequencies of the genotypes in the adults and absolute fitness W_{ij} and relative fitness w_{ij} both equal to $F_{ij} \cdot v(F_{ij}N/M)$.

The recurrence Eqs. (4) are written in adults. It is not possible to write model (4) in zygotes. Zygotes are not independent individuals in this life history. Adults are independent individuals. The recurrence equations stand in the earliest stage in the life cycle that consists of independent individuals.

Absolute fitness, and relative fitness, has the form $W_{ij} = F_{ij}\, v(F_{ij}N/M)$. This contrasts with the form $W_{ij} = v(n/M) \cdot F_{ij}$ of fitness in a life history where the zygotes are independent individuals. Absolute fitness $W = F \cdot v(FN/M)$ is closely related to the survival curve. If the viability function is such that the survival curve $n \cdot v(n/M)$ has a maximum, absolute fitness $W = F \cdot v(FN/M)$ has a maximum for the same viability function, since

$$\frac{\partial\, n \cdot v(n/M)}{\partial n} = \frac{\partial\, F \cdot v(NF/M)}{\partial F} \,. \tag{5}$$

Let the maximum of the survival curve be reached at n' for a certain amount of food M. The maximum of the absolute fitness is then reached at an F' such that $F'N = n'$ for all fixed N. This means that the optimum fecundity leading to the maximum fitness decreases with increasing adult number (Fig. 3). The evolutionary stable optimum fecundity is reached at the maximum number of adults possible. It is the optimum fecundity leading to the lowest maximum fitness, a maximum fitness of 1 (Fig. 4). In Fig. 4, let us start at a fecundity F_1 leading to equilibrium number K_1. The absolute fitness at $N = K_1$ is the curve labelled as such. At the equilibrium number K_1 the optimum fecundity is not F_1, but F'. Any mutation in fecundity leading to a higher fitness, i.e., any mutation leading to a fecundity value between F_1 and $F = 1$ will be selected for. Selection will lead to higher adult number, $N = K_2$. The absolute fitness will change to the curve labelled $N = K_2$. As long as the absolute fitness curve $W = F \cdot v(NF/M)$ is twice intersected by $W = 1$, selection for an intermediate fecundity value is possible. Only when the absolute fitness curve $W = F \cdot v(FN/M)$ is tangent to $W = 1$ is no further selection possible. This is at the maximum number of adults N^* from any number of zygotes. The amount of food M has no influence on

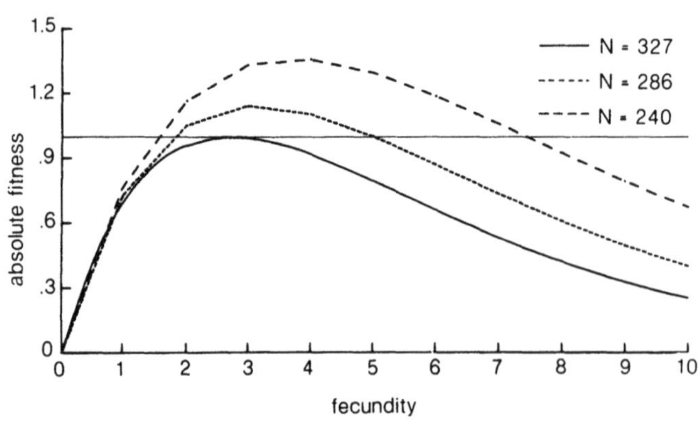

Fig. 3. Absolute fitness $F \cdot v(FN/M)$ for three values of N as a function of fecundity F. This absolute fitness decreases with increasing N. ($v = \exp[-c\, 2\, F/(M/\frac{1}{2}N)]$, $c = 4.5$, $M = 4{,}000$)

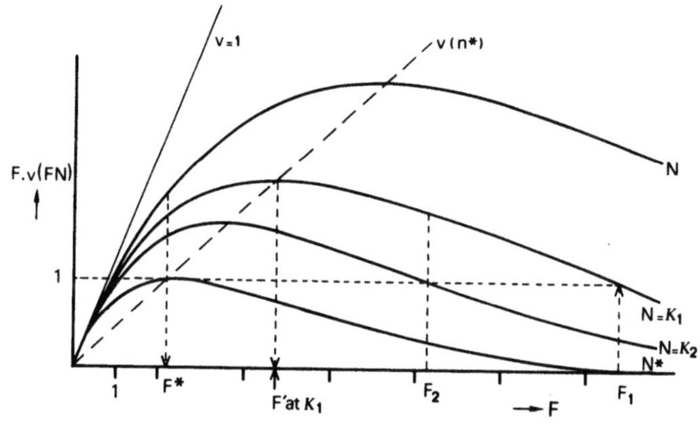

Fig. 4. Evolution of optimal fecundity with a fitness form F • v(FN/M). The evolutionary stable optimum fecundity is found when the maximum of F • v(FN/M) equals 1. This is at the absolute maximum number of adults. For further explanation see text

the value of F^*; its influence is on N^* as with any viability function of the form v(FN/M) the equilibrium number N^* increases linearly with M. Seeking the maximum of F • v(FN^*/M) means therefore seeking the maximum of something that is not a function of M. That is, optimal fecundity F^* is not dependent upon the amount of food M for all viability functions with the density as function argument. It means that a higher food supply does not lead to larger clutch size, but to higher numbers of nesting adults.

At the absolute maximum number of adults the local stability indicator $b = 0$. This is the consequence of the relation between the survival curve and the local stability indicator. Natural selection on fecundity in this life history leads to local stability indicating monotone convergence to equilibrium number at the highest possible rate. Note this life history with numerical stability is a life history with sib competition.

Fitness Forms in Parallel

We have considered one life history where all juveniles are independent and compete all together for one food supply; and we have considered another life history where the juveniles compete within a nest for the food the parents bring. In the one case juveniles compete for one food supply of size M. In the second case juveniles compete for one of $\frac{1}{2}$ N food supply shares of size M/$\frac{1}{2}$ N. A further possibility is that a group of juveniles competes for food in one of h food patches of environmentally fixed size M/h. The viability within any one patch will depend upon the number of juveniles in the patch and the amount of food in the patch, but it will be the same viability function as before. The number of adults emerging from a patch will be a survival curve of the third type, with a maximum at an intermediate number of juveniles in the patch. This leaves to consider how juveniles arrive in a patch of juvenile food.

There are three distinct possibilities for arrival of juveniles in a patch of food. The adult females can distribute single eggs as independent units over the patches according to a probability distribution, the adult females can distribute clutches of eggs as independent units over the patches according to a probability distribution, and the adult females can try to monopolize a patch for their own clutch. When the females distribute eggs or clutches according to a probability distribution over the patches, the expected value of the number of zygotes or adults in the next generation is used as if this was the number according to deterministic dynamics. In neither case are the dynamics identical to the dynamics when all juveniles homogeneously compete for one food supply or to the dynamics when the juveniles compete only with their sibs for a pairs' share of the juvenile food supply.

If single eggs are distributed over patches according to a probability distribution, say a Poisson distribution, local density in the patches varies. Competition for food among juveniles is no longer homogeneous. As a consequence of the variance in density among the patches and any nonlinear survival curve, the average number of adults from a patch is not the same as the number of adults from an average patch. With the same viability in any one patch as before, the overall viability now differs from the deterministic dynamics (De Jong 1979, Chesson 1981). Selection on female fecundity maximizes fecundity. Selection on fecundity maximizes the equilibrium number of zygotes, within the dynamics based upon a probability distribution of eggs over patches. The local stability indicator b increases with fecundity and numerical stability cannot be maintained by natural selection. In these respects the situation is analogous to the deterministic dynamics.

If each female lays one clutch of eggs and the clutches of eggs are distributed over the patches according to a probability distribution, local density in the patches varies. If female fecundity, i.e. clutch size, is different between the different genotypes, the variance in density over the patches has two sources, the probability distribution of dispersal of clutches and the clutch sizes. Competition for food among juveniles is nonhomogeneous. In any one patch juveniles are present from a few clutches. Part of the competition in a patch is between sibs, part is between unrelated juveniles. Let the 2 F sibs from one clutch and X juveniles from a number of other clutches be present in a patch. The number of adults emerging from the patch has the form $F \cdot v[(2F + X)/(M/h)]$. This is a form of absolute fitness that is a compromise between the two basic fitness forms described. If a patch is small, the number of other juveniles X will be relatively small and sib competition will be relatively high. The form of fitness approaches the form for pure sib competition $F \cdot v(FN/M)$. If a patch is large, the number of other juveniles X will be relatively large and sib competition will be relatively low. The form of fitness approaches the form for competition between independent juveniles, $F \cdot v(n/M)$.

In this life history of random dispersal of clutches of eggs over patches, selection on fecundity leads to an intermediate optimum fecundity value that operationalizes the maximum in the survival curve. The optimum fecundity is such that if one female laid eggs in a patch, she would have the maximum number of young adult offspring from the patch. This is therefore the same optimum fecundity as in the life history with monopolization of a patch for her own clutch by a female. Often, more than one female will lay a clutch of eggs in a patch if clutches are randomly distributed by all females. But the optimum fecundity is not affected by the difference in behavior. The optimum fecundity is related to the patch size. The optimum fecundity increases with the size of the patch M/h; this means too that the optimum fecundity increases with the number of unrelated juveniles a larva competes with. The female fecundity evolves therefore to a compromise between the fecundity in a situation with pure sib competition and the fecundity in a situation without any sib competition.

This same sort of compromise as in fecundity is evident in the equilibrium number. The equilibrium number reached under evolution of fecundity is neither the maximum number of zygotes nor the maximum number of adults (Fig. 5). If patches are large the equilibrium number of zygotes is relatively high and the equilibrium number of adults relatively low. If patches are small the equilibrium number of zygotes is relatively low and the equilibrium number of adults is relatively high. Again, this same sort of compromise is evident in stability. If patches are small and sib competition is relatively strong, the local stability indicator b is lower and the population is more stable than when patches are large and sib competition is relatively low. Hence, for stability it is not so much important whether the juveniles interact homogeneously, but at what stage they act as independent individuals. The basic situation is one of compromise between sib competition and competition with random individuals. The fecundity, the equilibrium number, and the stability are all compromises between consequences of the two basic fitness forms.

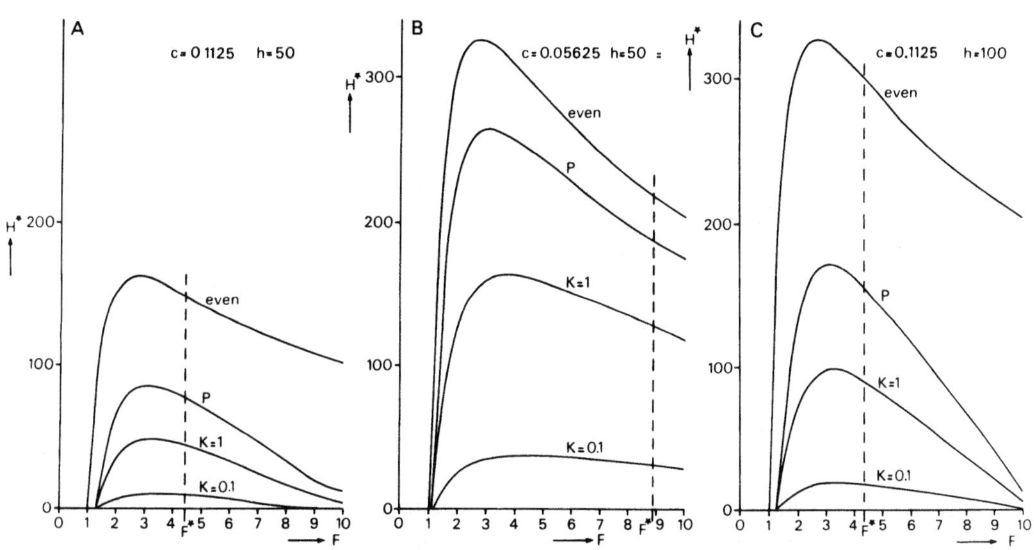

Fig. 5A–C. Equilibrium number of adults (marked H^* for N^*) as a function of per female fecundity. Fecundity evolution is toward $2F^* = 1/c$, with c as indicated in the figure. The viability function in each patch is $v = \exp[-cX/(M/h)]$ with M = 4,000, and with **(A)** c = 9, h = 50; **(B)** c = 4.5, h = 50; **(C)** c = 4.5, h = 100, and therefore $2F^* = M/hc$. The equilibrium number of adults at the optimum fecundity is a compromise between maximizing number of juveniles and number of adults. Four distributions of clutches over patches are indicated: deterministic even distribution and probabilistic Poisson (*P*) distribution and negative binomial distributions with parameter k = 1 and k = 0.1

Formally, the compromise is the result of a constant relative fitness and a frequency dependent absolute fitness. The census point has to be adults; emerging adults are the youngest independent individuals. In fact, it is impossible to write recurrence equations in zygotes. With the exponential viability function $v(X) = \exp\{-cX\}$ and a Poisson distribution of clutches over patches and only genotypic differences in fecundity, an analytical solution is possible. The recurrence equations in the expected numbers are according to model (4), with absolute fitnesses

$$W_{ij} = F_{ij} \cdot \exp\{-(c \cdot 2F_{ij})/(M/h\} \cdot \exp\{-(\tfrac{1}{2}N_t/h) \cdot (1 - \sum_{kl} x_{kl} \cdot \exp[-(c \cdot 2F_{kl})/(M/h)$$

The absolute fitnesses of the genotypes consist of three factors. The first is the female fecundity of the genotype. The second factor is the effect on the viability in a patch from one clutch laid in that patch by that genotype. The first two factors together give the effective female fecundity of a genotype in terms of young adults emerging from a patch if only that genotype laid one clutch in the patch. The third factor is the effect on viability over all patches caused by the total number of females $\frac{1}{2}N_t$, the dispersal distribution over the patches, and the average effect on viability in a patch by the laying of one clutch, $\sum_{kl} x_{kl} \cdot \exp\{-(c \cdot 2F_{kl})/(M/h)\}$. The relative fitness of the genotypes consists only of the first two factors. Relative fitness is a constant. Optimum fecundity is found at the highest relative fitness. The dispersal distribution causes the third factor to be frequency as well as density dependent. The absolute fitness of a genotype when it is the only genotype present is no longer the absolute fitness of a genotype when other genotypes are present. The case is not Roughgarden's but Prout's (1980): constant relative fitnesses followed by a density dependent factor that is identical for the three genotypes leads to frequency dependent absolute fitnesses. No maximization of any stage can occur. Nor is a weighted sum of the number in several life stages maximized, as is the case in purely density dependent selection (Iwasa and Teramoto 1980).

Fitness Forms in Series

The two basic fitness forms can be present consecutively in time in a life cycle. This means that in the first part of the life cycle sib competition occurs, but that another mortality factor is present between young adults emerging from the sib competition and egg laying adults. The fitness form is $F \cdot v(F) \cdot s$, where v refers to the viability during sib competition and s refers to the subsequent viability from young adult to egg laying adult.

The first part in such a life cycle is sib competition. The sib competition can be for an environmentally fixed food supply, as a monopolized patch of size M/h or for a pair's share of the general juvenile food supply $M/\frac{1}{2}N$. This has consequences for the optimal fecundity. After the evolution of the optimal fecundity the equilibrium number will be the same, the maximum number of young adults from the total juvenile food supply M. In both cases the census point appropriate to the fitness form is the number of emerging adults, the first stage in the life cycle individuals are independent entities. Both cases formally obey model (4). In both cases we have maximization of the number at the first stage that individuals are independent if the viability s from emerging young adult to egg-laying adult is the same for all genotypes. In both cases we have a numerically stable population at a value of the local stability indicator $b = 0$, indicating convergence.

Whatever the relative size of the viabilities s and v, as long as $F \cdot v(F)$ has a maximum, the population is stable in a stable environment. Stability therefore depends upon sib competition at the start of the life cycle. Only sib competition can make the shape of the survival curve operational.

If a pair gets a food share $M/\frac{1}{2}N$ for its nest, the fitness form is $W = s \cdot F \cdot v(F)$, if we write from census point to census point. When survival between emerging and egg laying adult is complete, $s = 1$ and the fitness form of sib competition $F \cdot v(F)$ comes back. When survival between egg and emerging adult is complete, $v(F) = 1$ and the fitness form of random competition vF comes back as sF. Fecundity will evolve towards F^*/s, where F^* is the optimal fecundity without any added adult mortality. It means that winter mortality in birds, for instance, causes higher clutch size. The causal sequence is from mortality s to optimum fecundity F^*/s. This is not so for the other case of total sib competition, monopolization of a food patch of environmentally given size by a female for her own clutch. Fecundity is optimized to patch size and the effective mortality between emerging and egg laying adults follows from that. In both cases a comparison between species or populations would lead us to suppose a trade-off existed between adult mortality and fecundity, as in both cases higher fecundity is accompanied by higher adult mortality - either as cause or as consequence. In reality, what we would be seeing is a compensation by density dependent mortality of a fixed factor in another part of the life cycle. There is no genetic trade-off, in the sense that selection on fecundity induces selection on viability. Only selection on fecundity is present, in the one case reacting on a fixed adult mortality s, in the other case reacting on a food patch of fixed size.

Discussion

Charlesworth (1972, 1981) introduced the concept of the critical age group. If only one stage in the life cycle is subject to density dependent selection and no other stage to any selection at all, the number in the age group experiencing density dependence is maximized. This age group is the critical age group. In the four situations dealt with here, this is immediately true for the first; and after some formal redefinition true for the second and fourth. In the first situation, i.e. all juveniles competing randomly with all other juveniles for the total amount of food, viability depends upon the number of zygotes present and the number of zygotes is maximized. Zygotes are the critical age group. In the second situation, i.e. juveniles compete only with their sibs in a nest, the density dependence acts on the juveniles. Yet it is the number of adults that is maximized, not the number of zygotes at the start of the density dependence. By a formal redefinition the adults can be made the critical age group. Juveniles are not independent individuals. The number of independent young of a pair is given by the fitness F • v(NF/M). Therefore, reckoning only from independent individual to independent individual, one has a density dependent fecundity; and the density this fecundity depends on is the density of adults. The adults are now formally the age group density dependent selection works on and, therefore, the critical age group; although biologically density dependence works on the juveniles.

The same redefinition, from a biologically density dependent viability to a formally density dependent fecundity, makes the egg laying adults the critical age group in the fourth situation dealt with, the case of additional mortality between emerging young adult and egg laying adult. For maximization to occur this additional mortality should be the same for all genotypes. The critical age group and the first age group with independent individuals are no longer identical, as in the first and second situation dealt with. Identical mortality for all genotypes between fledging young adult and egg laying adult means that after selection both the number of fledging young adults and the number of egg laying adults is the maximum that can be reached. The mechanism of sib competition leads to maximization of the number of individuals in the first stage of independence. This shows that the number of emerging young adults is really maximized and that the maximization of the number of egg laying adults is a consequence of that.

In the third situation, that of environmentally determined patches of food in which juveniles compete, a compromise is reached between maximization of the equilibrium number of zygotes and the equilibrium number of adults. In this situation, two differences with the first model exist: sib competition and nonhomogeneous interactions caused by a probability distribution of clutches over patches of food. It is the compromise in sib competition that causes the compromise in equilibrium number and stability. In the absence of sib competition, natural selection leads to the maximization of equilibrium number of zygotes. Nonhomogeneous competition has no influence on the latter result.

In ecology, the question of the origin of numerical stability has been hotly debated. Wynne-Edwards proposed group selection as a mechanism for the regulation of numbers; and it should be noted that regulation of numbers at that time and in that ecological context meant numerical stability rather than the existence of negatively density dependent processes. Lack denied group processes were necessary for regulation of numbers. He showed that in some species the clutch size was optimized and argued that this was the basis of numerical stability. Since then many group-selection models have appeared. Very little model work has, however, been devoted to the elaboration of Lack's ideas on the natural regulation of animal numbers. Perhaps the only one is Goodman (1974) and his life table analysis is not a model of natural selection.

In a model representing the evolution of clutch size according to Lack, the viability of the juveniles in the nest as determined by sib competition is a function of the initial number of eggs laid. Lack showed that in nature the survival curve often has a maximum. Such a maximum in the survival curve is indeed the distinguishing feature in a model of the evolution of fecundity that leads to optimization, not maximization of fecundity. Sib competition makes the survival curve the base line for natural selection, instead of the number of zygotes. Whatever the form of the survival curve, selection on fecundity with pure sib competition leads to numerical stability. Only with a survival curve that has a maximum at intermediate number of zygotes does selection on fecundity with pure sib competition lead to an intermediate optimum fecundity. Other shapes of the survival curve too lead to numerical stability with selection on fecundity in a situation of sib competition, but to maximization of number of eggs. The essential requisite for the evolution of numerical stability is therefore sib competition, not the existence of an intermediate optimum fecundity.

Acknowledgements. With many thanks to Mr. D. Smit, who provided the images of the animals in the several life histories and prepared the figures and to Mr. P. Brouwer who prepared the photographs.

References

Anderson WW (1971) Genetic equilibrium and population growth under density regulated selection. Am Nat 105:489–498

Charlesworth B (1971) Selection in density regulated populations. Ecology 52:469–474

Charlesworth B (1972) Selection in populations with overlapping generations. III. Conditions for genetic equilibrium. Theor Popul Biol 3:377–395

Charlesworth B (1981) Evolution in age structured populations. Univ Press, Cambridge

Chesson PL (1981) Models for spatially distributed populations: the effect of within patch variability. Theor Popul Biol 19:288–325

Clarke BC (1972) Density dependent selection. Am Nat 106:1–13

Goodman D (1974) Natural selection and a cost of reproductive effort. Am Nat 108:247–268

Heckel DG, Roughgarden J (1980) A species near its equilibrium size in a fluctuating environment can evolve a lower intrinsic rate of increase. Proc Natl Acad Sci USA 77:7497–7500

Iwasa Y, Teramoto E (1980) A criterion of life history evolution based on density dependent selection. J Theor Biol 84:545–566

Jong de G (1979) The influence of the distribution of juveniles over patches of food on the dynamics of a population. Neth J Zool 29:33–51

Jong de G (1982) Fecundity selection and maximization of equilibrium number. Neth J Zool 32:572–585

Léon JA, Charlesworth B (1978) Ecological versions of Fisher's fundamental theorem of natural selection. Ecology 59:457–464

MacArthur RH (1962) Some generalized theorems of natural selection. Proc Natl Acad Sci USA 48:1893–1897

MacArthur RH, Whilson EO (1967) The theory of island biogeograph. Princeton University Press, Princeton

Prout T (1980) Some relationships between density-independent selection and density dependent population growth. Evol Biol 13:1–68

Roughgarden J (1971) Density dependent natural selection. Ecology 52:453–468

Roughgarden J (1976) Resource partitioning among competing species – a coevolutionary approach. Theor Popul Biol 9:388–424

Roughgarden J (1979) Theory of population genetics and evolutionary ecology: an introduction. Macmillan, New York

Thomas WR, Pomerantz MJ, Gilpin ME (1980) Chaos, asymmetric growth and group selection for dynamical stability. Ecology 61:1312–1320

Turelli M, Petry D (1980) Density dependent selection in a random environment: an evolutionary process that can maintain stable population dynamics. Proc Natl Acad Sci USA 77:7501–7505

: its biological and conceptual aspects. Prentice Hall, NJ

Selection in a Cyclically Changing Environment

K.P. HADELER [1]

Introduction

The classical selection model of population genetics describes a diploid population of infinite size. The individuals are assumed genetically identical with the exception of a single autosomal locus with two alleles A and a. Viabilities are attributed to the genotypes AA, Aa, aa. The state of the population can be described by the frequency of the allele A. In the case of separated generations the model assumes the form of a difference equation, for overlapping generations one obtains a differential equation (see Crow and Kimura 1975, Edwards 1977, Hadeler 1973, 1974, Nagylaki 1977). It is well known that in discrete time as well as in continuous time there are essentially four qualitatively different cases: if the heterozygotes are intermediate, then either the allele A or the allele a is superior; if both alleles are present initially then the frequency of the superior allele increases monotonely and converges to 1. If the heterozygotes are superior then there is a single polymorphism; every population in which both alleles are present converges to that polymorphism. If the heterozygotes are inferior, then the two pure states are locally stable and a unique unstable polymorphism acts as a threshold.

Several authors have studied similar models in a changing environment. Then the viabilities depend in a deterministic way on the time variable or they represent random variables (Haldane and Jayakar 1963, Hoekstra 1975, Karlin and Liberman 1974, 1975, Nagylaki 1975, 1977, Cornette 1981, Gillespie 1973). Among the purely deterministic models an interesting case is the cyclically changing environment. Again, in the most simple case the cycle has length two, i.e., the environment is alternating between two states (e.g., representing summer and winter). For this case Hoekstra (1975) and Karlin and Liberman (1975) have given general results and various examples. We carry Hoekstra's approach somewhat further.

While in the classical models stationary polymorphisms are distinguished solutions, in a periodic environment the role of such polymorphisms is assumed by nontrivial (i.e., $\neq 0,1$) periodic solutions. The existence and stability of such solutions is of particular interest: it can be shown that in our models every population tends asymptotically to a periodic polymorphism or to fixation.

First, we shall review the conditions for the existence of periodic polymorphisms in the model with discrete time and cycle length two. Then we show that a bifurcation

1 Lehrstuhl für Biomathematik, Universität Tübingen, 7400 Tübingen, FRG

Population Biology and Evolution
Ed. by K. Wöhrmann and V. Loeschcke

of a stable polymorphism is a quite general phenomenon. In fact we show the following (Proposition 1): suppose in a cyclic model with two environments one of the two environments supports a stable polymorphism. Then we can choose the other environment such that it also supports a stable polymorphism and that the full cyclic model has two distinct stable periodic polymorphisms and an intermediate unstable periodic polymorphism.

In the continuous time case we first consider a situation in which the population stays in two alternating constant environments for fixed times h_1 and h_2, such that $h_1 + h_2$ is the prescribed period. In contrast to the case with discrete time one can show (Theorem 2): in addition to the pure states there is at most one periodic solution or every solution is periodic. Finally, we provide some conditions for periodic polymorphisms in the general continuous time case (Theorems 3 and 4).

The Classical Selection Model

The viabilities $\sigma, \rho, \tau > 0$ of the genotypes AA, Aa, aa enter the model as parameters. The state of the population is given by the frequency $u \in [0,1]$ of the allele A.

In the case of discrete time or separated generations, the evolution of the frequency of the allele A in successive generations is described by the difference equations

$$u_{t+1} = g(u_t), \qquad t = 0,1,2,\dots, \tag{1}$$

where the function $g: [0,1] \to [0,1]$ is given by

$$g(u) = \frac{\sigma u^2 + \rho u(1-u)}{\sigma u^2 + 2\rho u(1-u) + \tau(1-u)^2}. \tag{2}$$

In the case of continuous time or overlapping generations, the density u is described by the differential equation

$$\dot{u}(t) = f(u), \tag{3}$$

where $f: [0,1] \to R$ is given by

$$f(u) = u(1-u)[\rho - \tau - (2\rho - \sigma - \tau)u]. \tag{4}$$

Apparently in the case of continuous time only the relative viabilities of the homozygotes with respect to the heterozygotes are important. We put

$$\rho - \tau = a, \quad \rho - \sigma = b \tag{5}$$

and write

$$f(u) = u(1-u)[a - (a+b)u]. \tag{6}$$

In both cases the following is true: $u = 0$ and $u = 1$ are stationary points and there is at most one further stationary point

$$\bar{u} = \frac{\rho - \tau}{2\rho - \sigma - \tau} \quad \left(= \frac{a}{a+b}\right) \tag{7}$$

provided this number falls into the interval (0,1). This is the case if either $\sigma > \rho, \tau > \rho$, or $\sigma < \rho, \tau < \rho$.

Case 1: $\sigma < \rho, \tau < \rho$. Heterozygotes are superior, $\bar{u}$ is a stable polymorphism; $u = 0$ and $u = 1$ are unstable stationary points.

Case 2: $\sigma > \rho, \tau > \rho$. Then heterozygotes are inferior, $\bar{u}$ is a threshold; $u = 0$ and $u = 1$ are stable stationary points.

Case 3: $\sigma > \rho > \tau$ (or Case 4: $\sigma < \rho < \tau$). Heterozygotes are intermediate. Then $u = 0$ and $u = 1$ are the only stationary points, $u = 0$ is unstable, $u = 1$ is stable (or $u = 0$ stable, $u = 1$ unstable in Case 4).

For discrete time as well as for continuous time the mean viability

$$\phi(u) = \sigma u^2 + 2\rho u(1-u) + \tau(1-u)^2 \tag{8}$$

is a Lyapunov function, along trajectories $\phi(u)$ is nondecreasing.

One important fact is that the function g is strictly monotone for arbitrary choices of the parameters σ, ρ, τ. In fact, a simple calculation shows

$$g'(u) = \frac{\sigma\rho u^2 + 2\sigma\tau u(1-u) + \rho\tau(1-u)^2}{\phi^2(u)}. \tag{9}$$

From Eq. (9) follows that the inverse mapping $g^{-1}: [0,1] \to [0,1]$ exists and is again a differentiable function. The function g^{-1} cannot be represented in the form of Eq. (2). An explicit representation is

$$g^{-1}(v) = \frac{1}{2}\frac{\rho + 2v(\tau-\rho) \pm \sqrt{\rho^2 + 4v(1-v)(\sigma\tau-\rho^2)}}{v(\sigma+\tau-2\rho)-(\sigma-\rho)}. \tag{10}$$

Here the + sign has to be chosen if the denominator $v(\sigma+\tau-2\rho)-(\sigma-\tau)$ is positive, the − sign if it is negative. In the limit case, where the denominator vanishes

$$g^{-1}\left(\frac{\sigma-\rho}{\sigma+\tau-2\rho}\right) = \frac{\tau(\sigma-\rho)}{\tau(\sigma-\rho)+\sigma(\tau-\rho)}.$$

On the other hand, if we solve Eq. (3), Eq. (4) backward in time, we obtain a selection model of the same form, but with A and a interchanged.

Environment with Two Seasons

Hoekstra (1975) has given sufficient conditions for existence and stability of periodic polymorphisms.

Suppose $g: [0,1] \to [0,1]$ is *any* continuously differentiable, nondecreasing function with $g(0) = 0$, $g(1) = 1$, and $g'(u) > 0$ for $0 < u < 1$. Then the difference equation

$$u_{t+1} = g(u_t), \quad t = 0,1,2,... \tag{11}$$

has the following properties:

1. The stationary solutions are the zeros of $g(u) - u = 0$.
2. If a stationary solution $\bar{u}$ is nondegenerate, i.e., $g'(\bar{u}) \neq 1$, then the stability is determined by $g'(\bar{u})$: stable for $g'(\bar{u}) < 1$, unstable for $g'(\bar{u}) > 1$.

3. Every solution of Eq. (11) converges to a stationary solution.
4. Once the graph of g is known, the solution of Eq. (11) kann be easily constructed graphically (see Fig. 1).

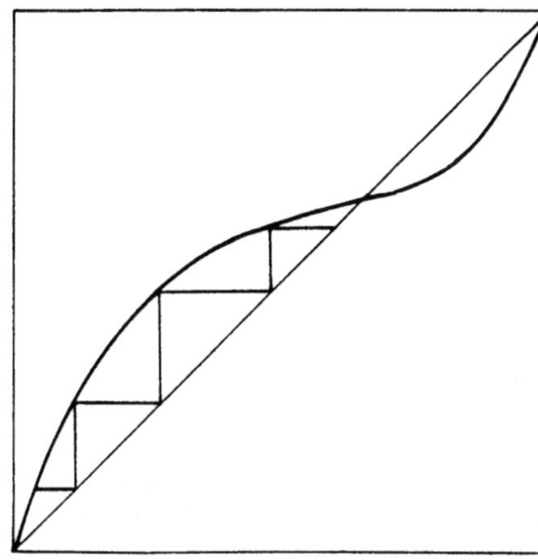

Fig. 1. Selection iteration (case of superior heterozygotes)

Now suppose that two functions g_1 and g_2 with the said properties represent alternating environments. Then the frequency of the allele A in successive generations is described by the equations.

$$\begin{aligned} u_{2k+1} &= g_1(u_{2k}) \\ & \qquad\qquad k = 0,1,2,\ldots \ . \\ u_{2k+2} &= g_2(u_{2k+1}) \end{aligned} \tag{12}$$

Then it is sufficient to investigate the equation

$$u_{t+1} = \bar{g}(u_t) \qquad t = 0,1,2, \tag{13}$$

where $\bar{g} = g_2 \circ g_1$ is the composition of the functions g_2 and g_1. The function $\bar{g}$ has the required properties also. Equation (13) describes the frequency as if the population is observed only every second generation. Fixed points of Eq. (13) correspond to solutions of period 2 in Eq. (12), and vice versa. If a fixed point of Eq. (13) is stable (or unstable) then the corresponding 2-periodic solution of Eq. (12) is stable (respectively, unstable).

Even if g_1, g_2 are rather simple functions, the expression for $\bar{g}$ may be quite cumbersome. Therefore, it is worthwhile to provide a simple graphical construction for the iterates u_t. One such construction has been given by Hoekstra (1975) (see Fig. 2): Draw the graphs of g_1, g_2. Start at u_0, find $g_1(u_0) = u_1$, reflect at the diagonal, find $g_2(u_1) = u_2$ etc. In this construction "spirals" typically appear and a stationary state of Eq. (13) appears as a solution of period 2, as a "box".

A simpler picture is obtained if we draw the graph g_2^{-1} instead (note that g_2 is monotone) which is the mirror image of the graph of g_2 with respect to the diagonal (Hoekstra 1975) (Fig. 3).

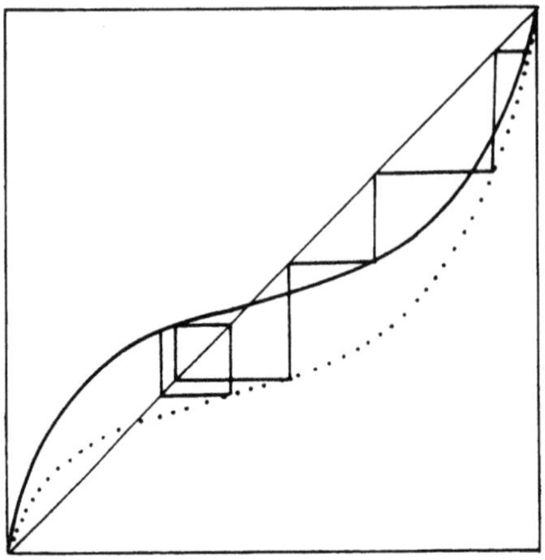

Fig. 2.

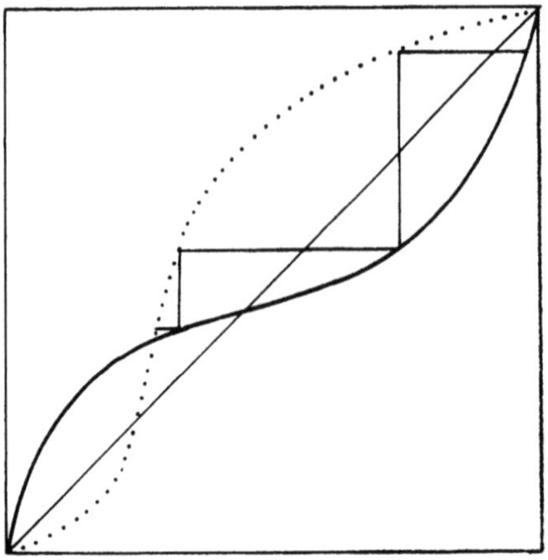

Fig. 3.

Fig. 2. Selection in a cyclic environment. The two curves represent the selection regimes. The interaction appears as a zigzag between each of the curves and the median

Fig. 3. Selection in a cyclic environment. The *dotted curve* represents the inverse of the second selection regime. The iteration appears as a zigzag between the two curves

The iteration is now represented by a zigzag line between the graphs of g_1 and g_2^{-1} and stationary solutions appear as intersections of the two graphs. From the graphical representation it is obvious that a stationary solution is stable if at the intersection the graph of g_2^{-1} is steeper than that of g_1. This condition just says that $(g_2^{-1})' > g_1'$, i.e., $(g_2')^{-1} > g_1'$, $g_2'g_1' = (g_2 \circ g_1)' < 1$, as it should be.

Hoekstra's sufficient conditions for the existence of a stable polymorphism result from the observation that from the four numbers $g_1'(0)$, $g_1'(1)$, $g_2'(0)$, $g_2'(1)$ one can draw some information how the graphs of g_1 and g_2^{-1} intersect. These conditions read as follows: if $g_2'(0)g_1'(0) > 1$ and $g_2'(1)g_1'(1) > 1$, then in Eq. (13) there is at least one stable fixed point corresponding to a stable polymorphism of period 2 in Eq. (12) [or there is an attracting continuum of fixed points].

However, this stable equilibrium may undergo a bifurcation, which typically will lead to two stable fixed points and an unstable fixed point between them.

Similarly, if $g_2'(0)g_1'(0) < 1$ and $g_2'(1)g_1'(1) < 1$, then there is at least one unstable polymorphism which, however, may bifurcate.

Now consider the special case, where g_1 and g_2 are given in the form of Eq. (2) with fitness parameters σ_1, ρ_1, τ_1 and σ_2, ρ_2, τ_2, respectively. With $\bar{g} = g_2 \circ g_1$ the equation $\bar{g}(u) = u$ is equivalent to a polynomial equation of degree 5. After the two trivial zeros $u = 0$ and $u = 1$ have been factored out, a polynomial of degree 3 remains, which indicates that there may be three solutions in (0,1).

We give a short account of the various possible cases. We have $g'(0) = \rho/\tau$, $g'(1) = \rho/\sigma$. Thus, Hoekstra's sufficient conditions are

$$\rho_1\rho_2 > \sigma_1\sigma_2, \quad \rho_1\rho_2 > \tau_1\tau_2 \tag{14}$$

for a stable periodic solution.

Also it is immediately clear that if in both seasons the environment is in the same case (case 3 or 4) then the model (12) has the corresponding behavior, $u \to 1$ in case 3, $u \to 0$ in case 4.

More complicated situations can arise. In this discrete time model the appearance of more than one polymorphism is a quite common feature although we have not found numerical evidence in the literature. Suppose one of the two environments supports a stable polymorphism. Then we can choose the other environment such that it also supports a stable polymorphism and that the full cyclic model has two distinct stable polymorphisms.

In mathematical terms we have to show:

Proposition 1. For given σ_1, ρ_1, τ_1 with $\tau_1 < \sigma_1 < \rho_1 = 1$ one can choose σ_2, ρ_2, τ_2 such that $\sigma_2 < \tau_2 < \rho_2 = 1$ and such that the equation $g_2[g_1(u)] - u = 0$ has three zeros in (0,1).

Proof: Insert the expression

$$\tilde{u} = g_1(u) = \frac{\sigma_1 u^2 + u(1-u)}{\sigma_1 u^2 + 2u(1-u) + \tau_1(1-u)^2}$$

into

$$u = g_2(\tilde{u}) = \frac{\sigma_2 \tilde{u}^2 + \tilde{u}(1-\tilde{u})}{\sigma_2 \tilde{u}^2 + 2\tilde{u}(1-\tilde{u}) + \tau_2(1-\tilde{u})^2}$$

and multiply by denominators. This manipulation gives a polynomial of degree 5

$$\begin{aligned}&\sigma_2(1-u)[\sigma_1 u^2 + u(1-u)]^2\\ &\quad + [\sigma_1 u^2 + u(1-u)][u(1-u) + \tau_1(1-u)^2](1-2u)\\ &\quad - \tau_2 u[u(1-u) + \tau_1(1-u)^2]^2 = 0\,.\end{aligned}$$

The zeros $u = 0$ and $u = 1$ can be factored out to give

$$\begin{aligned}&\sigma_2 u[\sigma_1 u + (1-u)]^2 + [\sigma_1 u + (1-u)][u + \tau_1(1-u)](1-2u)\\ &\quad - \tau_2(1-u)[u + \tau_1(1-u)]^2 = 0\,.\end{aligned}$$

The transformation $v = u/(1-u)$ carries the interval (0,1) into the interval $(0, \infty)$ and the polynomial into

$$Q(v) = \sigma_2 v(\sigma_1 v + 1)^2 + (1-v)(v+\tau_1)(\sigma_1 v + 1) - \tau_2(v+\tau_1)^2\,.$$

Now we evaluate Q at $v = 0$, $v = 1$, $\bar{v} = (1-\tau_1)/(1-\sigma_1) > 1$, $v = +\infty$.

$$\begin{aligned}Q(0) &= \tau_1(1-\tau_2\tau_1) \qquad Q(\infty) \sim \sigma_1(\sigma_2\sigma_1 - 1)v^3\,,\\ Q(1) &= \sigma_2(\sigma_1+1)^2 - \tau_2(\tau_1+1)^2\,,\\ Q(\bar{v}) &= (\bar{v}+\tau_1)^2[(\sigma_2-1)\bar{v} + 1 - \tau_2]\,.\end{aligned}$$

Then we choose τ_2 and σ_2 such that $\tau_2 < 1$, $\sigma_2 < 1$ (then $\tau_2\tau_1 < 1$, $\sigma_2\sigma_1 < 1$),

$$\frac{\sigma_2}{\tau_2} < \left(\frac{\tau_1+1}{\sigma_1+1}\right)^2$$

$$\tau_2 < 1 + (\sigma_2 - 1)\bar{v}\,.$$

It is easy to see that in the σ_2, τ_2 plane there is a triangle where these conditions are met.

Then $Q(0) > 0$, $Q(1) < 0$, $Q(\bar{v}) > 0$, $Q(\infty) < 0$. Thus, there are three sign changes and thus three simple zeros. There is at least one polymorphism, since the conditions of Eq. (14) are satisfied. There are two stable and one unstable polymorphisms because stable and unstable fixed points alternate (i.e., because of the invariance of the topological degree).

Continuous Time Models

Now we address the case of continuous time. For the moment we consider a general equation

$$\dot{u} = f(u)\,, \tag{16}$$

where f: $[0,1] \to \mathbb{R}$ is continuously differentiable and satisfies $f(0) = f(1) = 0$. Assume (for simplicity) that f vanishes only at isolated points and that the derivative does not vanish at such points (thus, f has only "simple" zeros).

Next consider the "solution operator" $G(t, u_0)$: for $u_0 \in [0,1]$ and $t \in \mathbb{R}$ let $G(t, u_0)$ be the solution of Eq. (14) with initial datum $u(0) = u_0$, evaluated at t. The solution operator $G(t, u)$ has the following obvious properties.

1. G: $\mathbb{R} \times [0,1] \to \mathbb{R}$ is differentiable in both variables.
2. As a function of t, $G(t, u)$ satisfies the differential equation

$$\frac{\partial G(t,u)}{\partial t} = f[G(t,u)]\,.$$

3. $G(t, u)$ satisfies the initial condition

$$G(0,u) = u\,.$$

4. $u = 0$ and $u = 1$ are stationary,

$$G(t,0) = 0, \qquad G(t,1) = 1, \qquad \text{for all } t.$$

5. From the uniqueness of solutions of differential equations it follows

$$\frac{\partial G(t,u)}{\partial u} \geqslant 0\,.$$

6. By separation of variables in Eq. (16) it follows

$$\frac{\partial G(t,u)}{\partial u} = \begin{cases} \dfrac{f[G(t,u)]}{f(u)} & \text{if } f(u) \neq 0 \qquad (17) \\[2ex] e^{tf'(u)} & \text{if } f(u) = 0\,. \qquad (18) \end{cases}$$

7. The solutions approach stationary points:

$$\lim_{t\to+\infty} G(t,u) \in \mathcal{N}(f),$$

where $\mathcal{N}(f)$ is the set of zeros of the function f. Now we establish a connection between the differential Eq. (16) and the difference Eq. (11). Let $h > 0$. Let $u(t)$ be a solution of the differential Eq. (16) describing the evolution of the population in continuous time. Then we observe the population at discrete points kh, $k = 0,1,2,\dots$. Denote the sequence of observed values by

$$u_k = u(kh), \qquad k = 0,1,2,\dots \; . \tag{19}$$

Then the sequence u_k satisfies the difference equation

$$u_{k+1} = g(u_k), \tag{20}$$

where the function g is defined by

$$g(u) = G(h,u). \tag{21}$$

Here the function g depends on the parameter h which measures the frequency of observations.

The function g has all properties required for Eq. (11). Thus, a continuous time selection model, when observed at discrete equidistant times, behaves like a discrete time model.

In particular, if the function f is given by Eq. (4) then the function g, as given by Eq. (21), in each of the four cases, looks qualitatively like the function given by Eq. (2). However, it cannot be represented in the form of Eq. (2), in fact a representation in terms of elementary functions cannot be given.

The function g constructed in Eq. (21) has a further property, namely, the inverse function g^{-1} has an explicit representation

$$g^{-1}(u) = G(-h,u). \tag{22}$$

The function g^{-1} is obtained from the differential equation by following the solutions for negative t or the solution of $\dot{u} = -f(u)$ for positive t.

In particular for the function f given by Eq. (4) the function $-f$ corresponds to selection with relative fitness coefficients $-a$ and $-b$.

In the following we shall frequently use the following fact: as far as the function g is concerned, a change in the duration h into $\tilde{h}$ is equivalent to a change in the viabilities $\sigma \to \tilde{h}\sigma/h$, $\rho \to \tilde{h}\rho/h$, $\tau \to \tilde{h}\tau/h$. In this sense frequent observation is equivalent with slow selection.

Now consider two functions f_1, f_2 which represent two environments. Let the population stay some time $h_1 > 0$ in the environment given by f_1 and during a time interval h_2 in environment f_2 alternatingly. With the functions f_1, f_2 we associate the solution operators G_1, G_2 according to Eq. (16)

$$g_1(u) = G_1(h_1,u), \qquad g_2(u) = G_2(h_2,u).$$

Then we can apply all arguments related to Eqs. (12) and (13).

From the conditions introduced earlier one derives easily: if

$$h_1 f_1'(0) + h_2 f_2'(0) > 0, \qquad h_1 f_1'(1) + h_2 f_2'(1) > 0 \tag{23}$$

then there is a stable polymorphic periodic solution. If the inequality signs are both reversed then there is an unstable polymorphic solution.

In the special case wehre the functions f_1, f_2 have the form of Eq. (6)

$$f_i(u) = u(1-u)\,[a_i - (a_i + b_i)u], \qquad i = 1,2 \tag{24}$$

these conditions of Eq. (23) are

$$h_1 a_1 + h_2 a_2 > 0, \qquad h_1 b_1 + h_2 b_2 > 0\,. \tag{25}$$

Here in contrast to the cyclic selection model with discrete time we can exclude the possibility that several nontrivial periodic solutions exist.

Theorem 2. Suppose a cyclically changing environment, where the population stays alternatingly in environments f_1 and f_2 of the form of Eq. (24) during intervals of length h_1 and h_2, respectively. Then in addition to $u = 0, u = 1$, there is at most one periodic solution or every solution is periodic.

Proof. Suppose there are numbers $\alpha, \beta, |\alpha| + |\beta| \neq 0$, such that $\alpha f_1(u) + \beta f_2(u) \equiv 0$. We can assume $\beta = -1$, thus $f_2 = \alpha f_1$. Then $G_2(h,u) = G_1(\alpha h, u)$ and, thus,

$$\begin{aligned} g_2 \circ g_1 &= G_2(h_2,\cdot) \circ G_1(h_1,\cdot) \\ &= G_1(\alpha h_2,\cdot) \circ G_1(h_1,\cdot) \\ &= G_1(\alpha h_2 + h_1,\cdot)\,. \end{aligned}$$

Thus, the result is the same as solving the equation $\dot{u} = f_1(u)$ with time step $\alpha h_2 + h_1$. In this case the assertion is true.

Now assume that f_1, f_2 are linearly independent. We have to show that for any $h_1, h_2 > 0$ the equation

$$G_1(h_1, u) = G_2(-h_2, u)$$

has at most one solution in (0,1).

Define a function φ: $\mathbb{R}^2 \to [0,1]$ by

$$\begin{aligned} \varphi(h_1,h_2,u) &= \frac{G_1(h_1,u) - G_2(-h_2,u)}{u(1-u)}, \qquad u \in (0,1) \\ \varphi(h_1,h_2,0) &= h_1 f_1'(0) + h_2 f_2'(0)\,, \\ \varphi(h_1,h_2,1) &= h_1 f_1'(1) + h_2 f_2'(1)\,. \end{aligned} \tag{26}$$

Assume for the moment that $f_1'(0) \neq 0$, $f_1'(1) \neq 0$.

Keep h_1 fixed and let h_2 vary. For $h_2 = 0$ the function $\varphi(h_1, h_2, \cdot)$ either has no zero in (0,1) or one zero, which is simple. If for some $h_2 = \bar{h}_2 > 0$ two or more zeros occur in (0,1) then, by the continuous dependence of φ on h_2, we have the following alternative.

Case 1: There is a number $h_2 \in (0, \bar{h}_2)$, such that $\varphi(h_1, h_2, u)$ has a double zero in (0,1).

Then the function $G_1(h_1, u) - G_2(-h_2, u)$ has a double zero $\bar{u}$ in (0,1). By appropriately redefining f_2 into $\bar{f}_2 = (-h_2/h_1) f_2$ and G_2 into $\bar{G}_2$, we can assume $-h_2 =$

$h_1 = h$. Then we have

$$G_1(h,\bar{u}) = \bar{G}_2(h,\bar{u}), \tag{27}$$

$$\frac{\partial}{\partial u} G_1(h,\bar{u}) = \frac{\partial}{\partial u} \bar{G}_2(h,\bar{u}). \tag{28}$$

Denote the left-hand side in Eq. (27) by $\bar{v}$. Then from Eq. (17) follows

$$\frac{f_1(\bar{v})}{f_1(\bar{u})} = \frac{\bar{f}_2(\bar{v})}{\bar{f}_2(\bar{u})}$$

or

$$\frac{f_1(\bar{v})}{\bar{f}_2(\bar{v})} = \frac{f_1(\bar{u})}{\bar{f}_2(\bar{u})}. \tag{29}$$

Denote the left-hand side of Eq. (29) by α. By Eq. (24) the functions $f_1, \bar{f}_2$ have the form

$$\begin{aligned} f_1(u) &= u(1-u)\,[a_1 - (a_1 + b_1)u] \\ \bar{f}_2(u) &= u(1-u)\,[a_2 - (a_2 + b_2)u] \end{aligned} \tag{30}$$

with some constants a_i, b_i.

Thus,

$$a_1 - (a_1 + b_1)\bar{v} = \alpha\,[a_2 - (a_2 + b_2)\bar{v}]$$

$$a_1 - (a_1 + b_1)\bar{u} = \alpha\,[a_2 - (a_2 + b_2)\bar{u}].$$

If $\bar{v} \neq \bar{u}$ it follows that $a_1 = \alpha a_2$, $b_1 = \alpha b_2$, or

$$f_1(u) \equiv \alpha \bar{f}_2(u).$$

If $\bar{v} = \bar{u}$ then $\bar{u}$ is a fixed point of $G_1(h,\cdot)$ and of $\bar{G}_2(h,\cdot)$. Thus, $\bar{u}$ is a zero of f_1 and of $\bar{f}_2$,

$$\bar{u} = \frac{a_1}{a_1 + b_1} = \frac{a_2}{a_2 + b_2}$$

and

$$f_1(u) = (a_1 + b_1)\,u\,(1-u)\,(\bar{u} - u)$$

$$\bar{f}_2(u) = (a_2 + b_2)\,u\,(1-u)\,(\bar{u} - u).$$

In both cases f_1 and $\bar{f}_2$ appear linearly dependent contrary to the assumption.

Case 2: There is a number $h_2 \in (0,\bar{h}_2)$ such that $\varphi(h_1,h_2,\bar{u}) = 0$ for some $\bar{u} \in (0,1)$ and also $\varphi(h_1,h_2,0) = 0$ or $\varphi(h_1,h_2,1) = 0$. We assume $\varphi(h_1,h_2,0) = 0$. The other case is treated similarly. Then the function $G_1(h_1,u) - G_2(-h_2,u)$ has a double zero at $\bar{u} = 0$, at least a simple zero at $u = 1$, and also a zero at $\bar{u} \in (0,1)$. As before, by appropriately redefining f_2, one can assume $-h_2 = h_1 = h$. Assume that $f_1, \bar{f}_2$ have the explicit representations of Eq. (30).

From Eq. (18) we infer $\exp[hf_1'(0)] = \exp[h\bar{f}_2'(0)]$, thus, $a_1 = a_2$. Assume $b_1 \leqslant b_2$. Then $f_1(u) \leqslant \bar{f}_2(u)$ for $0 \leqslant u \leqslant 1$. But from the equality $G_1(h,\bar{u}) = \bar{G}_2(h,\bar{u})$

follows $f_1(u) = \bar{f}_2(u)$ for $0 \leqslant u \leqslant \bar{u}$, thus, $b_1 = b_2$. Similarly from $b_1 \geqslant b_2$ it follows that $b_1 = b_2$.

Finally, if $f_1'(0) \neq 0$, $f_1'(1) \neq 0$ is not satisfied, one can either find an $a \in \mathbb{R}$ such that $(f_1 + a\bar{f}_2)'(0) \neq 0$, $(f_1 + a\bar{f}_2)'(1) \neq 0$ or $f_1, \bar{f}_2$ are dependent contrary to the assumption.

General Periodic Viabilities

In fact in Theorem 2 we have treated a nonlinear differential equation with a (discontinuous) periodic right-hand side. This concept can be easily generalized. Assume that in Eq. (16) the fitness coefficients are continuous functions of the time t, periodic with the same period. We can assume that this common period is equal to 1. Then the differential equation is

$$\dot{u}(t) = f[t, u(t)], \tag{31}$$

where

$$f(t,u) = u(1-u)[a(t) - (a(t) + b(t))u]$$

and

$$a(t+1) = a(t), \quad b(t+1) = b(t) \tag{33}$$

are the periodic relative fitness coefficients of the homozygotes relative to the heterozygotes. The relatively simple case where $a(t) = ab(t)$ with a constant has been treated by Nagylaki (1975).

Let $G(t, u)$ be the evolution operator of Eq. (31) [defined exactly as in Eq. (16)]. The derivative $dG(1,u)/du$ at u_0 is obtained as follows:

First solve Eq. (31) with initial datum $u(0) = u_0$. Then solve the linear equation

$$\dot{w}(t) = f_u[t, u(t)]w(t), \quad w(0) = 1$$

and evaluate the solution at $t = 1$. Thus, one finds

$$\left.\frac{\partial G(1,u)}{\partial u}\right|_{u=u_0} = e^{\int_0^1 f_u[t,u(t)]dt},$$

in particular

$$\left.\frac{\partial G(1,u)}{\partial u}\right|_{u=0} = e^{\int_0^1 a(t)dt}, \quad \left.\frac{\partial G(1,u)}{\partial u}\right|_{u=1} = e^{\int_0^1 b(t)dt}.$$

Thus, we find

Theorem 3: If

$$\int_0^1 a(t)dt > 0, \quad \int_0^1 b(t)dt > 0, \tag{34}$$

then Eq. (31) has at least one polymorphic periodic solution, i.e., a solution with values in (0,1). The same is true, if both inequality signs in Eq. (34) are reversed.

In general, i.e., if the periodic solutions are isolated, then Eq. (34) guarantees the existence of at least one stable polymorphic periodic solution. Similarly, if in condition Eq. (34) the inequality signs are reversed then there is at least one unstable solution of this type.

Theorem 4. Suppose $a(t) + b(t) > 0$ for all t [or $a(t) + b(t) < 0$ for all t]. Then Eq. (31) has at most one periodic solution in addition to $u = 0$ and $u = 1$.

Proof: The substitution [a similar substitution has been applied by Coleman (1981) to an ecological model]

$$u = \frac{1}{1 + e^{-v}} \tag{35}$$

carries the interval (0,1) into the real axis and the Eq. (26) into

$$\dot{v} = \tilde{f}(t, v) , \tag{36}$$

where

$$f(t, v) = a(t) - [a(t) + b(t)] \frac{e^v}{e^v + 1} . \tag{37}$$

If Eq. (31) has a periodic solution in (0,1), then Eq. (36) has a bounded periodic solution and conversely. Let $\tilde{G}(t, v)$ be the solution of Eq. (36) with initial condition $v(0) = v_0$, evaluated at t. Consider $\tilde{G}(1,\cdot)$. The derivative $d\tilde{G}(1,v)/dv$ at v_0 is obtained as before. Thus,

$$\frac{\partial \tilde{G}(1, v)}{\partial v} = \exp \int_0^1 \tilde{f}_v \, [t, v(t)] dt . \tag{38}$$

Let $a(t) + b(t) \geqslant K > 0$. Then

$$\tilde{f}_v(t, v) = - [a(t) + b(t)] \frac{e^v}{(e^v + 1)^2} \leqslant - K \frac{e^v}{(e^v + 1)} .$$

Thus, $\tilde{G}(1, v)$ is a contraction in every bounded domain.

Discussion

We have shown that the selection model with overlapping generations, when observed at discrete intervals, behaves qualitatively like the model with separated generations. If in the continuous time model the environment is cyclically alternating between two states, with fixed durations, then there is at most one periodic solution in addition to the trivial constant solutions (Theorem 2). This unique result is valid independently of the values of the viabilities. Sufficient conditions for the existence and stability of such polymorphic periodic solutions are provided.

If in the continuous time model the environment is an arbitrary periodic function then there is at most one polymorphic periodic solution if the viability of the heterozygotes is always less or always greater than the arithmetic mean of the viabilities of the homozygotes (Theorem 3). There is at least one polymorphic periodic solution if the averaged viabilities of the homozygotes are both smaller (or both greater) than the averaged viability of the heterozygotes (Theorem 4).

References

Coleman BD (1981) On optimal intrinsic growth rates for populations in periodically changing environments. J Math Biol 12:343–354

Cornette JL (1981) Deterministic genetic models in varying environments. J Math Biol 12:173 to 186

Crow JF, Kimura M (1975) Introduction to population genetics. Harper & Row, New York

Edwards AWF (1977) Foundations of mathematical genetics. Univ Press, Cambridge

Gillespie J (1973) Polymorphisms in random environments. Theor Popul Biol 4:193–195

Hadeler KP (1973) Selektionsmodelle in der Populationsgenetik. Meth Verf Math 9:137–160

Hadeler KP (1974) Mathematik für Biologen. Springer, Berlin Heidelberg New York

Haldane JBS, Jayakar JD (1963) Polymorphism due to selection of varying direction. J Genet 58:237–242

Hoekstra RF (1975) A deterministic model of cyclic selection. Genet Res 25:1–15

Karlin S, Liberman U (1974) Random temporal variation in selection intensities: Case of large population size. Theor Pop Biol 6:355–382

Karlin S, Liberman U (1975) Random temporal variation in selection intensities: One-locus two allele model. J Math Biol 2:1–17

Nagylaki T (1975) Polymorphisms in cyclically varying environments. Heredity 35:67–74

Nagylaki T (1977) Selection in one- or two-locus systems. Lecture notes in biomathematics, vol 15. Springer, Berlin Heidelberg New York

Evolution in a Temporally Varying Environment: Density and Composition Dependent Genotypic Fitnesses*

F.B. CHRISTIANSEN[1]

Introduction

It is difficult to imagine a biotic environment remaining unchanged for a sufficiently long time to allow significant evolutionary change. Nevertheless, our understanding of the process of genetic evolution by natural selection is founded primarily on considerations of deterministic processes in a constant environment. This discord is avoided if time-homogeneous deterministic models are considered as a description of the average evolutionary change in the population. This assertion is corroborated by results of population genetic models in which selection varies through time. Haldane and Jayakar (1963) and Gillespie (1973) showed that the condition for initial increase of a rare allele is the same in a varying environment as in a constant environment when the genotypic fitnesses are calculated as the geometric average fitness through time. Thus, the establishment of a new allele in the population is fully described by the average selection over time and the condition for protected polymorphism of the allele and its alternative is also formulated in terms of time-average genotypic fitnesses. However, directional selection in time-average fitnesses does not exclude the existence of stable polymorphism (Karlin and Liberman 1974).

These simple results on the initial increase of an allele are at variance with ideas on the influence of varying environment on evolution, when genotypic fitness is influenced by the density and composition of the population. The concepts of r and K selection (MacArthur 1962, MacArthur and Wilson 1967) are founded on the distinction between varying environments and environments that are stable through time. Recently Heckel and Roughgarden (1980) and Turelli and Petry (1980) considered evolution by density dependent selection in a varying environment and obtained results that could not be inferred from constant environment models, so the immediate expectations from the population genetical results are violated. Here I will attempt to elucidate the apparent conflict between the results obtained in population genetics and those obtained in evolutionary ecology. I will consider the model of Poulson (1979, Christiansen and Fenchel 1977) and a generalization of this model (Iwasa and Teramoto 1980) in varying environments. The model is a discrete time version of the logistic model without any time lag in the density or composition dependent effects; its simplest ecological version is formally equivalent to the hyperbolic model presented in Turelli and Petry (1980). However, the parameterization of Poulsen's model is in

* Research supported in part by NIH grant CM 28016

1 Institute of Ecology and Genetics, University of Aarhus, 8000 Aarhus C, Denmark

Population Biology and Evolution
Ed. by K. Wöhrmann and V. Loeschcke

terms of fecundity and survival of the individuals in the population, whereas the parameterization used in Heckel and Roughgarden (1980) and Turelli and Petry (1980) is in terms of descriptive parameters of population growth. Therefore, the reference to classical population genetic result is more natural in the Poulsen model allowing a discussion of the apparent discrepancy between the results of population genetics and of evolutionary ecology.

Evolution in a Constant Environment

Consider a population of individuals with nonoverlapping generations and census the population immediately prior to reproduction. Let x denote the population size at census and let these individuals each produce B offspring, so that the initial population size is $u(0) = Bx$. While the offspring are aging they are subject to density independent and density dependent death, so the population size $u(t)$ at age t changes according to

$$du/dt = -(d + cu)u, \quad 0 \leqslant t \leqslant T, \tag{1}$$

where d and c are constants describing the density independent and density dependent death rates and T is the age of maturity (Poulsen 1979). The size of the offspring population at census then becomes $x' = u(T)$ or

$$x' = DBx/[1 + \gamma(1 - D)Bx], \tag{2}$$

where $D = \exp(-dT)$ is the survival during the growth phase in a sparse population and $\gamma = c/d$. The population grows for x small when $DB > 1$; and in this case it converges to the globally stable equilibrium

$$\hat{x} = (DB - 1)/[\gamma(1 - D)B] \tag{3}$$

(Poulsen 1979). A generalization of the model (Iwasa and Teramoto 1980) allows the density independent death rate in Eq. (1) to vary with age, $d(t)$, and the density dependent death rate coefficient to vary with population density and age, $c[u(t),t]$. [These functions are positive and continuously differentiable in the closed interval $(0,T)$.] In the general model an explicit recurrence equation like Eq. (2) has not been found, but it will be useful to consider the implicit form:

$$x' = DBx \exp\left\{-\int_0^T c[u(t),t]\,u(t)\,dt\right\} \tag{4}$$

(Iwasa and Teramoto 1980, Appendix C), where $D = \exp[-\int_0^T d(t)\,dt]$. The population grows for x small when $DB > 1$; it then converges to a globally stable equilibrium under the reasonable assumptions that the density dependent death rate increases with population size, namely, $(\partial/\partial u)\,[c(u,t)u] > 0$ and that the growth of the population is limited. At equilibrium Eq. (4) provides:

$$\log(DB) = \int_0^T c[\hat{u}(t),t]\,\hat{u}(t)\,dt, \tag{5}$$

which expresses that equilibrium is reached when the density independent growth rate, log(DB), is balanced by the density dependent death rate.

Suppose the population is initially monomorphic for an allele a at an autosomal locus and consider the fate of an introduced rare allele A. Let the three genotypes AA, Aa, and aa be indexed 1, 2, and 3 and let B_i and d_i be the fecundity and density independent death rate of the genotypes i = 1,2,3. Further, let $c_{ij}(\mathbf{u}, t)u_j$ be the density dependent death rate inflicted on genotype i by the presence of u_j individuals of genotype j, where $\mathbf{u} = (u_1, u_2, u_3)$, i.e., the death rate of genotype i at age t is

$$d_i(t) + \sum_{j=1}^{3} c_{ij}[u_1(t), u_2(t), u_3(t), t]\, u_j(t)\,. \tag{6}$$

Assuming reproduction by random mating, the fitness w_2 of genotype Aa is approximately $w_2 = (x_2'/x_2)/(x_3'/x_3)$ relative to that of genotype aa ($w_3 = 1$) when allele A is rare. The rare allele A increases in frequency when $w_2 > 1$ and it decreases in frequency when $w_2 < 1$. For A sufficiently rare, the change in the frequency of A is so slow that we may assume that the population is at ecological equilibrium, i.e., $x_3' = x_3$. The condition for A to increase in the population is (by neglecting higher order terms in the frequency of A) given as $x_2' > x_2$ or from an equation similar to Eq. (4):

$$\log(D_2 B_2) > \int_0^T c_{23}\,[0, 0, \hat{u}_3(t), t]\, \hat{u}_3(t)\, dt\,, \tag{7}$$

where $\hat{u}_3(t)$ is the equilibrium density of genotype aa at age t. The opposite of condition (7) is the condition that the frequency of allele A decreases toward zero when A is initially rare.

In an equilibrium population monomorphic for allele a, Eq. (5) becomes

$$\log(D_3 B_3) = \int_0^T c_{33}\,[0, 0, \hat{u}_3(t), t]\, \hat{u}_3(t)\, dt\,. \tag{8}$$

Therefore, in the case $c_{ij}(\mathbf{u}, t) = a_{ij}c(u, t)$ the condition (7) for initial increase of A simplifies by Eq. (8) to the condition given by Poulsen (1979), namely,

$$(1/a_{23}) \log(D_2 B_2) > (1/a_{33}) \log(D_3 B_3)\,. \tag{9}$$

Thus, as long as the genotypes do not differ in the functional form of their density dependent death rate, the condition for initial increase of a new allele does not depend on this functional form.

Evolution in a Varying Environment

Suppose now that the environment varies cyclically with a period of n generations. Let $B^{(k)}$, $d^{(k)}(t)$, and $c^{(k)}(u, t)$ be the environmental parameters in generation k of the cycle, k = 1,2,...,n. The population size $x^{(k)}$ in generation k is then given by

$$x^{(k)} = x^{(k-1)} D^{(k)} B^{(k)} \exp\left\{- \int_0^T c^{(k)}\,[u^{(k)}(t), t]\, u^{(k)}(t)\, dt\right\}\,. \tag{10}$$

At equilibrium $x^{(n)} = x^{(0)}$ so the relation corresponding to Eq. (5) is

$$\sum_{k=1}^{n} \log(D^{(k)} B^{(k)}) = \sum_{k=1}^{n} \int_0^T c^{(k)} [u^{(k)}(t), t]\, u^{(k)}(t)\, dt\,. \tag{11}$$

Thus, at equilibrium the mean density independent growth rate equals the mean density dependent death rate.

The condition for initial increase of allele A in a population monomorphic for a corresponds to condition (7) (Haldane and Jayakar 1963, Hoekstra 1975, Cornette 1981):

$$\log(\bar{D}_2 \bar{B}_2) > \int_0^T (1/n) \sum_{k=1}^{n} c_{23}^{(k)} [0, 0, \hat{u}_3^{(k)}(t), t]\, \hat{u}_3^{(k)}(t)\, dt\,, \tag{12}$$

where $\bar{D}_i$ and $\bar{B}_i$ are the geometric means over generations of the paramters $D_i^{(k)}$ and $B_i^{(k)}$ and $\hat{u}_3^{(k)}(t)$ is the population size at age t in the k'th generation of the cycle at equilibrium in a population monomorphic for genotype aa. This equilibrium population size satisfies Eq. (11), but in general we need the equilibrium solution to evaluate condition (12). However, if the three genotypes have the same functional form of the density-dependent death rate, so only the strength of the death rate varies with the environment, *viz*, $c_{ij}^{(k)}(\mathbf{u}, t) = a_{ij}(t)\, c^{(k)}(u, t)$ with $u = u_1 + u_2 + u_3$ then condition (12) is

$$\log(\bar{D}_2 \bar{B}_2) > \int_0^T a_{23}(t) \left\{(1/n) \sum_{k=1}^{n} c^{(k)} [\hat{u}_3^{(k)}(t), t]\, \hat{u}_3^{(k)}(t)\right\} dt \tag{13}$$

subject to the constraint

$$\log(\bar{D}_3 \bar{B}_3) = \int_0^T a_{33}(t) \left\{(1/n) \sum_{k=1}^{n} c^{(k)} [\hat{u}_3^{(k)}(t), t]\, \hat{u}_3^{(k)}(t)\right\} dt\,. \tag{14}$$

The condition (12) for initial increase of A is similar to condition (7) of a constant environment. Moreover, we can define an average environment of the population which makes condition (13) a version of condition (7). The density independent parameters in this average environment are given by the geometric means $\bar{B}_i$ and $\bar{D}_i$ [$\bar{D}_i$ corresponds to the arithmetic mean $\bar{d}(t) = (1/n) \sum_{k=i}^{n} d^{(k)}(t)$]. The density dependent death rate coefficients $\bar{c}_{ij}$ are given in the above product form with the function $\bar{c}(u, t)u$ corresponding pointwise to the bracketed functions in Eqs. (13) and (14) at equilibrium. This function is constructed by extension from the definition

$$\bar{c}(\bar{u}, t) = (1/n) \sum_{k=1}^{n} c^{(k)} [u^{(k)}(t), t]\, u^{(k)}(t)/\bar{u}(t)\,,$$

where

$$\bar{u}(t) = [\sum_{k=1}^{n} u^{(k)}(t)]^{1/n}\,. \tag{15}$$

Thus, the condition for initial increase of A can be formulated as a condition in an average environment of the population.

When the age dependence of the density dependent death rate is common among genotypes, i.e., $a_{ij}(t) = a_{ij}$, condition (13) immediately reduces to condition (9) in a constant environment if the appropriate time averages are substituted.

Now consider a stochastic environment. Let B and d(t) be random variables and let the density dependent death rate coefficient be $\phi(t)c(u,t)$ where $\phi(t)$ is a random variable. Suppose that B is a discrete stochastic process and that d(t) and $\phi(t)$ are arrays of continuous time stochastic processes, all of a reasonably stationary nature. With these assumptions the population size x reaches a stationary distribution after some time independent of the initial population size. When the population size distribution is stationary, then the size is expected to neither decrease nor increase, so $E[\log(x'/x)] = 0$. Therefore, in a stationary population, using Eq. (4) we have

$$E[\log(DB)] = E\int_0^T c[u(t),t]\,\phi(t)u(t)\,dt\,, \tag{16}$$

which states again that the average density independent growth rate is balanced by the average density dependent death rate in an equilibrium population.

The condition for initial increase of allele A in a stationary population monomorphic for allele a is a condition corresponding to Eqs. (7) and (12) [Gillespie 1973, Karlin and Liberman 1974]:

$$E[\log(D_2B_2)] > E\int_0^T c_{23}[0,0,\hat{u}_3(t),t]\,\phi(t)\hat{u}_3(t)\,dt\,, \tag{17}$$

where $\hat{u}_3(t)$ is the population size at age t in a stationary population monomorphic for a. This condition is of the same form as condition (12) and therefore the same correspondence to the constant environment model is obtained. If $c_{ij}(\mathbf{u},t) = a_{ij}(t)\,c(u,t)$, then Eq. (17) reduces to

$$E[\log(D_2B_2)] > \int_0^T a_{23}(t)\,E\{c[\hat{u}_3(t),t]\,\phi(t)\,\hat{u}_3(t)\}\;dt \tag{18}$$

with

$$E[\log(D_3B_3)] = \int_0^T a_{33}(t)\,E\{c[\hat{u}_3(t),t]\,\phi(t)\,\hat{u}_3(t)\}\;dt\,. \tag{19}$$

Again, this condition corresponds to the condition for increase of a new allele with the appropriate average density independent environmental parameters and a density dependent death rate coefficient given in the above product from with c from equations similar to Eq. (15). Furthermore, if $a_{ij}(t) = a_{ij}$ condition (17) immediately by Eq. (19) becomes

$$E[\log(D_2B_2)](1/a_{23}) > E[\log(D_3B_3)](1/a_{33})\,, \tag{20}$$

which is condition (9) in terms of geometric average fitnesses.

Time Lag in the Density Response

The characterization of evolution in a varying environment may be extended to models incorporating a time lag in the density response within a generation. The per capita death rate $d(t) + c[u(t),t]\,u(t)$ of juveniles of age t is in Eq. (1) replaced by

$$d(t) + \int_0^t c[u(\tau),\tau]\,u(\tau)\,\psi(t,\tau)\,d\tau \qquad (21)$$

where $\psi(t,\cdot)$ is a probability density on the interval $[0,t]$:

$$\int_0^t \psi(t,\tau)\,d\tau = 1\,. \qquad (22)$$

Within this formulation we may also include the exponential model used by Turelli and Petry (1980) as the limit where all the mass of $\psi(t,\cdot)$ is at the origin, i.e., the per capita death rate at age t is given by

$$d + c\,u(0)\,. \qquad (23)$$

The other extreme, where all the mass of $\psi(t,\cdot)$ is at age t, restore the original Poulsen model with an age and density dependent coefficient of the density dependent death rate.

None of the arguments in the preceding sections are seriously altered by the more general form at the density dependent death rate given by Eq. (21). Therefore, we get the generalization of condition (18) for the increase at allele A in a population monomorphic aa as

$$E[\log(D_2B_2)] > \int_0^T\int_0^t a_{23}(\tau)E\,\{c[\hat{u}_3(\tau),\tau]\,\phi(\tau)\,\hat{u}_3(\tau)\}\;\;\psi(t_1\tau)\,d\tau\,dt \qquad (24)$$

with

$$E[\log(D_3B_3)] = \int_0^T\int_0^t a_{33}(\tau)E\,\{c[\hat{u}_3(\tau),\tau]\,\phi(\tau)\,\hat{u}_3(\tau)\}\;\;\psi(t_1\tau)\,d\tau\,dt \qquad (25)$$

at stationarity in the original monomorphic population. Again this condition corresponds to a condition in a constant environment and if $a_{ij}(t) = a_{ij}$, then condition (20) is recovered.

As an example we may consider the exponential model given by Eq. (23). The general condition (17) for initial increase in a stochastic environment now reads

$$E[\log(D_2B_2)] > TE\,[c_{23}B_3\hat{x}_3] \qquad (26)$$

with

$$E[\log(D_3B_3)] = TE\,[c_{33}B_3\hat{x}_3]\,. \qquad (27)$$

In the exponential model we therefore get that if

$$E[c_{23}B_3\hat{X}_3] = E\left[\frac{c_{23}}{c_{33}}\right]E\,[c_{33}B_3\hat{x}_3] \qquad (28)$$

then condition (26) corresponds to a condition for initial increase in an exponential model in a constant environment. The requirement that $c_{ij} = a_{ij}\,c$, where a_{ij} is a constant and c a random variable, is sufficient (but not necessary) to secure the fulfillment of condition (28).

In a constant environment, the time lag model Eq. (21) does not necessarily have a stable equilibrium density, as oscilliatory or chaotic behavior may occur (May 1974). In that situation the increase of the new allele A will occur in a varying environment, where the only source of variation is the deterministic variation in the population size, $u_3(t)$. The condition for increase of A is still given by condition (24) with averages now taken over the distribution of $u_3(t)$ prevailing during a long time interval after the introduction of the new allele into an old monomorphic population.

Discussion

In classical population genetics the condition for initial increase of a new allele in a temporally varying environment is formulated as a condition in a constant average environment. The general conclusion of this paper is that under wide conditions the same is true when the genotypic fitness values are allowed to depend on the density and composition of the population: if the age specific death rate of genotype i is given as

$$d_i(t) + \int_0^t \sum_{j=1}^3 a_{ij}(\tau)\, c\,[\mathbf{u}(\tau),\tau]\, u_j(\tau)\, \psi(t,\tau)\, d\tau\,, \tag{29}$$

then d_i and c, as well as the fecundity parameter B_i may vary through time. The correspondence between varying and constant environments may only break down (1) when the density dependent death rate coefficient, c, differs among genotypes in its dependence on the population size or (2) when the competition coefficients between genotypes, a_{ij}, vary between generations. Therefore, new phenomena are only expected in a varying environment if rather complicated differences exist between the density dependent death rates of the genotypes. Any difference among genotypes in density independent demographic parameters translates into differences between classical population genetic fitnesses when initial increase phenomena are considered. In a varying environment differences among genotypes in density dependent death and competition in a Lotka-Volterra type model are to be considered as differences in an appropriately defined constant environment when competitive relations between genotypes are constant over generations.

These results are derived from the Poulsen model and its generalization where an important simplifying assumption is made, *viz*, only the death rate is density dependent with fecundity being density independent. This assumption is partly a reflection of the lack of specification of the breeding stage, but qualitatively it is an assumption on the simplicity of the time lag effects. However, if density dependence of fecundity is modeled as a process, where reproductive potential of the individual is built-up over time in competition with the other individuals in the population, then very similar qualitative conclusions apply.

Of the recurrence equation models considered by Turelli and Petry (1980) the hyperbolic model is contained in the set of population dynamic models without time lag, whereas the others, the linear and the exponential, involve either time lags or destabilizing density response in the breeding stage. In any case, these biological phenomena are not explicit in the traditional formulation of either the linear or the exponential models, so the evolutionary phenomena discovered in Heckel and Roughgarden (1980) and Turelli and Petry (1980) can neither be specified nor evaluated in context of these biological phenomena. Further, the definition of individual fitnesses in the recurrence equation models are fundamentally different from the description of fitness in the Poulsen model. Even though the hyperbolic model is equivalent to the Poulsen model in describing the population dynamics of a monomorphic population, the fitnesses in the hyberbolic model are simple density dependent values, whereas natural selection is described in the Poulsen model as a process originating from the interaction of individuals with their environment. Therefore, the purely density dependent version of the Poulsen model will give rise to density and frequency dependent fitness values.

Probably the most astonishing result emerges in the linear model in a varying environment. Here the intrinsic rate of increase, r, can evolve to zero (Heckel and Roughgarden 1980). However, in that situation the parameter r has three different conflicting interpretations: the intrinsic rate of increase, the rate of convergence to ecological equilibrium, and a coefficient weighting the influence of environmental variation. The decrease of r parallels the result of Gillespie (1973) that selection favors the genotype which is influenced the least by the environmental variation. Thus, the evaluation of the results in terms of either of the first two interpretations of r is questionable. In addition, the result of Turelli and Petry (1980) that the parameter r is neutral when noise is introduced in the parameter K of their exponential model is changed in the time lag formulation (23). The K parameter becomes $r/(cB)$ and the assumption that $K_2 = K_3$ provides the admittedly rather artificially looking relation $c_2/c_3 = (r_2/r_3)/(B_2/B_3)$. Using this, the condition for initial increase of A is $B_2 > B_3$, under the assumption of Turelli and Petry that the parameters of the intrinsic rate of increase, r, are not influenced by the environmental fluctuations. Thus, with noise in the parameter K of the exponential model, higher fecundity will be favored. This argument shows that the evolutionary results depend crucially on the particular parametrization of the population dynamic model that provides the fitness description. The carrying capacity parameter K is not a natural parameter in the Poulsen model, it is just the equilibrium population size.

The models considered by Turelli and Petry (1980) are conceptually different from the Poulsen model. The recurrence equation models specify the population size at the census stage as a function of the size at census in the previous generation. The models are phenomenological in that they link observable stages and they are descriptive in a statistical sense in that they can be fitted to data. The Poulsen model, on the other hand, is an explanatory model in that it aims at a description of the events in the life history of the individual. Evolution is driven by differences in individual fitness which are manifested as differences in survival probability and fecundity. Therefore, an explanatory model is preferable for evolutionary arguments, especially for the evaluation of the impact of general phenomena like environmental variation. However, analytical complications set a limit to the amount of detail in the results that can be attained using the Poulsen model, so for the study of some problems, the simpler recurrence equation models are useful and may even be given an explanatory air (Christiansen and Fenchel 1977, Roughgarden 1979, Christiansen and

Loeschcke 1980). In any case care should be taken to formulate the evolutionary question in terms of individual fitness modified by biologically plausible phenotypic traits. For most considerations this task is far easier with the explanatory models.

The results of evolutionary convergence of the intrinsic rate of increase in a varying environment found by Turelli and Petry (1980) can all be derived in the Poulsen model in a constant environment (Christiansen and Fenchel 1977), although the result $r = \log(DBP) \to 0$ is biologically unrealistic. With genotypic variation in only the density independent parameters conditions (12), (17), and (24) immediately state that r increases during evolution. Genotypic variation in only the density dependent parameters leaves r unchanged and simultaneous variation in the two sets of parameters may result in any limit of $r > 0$ with suitable constraints on the parameters. The evolutionary change in the equilibrium population size at census may in general have any direction (Poulsen 1979). However, if the composition dependence of the death rate is abandoned, then the change in population size may be characterized in terms of the density dependent death rate (Iwasa and Teramoto 1980).

The theoretical correspondence between the condition for initial increase of an allele in a constant and a varying environment demonstrates that the evolutionary process under wide conditions is qualitatively equal in the two environments. However, this result is limited to the cases where the period in, e.g., a cyclical environment is short compared to the rate of increase from the minimum gene frequency of a rare allele, *viz*, $1/(2x)$. Stochastic loss of a rare allele must also be considered. Nevertheless, the present results delimit a class of genetic variation which may play a role in specific adaptations of an organism to environmental variation, namely, variation that affects the higher order terms of an individuals density response or that shapes the time lag effects. Unfortunately, neither of these kinds of variation can be easily visualized. If this kind of genetic variation is absent, the treatment of initial evolutionary increase of new alleles in a varying environment amounts to a calculation of appropriate time average genotypic fitnesses. However, these time averages may well favor phenotypes which mediate an adaptation to the environmental variation, e.g., homeostatic phenotypes (Gillespie 1973) and phenotypes that average out the variation (Cohen 1967), but whether these phenotypes express their adaptations through density independent or density dependent fitness parameters is to a large extent immaterial.

Acknowledgements. My interest of evolution in a varying environment arose during collaboration with T. Fenchel on problems of evolutionary ecology. The presentation benefited from discussions with A. Clark, M.W. Feldman, T. Fenchel, S.E. Graversen, V. Loeschcke, and E.T. Poulsen.

References

Christiansen FB, Fenchel TM (1977) Theories of populations in biological communities. Springer, Berlin Heidelberg New York

Christiansen FB, Loeschcke V (1980) Evolution and intraspecific exploitative competition I. One-locus theory for small additive gene effects. Theor Popul Biol 18:297–313

Cohen D (1967) Optimizing reproduction in a randomly varying environment. J Theor Biol 16: 1–14

Cornette JL (1982) Deterministic genetic models in a varying environment. J Math Biol 12:173 to 186

Gillespie J (1973) Polymorphism in random environments. Theor Popul Biol 4:193–195

Haldane JBS, Jayakar SD (1963) Polymorphism due to selection of varying direction. J Genet 58:237–242

Heckel DG, Roughgarden J (1980) A species near its equilibrium size in a fluctuating environment can evolve a lower intrinsic rate of increase. Proc Natl Acad Sci USA 77:7497–7500

Hoekstra RF (1975) A deterministic model of cyclical selection. Genet Res 25:1–15

Iwasa Y, Teramoto E (1980) A criterion of life history evolution based on density-dependent selection. J Theor Biol 84:545–566

Karlin S, Liberman U (1974) Random temporal variation in selection intensities: case of large population size. Theor Popul Biol 6:355–382

MacArthur RH, Wilson EO (1967) The theory of island biogeography. Univ Press, Princeton, NJ 48:1893–1897

MacArthur RH, Wilson EO (1967) The theory of Island biogeography. Univ Press, Princeton, NJ

May RM (1974) Biological populations with non-overlapping generations: stable points, stable cycles, and chaos. Science 186:645–647

Poulsen ET (1979) A model for population regulation with density- and frequency-dependent selection. J Math Biol 8:325–343

Roughgarden J (1979) Theory of population genetics and evolutionary ecology: an introduction. Macmillan, New York

Turelli M, Petry D (1980) Density-dependent selection in a random environment: An evolutionary process that can maintain stable population dynamics. Proc Natl Acad Sci USA 77:7501–7505

Genetic Heterogeneity and Ecological Factors

The Alcohol Dehydrogenase Polymorphism in *Drosophila melanogaster*, Facts and Problems

W. VAN DELDEN[1]

Introduction

The finding that most animal and plant species are highly polymorphic for allozyme variants (see reviews by Powell 1975, Nevo 1978, Brown 1979, Hamrick et al. 1979, and Avise and Aquadro 1982) has created a heated dispute concerning the significance and the maintenance of this kind of molecular variation, known as the selectionist-neutralist controversy. In the neutralist view nucleotide substitutions in the course of evolution are mainly due to fixation of selectively neutral variants (Kimura 1968, Kimura and Ohta 1971, Kimura 1977). Allozyme variation in present-day populations is considered as biochemical noise in the neutralist view and is maintained for the greater part by mutation and drift. Selectionists, on the contrary, claim that selection, and most probably some kind of balancing selection, is the force maintaining allozyme variation (Clarke 1970, Richmond 1970, Wills 1973).

Numerous attempts have been made to determine the nature, either neutral or selective, of the allozyme polymorphisms in nature. Many of these efforts involved rather indirect methods and included statistical tests by which observations from allozyme variation in nature were compared with theoretical expectations based on the neutrality hypothesis. These tests concerned expectations as to the effective number of alleles, means, variances, and distributions of heterozygosities and geographic allele frequency distributions. Often, however, these attempts gave ambiguous results. This was partly due to difficulties in the estimation of the parameters involved in natural populations, such as effective population size and gene flow, partly also because of theoretical objections to the tests applied.

A direct approach is, however, available which appears more promising and straightforward. This approach can be applied when the allozyme variants differ in in vitro biochemical properties, such as activity, heat stability, pH optimum, etc. When such differences lead to physiological differences among allozyme genotypes, fitness differences may occur, at least under relevant environmental conditions. This provides a methodological approach allowing predictions to be made, based on enzyme properties, leading to testable fitness differences. Biochemical differences among allozyme variants are not uncommon, for a sample of allozyme loci in man, e.g., it was shown that about 60% showed differences in kinetic and other properties (Harris 1976).

1 Institute of Genetics, Biology Center, University of Groningen, 9751 NN Haren (Gn), The Netherlands

Population Biology and Evolution
Ed. by K. Wöhrmann and V. Loeschcke

When the selectionist view is correct and selection is acting on allozyme polymorphisms, at least under appropriate environmental conditions, this is an important and relevant finding for population biologists. It has then to be taken into account that in addition to ecologically relevant quantitative genetic variation as, e.g., morphological variation, also allozyme variation may play a role in adaptation. Populations then appear even more genetically heterogeneous than before. In addition to the problem of increasing the level of complexity, allozyme variation has the advantage that compared to most quantitative genetic variation, the genetics is generally simple.

It may be asked what the available evidence is for the selectionist hypothesis. Especially in *Drosophila melanogaster* allozyme loci have been studied for the relation between biochemical and fitness differences. In a few well-studied cases unambiguous results were obtained where under stress conditions in the laboratory, selection operated on allozyme polymorphisms. Clear examples for selection are provided by the amylase polymorphism (De Jong and Scharloo 1976, Hoorn and Scharloo 1979), by the glucose-6-phosphate dehydrogenase, the 6-phosphogluconate dehydrogenase polymorphisms (Bijlsma and Van Delden 1977, Bijlsma and Kerver 1983), and the alcohol dehydrogenase polymorphism (Van Delden 1982). The alcohol dehydrogenase (ADH) polymorphism is probably the best studied allozyme polymorphism in *Drosophila* at the moment and will be considered here in more detail. In the following, first, a few relevant data, pointing at strong selection pressures, will be summarized; for a more extensive treatment the reader is referred to a recent review by Van Delden (1982) and the references cited there. Thereafter, a few recently encountered complications will be dealt with.

The Alcohol Dehydrogenase Polymorphism

The ADH locus in *D. melanogaster* is located on the second chromosome (2–50.1). Populations in nature are nearly always polymorphic for two electrophoretic alleles: the slow (S) and the fast (F) allele, as has been repeatedly reported for populations from different continents. Heat stability studies have shown additional variation as variants with lower or higher heat stability occur at low frequencies in some natural populations. The complete amino acid sequences of the S and F allozymes have been determined: they appear to differ by only a single amino acid substitution. ADH catalyzes the oxidation of alcohols to aldehydes and concurrently reduces NAD^+ to NADH. Ethanol and a variety of both primary and secondary higher alcohols can serve as substrates. Recently, it was reported (Heinstra et al. 1983) that ADH, in addition to the oxidation of ethanol to acetaldehyde, also catalyzes, at least in vitro, the conversion of acetaldehyde into acetate. As far as biochemical properties are concerned: larvae and adults homozygous for the F allele possess greater in vitro ADH activity than homozygotes for the S allele, as the heterozygotes show intermediate activities. In vitro studies further showed that ADH from SS homozygotes is more stable at high temperatures than ADH from FF and FS.

The ADH polymorphism thus fulfils the conditions for a straightforward relation between biochemical differences and fitness differences, assuming that differences in

ADH activity will lead to physiological differences as to the detoxifying properties of the allozyme variants, subsequently resulting in differential survival. Such a proposed relationship would also have ecological relevance as *D. melanogaster* uses fermenting fruits for oviposition, where ethanol is present due to yeast action. Also other alcohols are sometimes present in considerable quantities in decaying fruits. Laboratory experiments have confirmed the predictions based on the in vitro differences between the ADH variants. Considerable differences in survival on ethanol supplemented food, both in juveniles and adults, were found. SS homozygotes, the genotype with the lowest in vitro ADH activity, show a much higher mortality than FF homozygotes (the genotype with the highest ADH activity), while the heterozygotes generally exhibit an intermediate mortality (Fig. 1). The relation between ADH activity and survival was further confirmed by an experiment where different, electrophoretically identical (FF homozygotes) strains were exposed to ethanol (Kamping and Van Delden 1978). In this case a positive correlation between ADH activity and survival was also found. In view of these fitness differences it is not surprising to find that polymorphic populations kept on food supplemented with ethanol or other alcohols show a rapid decrease in frequency of the S allele, eventually leading to fixation of the F allele. As in the case of ADH activity an unambiguous relation is found with detoxifying ability and consequently with fitness, a less clear relationship is found to exist for temperature stability. In view of the higher in vitro stability of ADH of SS homozygotes, compared to that of the FF and FS genotypes, it has been argued that at high temperature SS should have an advantage. Evidence for this view has been derived from a latitudinal cline in allele frequencies observed both in the USA and in Australia, where the frequency of the S allele is found to be higher in the hotter areas (Johnson and Schaffer 1973, Oakeshott et al. 1982). Though Johnson and Powell (1974) reported from a laboratory experiment a higher S frequency for survivors of heat shock compared to controls in some of their populations, Van Delden and Kamping (1980) were unable to confirm their results. They found instead that heterozygotes have the highest survival at high temperatures, a finding confirmed by the results of Bijlsma-Meeles and Van Delden (1974) obtained in population extinction experiments. A further notable phenomenon met in polymorphic laboratory populations is the convergence of allele frequencies in populations started with different initial frequencies (Van

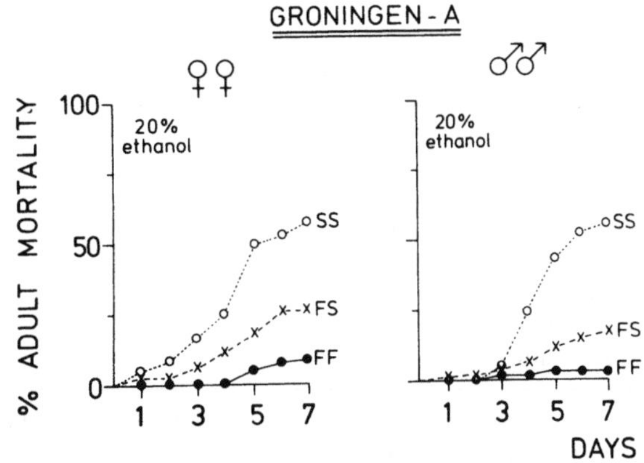

Fig. 1. Adult mortality of ADH genotypes of the Groningen-A population on ethanol supplemented food

Delden et al. 1978). These results, obtained for regular food and temperature conditions, point to the action of some kind of balancing selection.

It may be asked to what extend selection pressures are acting on the ADH polymorphism in natural populations of *D. melanogaster*, in view of the results obtained by laboratory experiments. In the following sections two aspects will be dealt with, one concerning the effects of - long-term - ethanol exposure and one concerning the relation of the ADH locus with other loci. In both cases a more complex situation than previously envisaged is met with.

Selection for Alcohol Tolerance

The action of selection on the ADH polymorphism in nature, if any, due to the presence of alcohols is expected to be strongest in those cases where the alcohol concentrations are the highest. On the species level this is demonstrated by the differential distribution of *Drosophila* species. *Drosophila* species differ greatly in their tolerance to ethanol and other alcohols (see Van Delden 1982 for review). Such species differences, correlated with ADH activity, are reflected in their distribution. Comparison of *D. melanogaster* and *D. simulans*, e.g., shows that the latter, a much less ethanol-tolerant species, is absent in maturation cellars of vineyards, where *D. melanogaster* abundantly occurs, while both species are found outside the cellars. Concerning the ADH polymorphism in *D. melanogaster*, it is to be expected that winery populations will possess higher F frequencies than surrounding populations not exposed to extreme ethanol concentrations. Table 1 lists the results of several workers who compared winery with nonwinery populations. Higher F frequencies are found indeed in two cases for Spanish and Canadian winery populations. In the other cases, however, no differences between inside and outside populations are found. In those cases where the populations were investigated for ethanol tolerance, on the other hand, it was demonstrated that tolerance to ethanol was higher in cellar than in noncellar populations, while in these cases no differences in allele frequencies were noticed.

The increased tolerance was genetically determined and this suggests that, in addition to the ADH locus, other loci may be involved in tolerance to alcohols.

Table 1. Comparison of wine cellar populations of *D. melanogaster* with surrounding populations for higher frequency of the ADH^F allele and higher ethanol tolerance in the former populations

Origin population	Higher ADH^F frequency	Higher tolerance	Reference
Spain	+[a]		Briscoe et al. (1975)
Canada	+	(+)	Hickey and McLean (1980)
Australia	–[b]	+	McKenzie and Parsons (1974)
Australia	–	+	McKenzie and McKechnie (1978)
USA-California	–		Marks et al. (1980)

[a] + indicates higher ADH^F frequency or ethanol tolerance for the wine cellar populations
[b] – indicates no difference

A series of laboratory experiments were set up in our laboratory, designed to elucidate the nature of the increase in tolerance after long-term exposure to alcohols. For this purpose, strains either homozygous for the S or the F allele, were kept on food supplemented with a particular alcohol. After many generations the strains were tested for increased tolerance, if any, to the alcohol involved (details in Van Delden and Kamping 1983, Kerver and Van Delden 1984). In addition tolerance to other alcohols was also tested in order to obtain information about cross-tolerance. The main purpose was to determine the contribution of the ADH locus in the process of adaptation to the toxic alcohols. In view of the crucial role of ADH in the differential survival of ADH genotypes, it is plausible to assume that increased tolerance after long-term exposure to alcohols can be brought about by augmented ADH activity. For the tolerance tests the alcohol strains were compared with control strains (kept on regular food and never before exposed to alcohols) for mortality, both in the egg-to-adult and the adult life phase, at several concentrations of the alcohol concerned (Fig. 2 provides an example for survival on hexanol supplemented food). Thus, LD_{50} values could be calculated. Table 2A gives the LD_{50}'s for egg-to-adult survival for strains kept for 90 generations on ethanol (SSE and FFE) compared with control

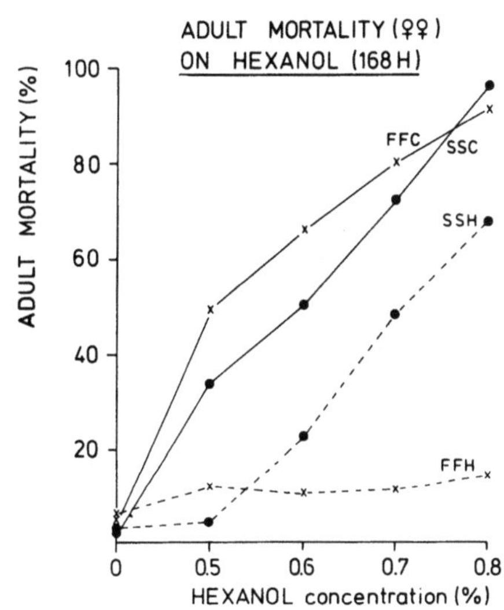

Fig. 2. Adult mortality of hexanol and control strains (females) on hexanol supplemented food, after 168 h exposure

Table 2A. Median lethal doses (LD_{50}) of ethanol and hexanol for egg-to-adult survival of ethanol and control strains

Strains	LD_{50} of ethanol (%)	LD_{50} of hexanol (%)
SSC	12.6 (11.9–13.3)[a]	0.34 (0.33–0.35)
SSE	18.4 (16.1–20.9)	0.27 (0.26–0.28)
FFC	16.8 (15.7–17.9)	0.39 (0.38–0.39)
FFE	20.2 (18.5–22.0)	0.42 (0.41–0.43)

[a] 95% confidence intervals given in parentheses

Table 2B. Median lethal doses (LD_{50}) of ethanol and hexanol for egg-to-adult survival of hexanol and control strains

Strains	LD_{50} of ethanol (%)	LD_{50} of hexanol (%)
SSC	12.9 (12.5–13.3)	0.34 (0.33–0.35)
SSH	9.7 (9.1–10.3)	0.39 (0.38–0.41)
FFC	14.1 (13.4–14.9)	0.39 (0.38–0.39)
FFH	18.0 (17.0–19.0)	0.47 (0.45–0.49)

strains (SSC and FFC). Both for SSE and FFE a considerable increase (46% and 20% respectively) in ethanol tolerance is observed. When the ethanol strains were tested on hexanol, increased tolerance was only found for the FFE strain; the SSE strain even possessed a significant lower LD_{50}. Hexanol was chosen because it had previously been found that polymorphic populations kept on hexanol supplemented medium show a very rapid increase in frequency of the F allele (Van Delden et al. 1975). Table 2B shows a separate experiment where hexanol strains (SSH and FFH) were compared with control strains both for tolerance to hexanol and for ethanol. The hexanol strains both show an increase in tolerance to hexanol, but cross-tolerance to ethanol is only found for the FFH strain, where as the SSH strain even shows a decrease in resistance to ethanol. The results show a clear-cut increase in the juvenile stage in tolerance to the particular alcohol the E and H strains have been exposed to for 90 generations. Thus, cross-tolerance to a "new" alcohol, which the strains never experienced, exists both for the FFE and FFH strains, possibly suggesting a common factor in the adaptation to alcohols in general. The SS strains, however, behave quite differently; on the alien alcohol even a significant decrease in tolerance is observed. This points to a different mechanism of adaptation in SS and FF strains, suggesting that in the former strains the mechanism providing increased tolerance to the "own" alcohol is different from that in the FF strains, even leading to a disadvantage at the alien alcohol.

Though the juvenile stage in *Drosophila* is probably the most vulnerable to the toxic effects of alcohols, the adult mortalities were also determined in separate tests. Table 3A gives the LD_{50}'s for adult survival for ethanol, hexanol, and control strains when tested on ethanol supplemented food. An increased tolerance to ethanol is found in the ethanol strains, except for FFE ♀♀. Also the hexanol strains have a higher tolerance to ethanol, except for SSH ♂♂. When the same strains are tested on hexanol supplemented food it appears that the adult tolerance is consistently increased in the hexanol strains (Table 3B). For the ethanol strains only a moderate, though significant, increase is found for the SSE strain, while the FFE strain even shows a decreased tolerance when compared with controls. The general picture, emerging from these results, is that *D. melanogaster* strains, after long-term exposure to a particular alcohol, are able to develop a, genetically based, increased tolerance to that alcohol. In a number of cases the increased tolerance is nonspecific and also holds for alcohols different from the one the strain was confronted with for nearly a hundred generations. Strain specific effects are likely, while tolerance in juveniles and adults probably depends on different mechanisms.

Table 3 A. Median lethal doses (LD_{50}) of ethanol for adult survival of ethanol, hexanol, and control strains

	LD_{50} (%) of ethanol	
Strain	♀♀	♂♂
SSC	24.5 (23.4–25.5)[a]	23.3 (22.7–24.0)
SSE	29.4 (28.5–30.4)	24.9 (24.3–25.6)
FFC	28.6 (27.7–29.6)	28.1 (27.1–29.1)
FFE	28.6 (27.6–29.5)	31.7 (30.4–33.1)
SSC	25.7 (24.8–26.7)	25.6 (24.7–26.4)
SSH	28.7 (27.7–29.8)	26.4 (25.7–27.1)
FFC	33.1 (32.0–34.3)	29.0 (28.1–29.9)
FFH	38.3 (36.1–40.5)	38.5 (36.5–40.7)

[a] 95% confidence intervals given in parentheses

Table 3 B. Median lethal doses (LD_{50}) of hexanol for adult survival of ethanol, hexanol, and control strains

	LD_{50} (%) of hexanol	
Strain	♀♀	♂♂
SSC	0.74 (0.72–0.76)	0.60 (0.58–0.61)
SSE	0.81 (0.77–0.85)	0.63 (0.61–0.66)
FFC	0.68 (0.66–0.70)	0.71 (0.70–0.73)
FFE	0.62 (0.60–0.64)	0.59 (0.57–0.60)
SSC	0.60 (0.58–0.63)	0.47 (0.45–0.49)
SSH	0.73 (0.71–0.76)	0.64 (0.62–0.66)
FFC	0.56 (0.54–0.59)	0.58 (0.56–0.60)
FFH	$\gg 1$[b]	$\gg 1$[b]

[b] Unrealistic high LD_{50}'s as result of very low mortalities at the chosen test concentrations

The role of ADH in the improved tolerance to alcohols was investigated by a comparison of in vitro ADH activities of crude extracts of adults, cultured on regular food without crowding. The ethanol strains, however, do not exhibit higher ADH activities than the control strains (Kerver, in preparation). This is contrary to the findings of McDonald et al. (1977) who found an increase in ADH activity in a strain selected for increased tolerance to ethanol. The selection procedure in this case, however, was different from ours as selection was only exercised on adults by exposure during 3 days to an aqueous solution of sucrose supplemented with ethanol. Kerver (unpublished) has performed chromosome substitutions between ethanol and control strains and found effects on ethanol tolerance for all three major chromosomes, though distinct differences existed between SSE and FFE strains in the relative contributions of the chromosomes. It, thus, appears that apart from the ADH locus, other loci also contribute to tolerance to alcohols. For the hexanol strains, however, higher levels of ADH activity in adults were found (Table 4) when compared with controls, except for

Table 4. Mean ADH activities in batches of 25 adults of the hexanol and control strains

Strain	ADH activity in ♀♀ $\Delta E\ mg^{-1} min^{-1} \times 10^3$ (s.e.)	ADH activity in ♂♂ $\Delta E\ mg^{-1} min^{-1} \times 10^3$ (s.e.)
SSC	31.0 (0.2)	62.3 (1.2)
SSH	35.5 (0.4)[a]	61.8 (0.7)
FFC	155.6 (2.4)	269.1 (5.0)
FFH	204.8 (3.3)[a]	318.4 (6.0)[a]

[a] Significant ($P < 0.001$) differences in ADH acitivity between hexanol and control strains

SSH ♂♂. It appears that at least in the case of hexanol adaptation the supposed relation with increased ADH activity can be shown. It should be stressed, however, that the level of ADH activity alone cannot be responsible for all the observed increases in tolerance. This follows from the data for adult survival on hexanol: SSH survives better than FFC, in spite of a lower ADH activity in the former strain. On ethanol, on the other hand, the opposite is true. It may be that survival on ethanol is more dependent on ADH activity than survival on hexanol. In the latter case the role of hexanol specific adaptation mechanisms will be more prominent. The interpretation of the impact of ADH activity in the improved tolerance to alcohols is obscured, however, by an additional finding, represented in Fig. 3. In those cases, namely, where an increase in ADH activity is found (the hexanol strains, except for SSH ♂♂), an accompanying increase in body weight is also observed. This rise in body weight may in itself be related with improved tolerance to alcohols, e.g., because heavier individuals do survive better. It has been shown that for a number of stresses like desiccation, anoxia, and insecticides survival in *Drosophila* and other insects was positively correlated with body size (Parsons 1973, Perry and Agosin 1974). A further complication is that it has been found that both the amount of ADH protein (Clarke et al. 1979) and ADH activity per unit weight (Kamping and Van Delden 1978, Van Dijk 1981) increase with increasing body weight (Fig. 4). In view of this positive correlation between ADH activity and body weight, it may be asked whether in the adapted strains selection has occurred for increased body weight with an accompanying,

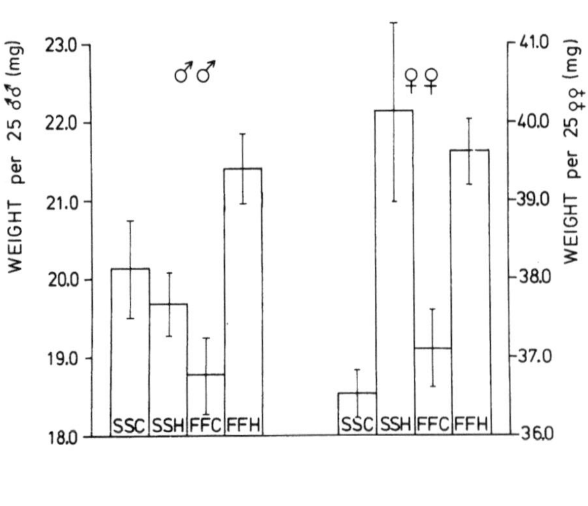

Fig. 3. Mean weights in batches of 25 adults of hexanol and control strains *Vertical bars* indicate 95% confidence intervals

Fig. 4. The relation between ADH activity ($\Delta E\ mg^{-1}\ min^{-1}$) and body weight (mg) of FF and SS genotypes

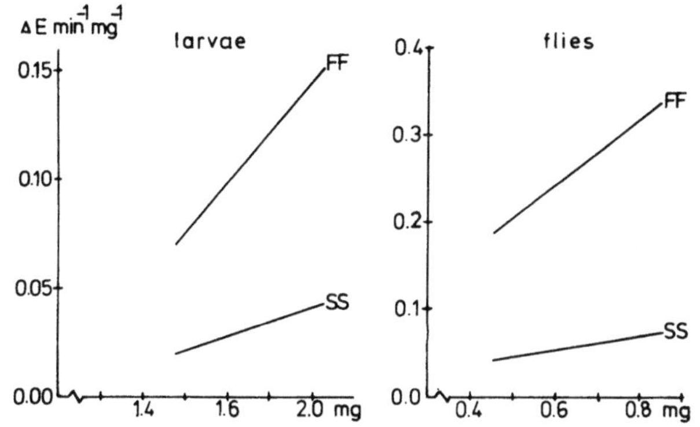

secondary, rise in ADH activity or that the opposite has occurred or that both selective forces interacted. At this moment the mechanism of the increased tolerance to alcohols and the contribution of ADH is not elucidated. It appears that in addition to the biochemical aspects playing such a prominent role in the ADH polymorphism, other factors, like body weight and developmental time, are of utmost importance. This stresses the need of a combined study of allozyme variation and "classical" quantitative characters. A further aspect of adaptation to alcohols, which will be the subject of further study, is the drastic increase in survival from egg-to-adult, obtained by exposure of eggs during a short period after laying. This, apparently maternal effect, first observed by Bijlsma-Meeles (1979), may be a key factor in the tolerance to alcohols, lessening the need for genetic adaptation.

Relationship Between ADH and αGPDH Loci

Another factor to be considered in relation with the significance and the maintenance of the ADH polymorphism is the relation, both in a structural and a functional sense, of the ADH locus with other loci. This aspect was encountered in an experiment where a seminatural population of *D. melanogaster* was founded in a part of a tropical greenhouse. This greenhouse population was started in 1972 with flies from the Groningen K-71 population (details in Bijlsma and Van Delden 1977) and was variable for several allozyme loci. Considerable fluctuations occur in the frequencies of the ADH alleles: the ADH S frequency varies between 0.50 (December 1976) and 0.16 (November 1982), but does not show a regular, consistent pattern as periods of fall in frequency alternate with periods of increasing frequency. Genetic drift is not responsible for this behavior, as the population always contained several thousands of flies. When the course of the ADH S frequency is considered in relation with the allele frequencies at the αGPDH locus (2–20.5) an even more complex picture arises (Fig. 5). In the 6-year period during which allele frequency changes for both loci combined were recorded, a tendency for a correlation in allele frequencies for both loci can be seen. A decrease in ADH S frequency is accompanied by an increase in αGPDH S frequency, as the αGPDH S frequency declines when the ADH S frequency goes up. The αGPDH S frequency consequently also shows considerable changes: from 0.35 in 1976 till 0.80 in

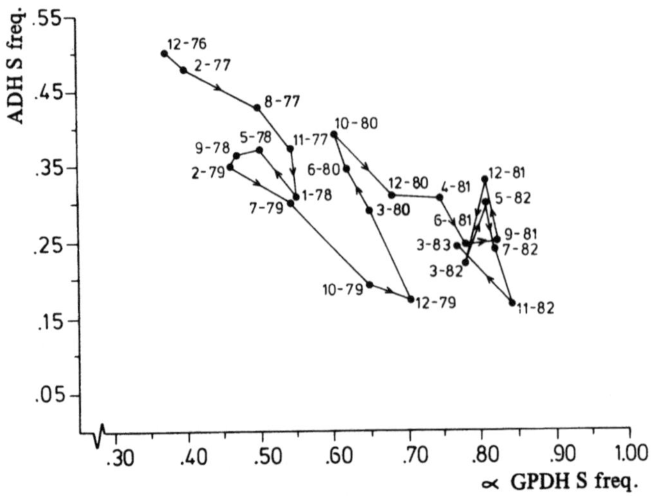

Fig. 5. Changes in ADH and αGPDH allele frequencies in the greenhouse population in time. *Numbers* in the figure indicate month-year

1983. These changes in allele frequencies, both for the ADH and the αGPDH locus raise the question whether selection is involved. In this respect another important finding is that when individuals are assayed both for their ADH and αGPDH genotype, considerable linkage disequilibrium is found. This is illustrated in Fig. 6 where linkage disequilibrium (D) is expressed as a fraction of the maximum linkage disequilibrium (D_{max}) (Hedrick et al. 1978). For many years significant linkage disequilibrium exists for the two loci concerned. It was only in 1982 that values dropped to near zero. Table 5 gives the genotypic composition of a representative sample of flies taken in July, 1979. Apparently some genotypes occur in excess (first letters indicate ADH alleles, last letters αGPDH alleles): $\frac{SF}{SF}, \frac{FF}{SS} (\frac{FS}{SF}), \frac{FS}{FS}, \frac{FF}{SF}$; while the other genotypes are present in numbers lower than expectation. A cytological examination of the salivary gland chromosomes showed the occurrence of a second chromosome inversion, identified as In(2L)t. In addition to this long left arm inversion, smaller inversions of the left arm of the second chromosome were also found. With the aid of a SM_5/Bla strain, individual second chromosomes were extracted from the population and subsequent-

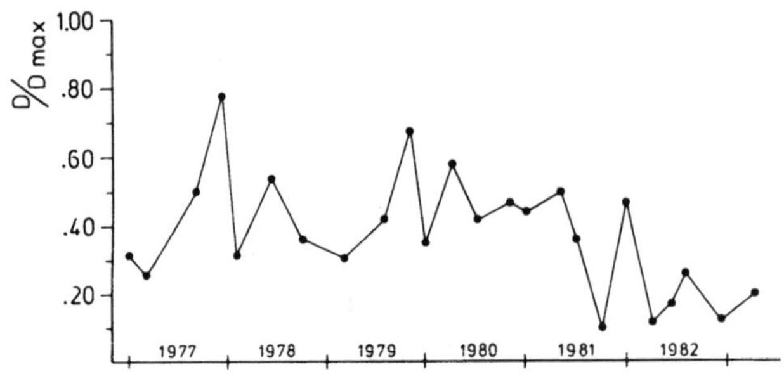

Fig. 6. Linkage disequilibrium between the ADH and αGPDH loci (expressed as D/D_{max}) in the greenhouse population in the course of time

Table 5. Genotypic composition of a representative sample of flies taken in July, 1979 from the greenhouse population

		αGPDH		
		FF	FS	SS
ADH	FF	3	17	27
	FS	17	24	5
	SS	5	2	–

ly tested for recombination between the ADH locus and the αGPDH locus using a dp b strain (Van Delden and Kamping, in preparation). This provides three groups of chromosomes: one with no recombination; one with the regular recombination fraction; and a third group with reduced recombination. Comparison with cytological results shows that these three groups are characterized by the long inversion, no inversion, and a small inversion, respectively. When the chromosomes are classified according to allozyme genotype as far as the ADH and αGPDH loci are concerned, it becomes clear that the long inversion only occurs in SF chromosomes (slow allele for ADH, fast allele for αGPDH) and not in the other three chromosome types. Table 6 gives representative data for a sample of 192 extracted chromosomes. The association of allozyme genotypes and long inversion genotype is diagrammatically given in Fig. 7. It is remarkable that in the Groningen-K-71 base population, from which an offshoot was used to found the greenhouse population, kept as a laboratory population at 25 °C, no linkage disequilibrium between the ADH and the αGPDH locus was found during the period 1977–1983. No inversions were found in this population.

The long inversion is apparently the cause for the linkage disequilibrium and it may be asked for what reason it was maintained for such a long period in the greenhouse population. One reason could be that the inversion itself possessed a selective advantage during several years, another that a specific combination of alleles of the ADH and the αGPDH loci is selectively advantageous. Concerning the latter possibility,

Table 6. Results of a test for recombination between the ADH and αGPDH loci

Chromosome type ADH-αGPDH	Number of chromosomes with:			Total number of chromosomes
	Regular recombination %	Reduced recombination %[a]	No recombination[b]	
FF	11	4	–	15
FS	100	29	–	129
SF	5	4	11	20
SS	24	4	–	28
Totals	140	41	11	192

[a] Reduced recombination % indicates the presence of a shorter inversion
[b] No recombination indicates the presence of the In(2L)t inversion

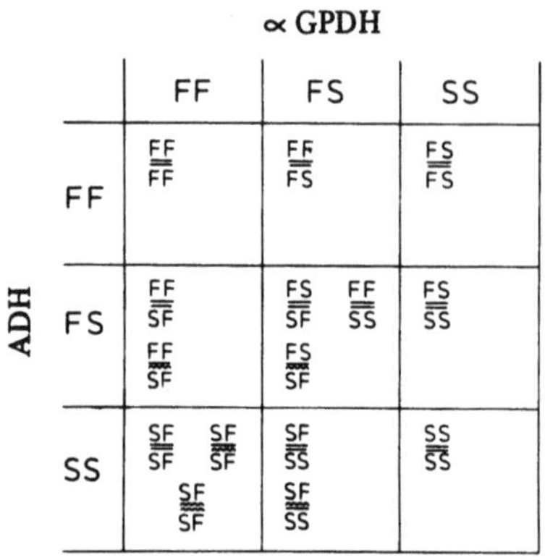

Fig. 7. ADH/αGPDH genotypes in the greenhouse population. *Wavy lines* indicate chromosomes with the In(2L)t inversion, *straight lines* represent chromosomes without this inversion

the results of Cavener and Clegg (1981) should be mentioned. They found an increase of the αGPDH^S frequency on ethanol supplemented food accompanied by a decrease in ADH^S frequency. They calculated two-locus joint fitness values which indicated a fitness interaction between the two loci. A role of αGPDH in ethanol metabolism in connection with the NADH/NAD$^+$ ratio possibly caused this interaction. Though in the greenhouse no stress ethanol concentrations are likely, it is possible that other conditions, prevailing in the greenhouse, will promote certain ADH/αGPDH allele combinations.

A more important factor may be the chromosome inversion polymorphism. Many authors have reported that natural populations of *D. melanogaster* are polymorphic for inversions, some of them being cosmopolitan (Mukai et al. 1974, Mukai and Voelker 1977, Mettler et al. 1977, Voelker et al. 1977, Inoue and Watanabe 1979, Knibb et al. 1981, Knibb 1982). It further appears that cosmopolitan inversions like In(2L)t have latitudinal clines in North America, Australia, and Asia: inversion frequency decreases with increasing latitude on both hemispheres. Inversion frequencies are found to be positively correlated with temperature (Knibb 1982). It is not unlikely that for *Drosophila melanogaster*, as far as inversion polymorphism is concerned, a situation exists similar to the well-studied third chromosome inversion polymorphisms in *Drosophila pseudoobscura* which are found to be subject to natural selection (Dobzhansky 1970). The geographic clines in inversion frequencies are accompanied by clines for ADH and αGPDH (Johnson and Schaffer 1973, Vigue and Johnson 1973, Oakeshott et al. 1982) as the frequencies of ADH^S and αGPDH^F decrease with latitude. Associations of ADH and αGPDH alleles with some inversions are not random, for In(2L)t, e.g., the frequencies of both the ADH^S and the αGPDH^F alleles in the inversion are high (Voelker et al. 1978, Inoue et al. 1981). Returning to the greenhouse population, it may be asked what caused the high frequencies of the inversions and the In(2L)t particularly, highly unusual for regions at higher latitudes? The reason may be found in the conditions met in this tropical greenhouse, where in summer, for Holland, exceptionally high temperatures (> 40 °C) occur together with high humidity. An occasional inversion, occurring in the founder population may have been selected for, which gave rise to the, relatively, high frequencies. It is then not surprising to find that the laboratory population from the same origin has no inversion, because it has been shown that inversions disappear under laboratory conditions as they are selected against (Inoue 1979).

In an attempt to trace the causes of the high frequency of the long inversion and the changes in inversion frequency and in allele frequencies at the ADH and αGPDH loci, an experiment was started in 1981 with flies from the greenhouse population. Three laboratory populations were started with large sample of flies derived from the greenhouse population. The three populations were kept in incubators set at 20 °C, 25 °C, and 29.5 °C. The course of the allele frequencies is given in Fig. 8. The effects of the various temperature regimes on allele frequencies are considerable: ADH^S frequency goes up at the highest temperature, goes down at the lowest temperature, and initially declines at 25 °C, but stabilizes then soon. The $\alpha GPDH^S$ frequencies at 25° and 29.5° decrease, but at 20 °C an increase, nearly to fixation, is found. Linkage disequilibrium between the ADH and αGPDH loci disappears in both populations kept at the lower temperatures within 20 generations. In the population kept at 29.5 °C, however, linkage disequilibrium, expressed as D/D_{max}, rises sharply from an initial value of about 0.30 to 0.50 and stabilizes there. Subsequent assays of long inversion frequencies show no inversions at 20 °C and 25 °C. At 29.5 °C, on the contrary, the inversion frequency has increased till 24.5%. All the inversions only contain ADH^S and $\alpha GPDH^F$ alleles, but among the ADH^S $\alpha GPDH^F$ chromosomes, the frequency of chromosomes carrying the inversion is as high as 83% (for comparison: see Table 6). High temperature consequently leads to changes in both allele frequencies and inversion frequencies and leads to considerable linkage disequilibrium. All these phenomena have been observed in the greenhouse population and are also probably associated with temperature or with temperature correlated factors. Not clear at present is the target of selection, either the inversion itself or particular combinations of allozyme

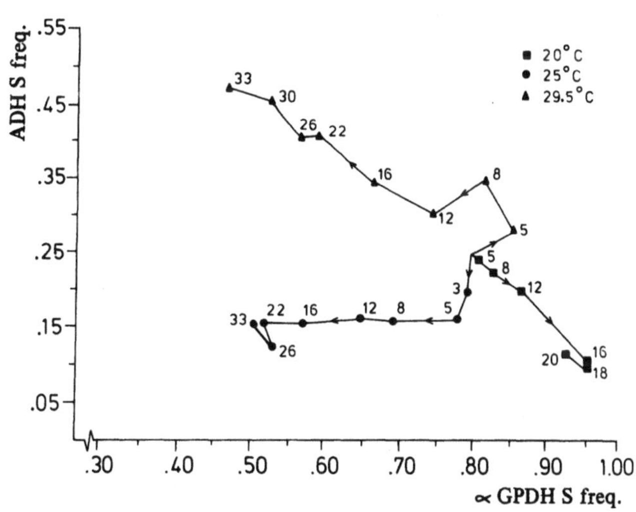

Fig. 8. Changes in ADH and αGPDH allele frequencies in three populations derived from the greenhouse population and kept at 20°, 25°, and 29.5 °C, respectively. *Numbers* in the figure indicate generations

alleles. Further experiments, now in progress with different types of chromosomes, with or without inversions, will help elucidate this matter.

What initially appeared as a mere case of allozyme polymorphism, thus appears to be associated with problems around inversion polymorphisms; further research will have to include both aspects. As in the case of selection for tolerance to alcohols, treated in a previous section where quantitative characters were involved, the study of the ADH polymorphism extends to other relevant parts of the genotype.

References

Avise JC, Aquadro CF (1982) A comparative summary of genetic distances in the Vertebrates: patterns and correlations. Evol Biol 15:151–185

Bijlsma R, Delden Van W (1977) Polymorphism at the G6pd and 6Pgd loci in *Drosophila melanogaster*. I. Evidence for selection in experimental populations. Genet Res 30:221–236

Bijlsma R, Kerver JWM (1983) The effect of DDT on the polymorphism at the G6pd and Pgd loci in *Drosophila melanogaster*. Genetics 103:447–464

Bijlsma-Meeles E (1979) Viability in *Drosophila melanogaster* in relation to age and ADH activity, of eggs transferred to ethanol food. Heredity 42:79–89

Bijlsma-Meeles E, Delden Van W (1974) Intra- and interpopulation selection concerning the alcohol dehydrogenase locus in *Drosophila melanogaster*. Nature (London) 247:369–371

Briscoe DE, Robertson A, Malpica JM (1975) Dominance at Adh locus in response of adult *Drosophila melanogaster* to environmental alcohol. Nature (London) 255:148–149

Brown AHD (1979) Enzyme polymorphism in plant populations. Theor Popul Biol 15:1–42

Cavener DR, Clegg MT (1981) Multigenic adaptation to ethanol in *Drosophila melanogaster*. Evolution 35:1–10

Clarke B (1970) Darwinian evolution of proteins. Science 168:1009–1011

Clarke B, Camfield RG, Galvin AM, Pits CR (1979) Environmental factors affecting the quantity of alcohol dehydrogenase in *Drosophila melanogaster*. Nature (London) 280:517–518

Delden Van W (1982) The alcohol dehydrogenase polymorphism in *Drosophila melanogaster*. Selection at an enzyme locus. Evol Biol 15:187–222

Delden Van W, Kamping A (1980) The alcohol dehydrogenase polymorphism of *Drosophila melanogaster*. IV. Survival at high temperature. Genetica 51:179–185

Delden Van W, Kamping A (1983) Adaptation to alcohols in relation to the alcohol dehydrogenase locus in *Drosophila melanogaster*. Entomol Exp Appl 33:97–102

Delden Van W, Kamping A, Dijk Van H (1975) Selection at the alcohol dehydrogenase locus in *Drosophila melanogaster*. Experientia 31:418–419

Delden Van W, Boerema AC, Kamping A (1978) The alcohol dehydrogenase polymorphism in populations of *Drosophila melanogaster*. I. Selection in different environments. Genetics 90: 161–191

Dijk Van H (1981) The relationship between ADH activity and body weight in *Drosophila melanogaster*. Drosophila Inf Serv 56:150–151

Dobzhansky T (1970) Genetics of the evolutionary process. Columbia Univ Press, New York

Hamrick JL, Linhart YB, Mitton JB (1979) Relationship between life history characteristics and electrophoretically detectable genetic variation in plants. Annu Rev Evol Syst 10:173–200

Harris H (1976) Molecular evolution: the neutralist-selectionist controversy. Fed Proc Fed Am Soc Exp Biol 35:2079–2082

Hedrick P, Jain S, Holden L (1978) Multilocus systems in evolution. Evol Biol 11:101–184

Heinstra PWH, Eisses KT, Schoonen WGEJ, Aben W, Winter De AJ, Horst Van der DJ, Marrewijk Van WJA, Beenakkers AMT, Scharloo W, Thörig GEW (1983) A dual function of alcohol dehydrogenase in *Drosophila*. Genetica 60:129–137

Hickey EA, McLean MD (1980) Selection for ethanol tolerance and ADH allozymes in natural populations of *Drosophila melanogaster*. Genet Res 36:11–15

Hoorn AJW, Scharloo W (1979) Selection on enzyme variants in *Drosophila*. Aquilo Ser Zool 20:41–48

Inoue Y (1979) The fate of polymorphic inversions of *Drosophila melanogaster* transferred to laboratory conditions. Jpn J Genet 54:83–96

Inoue Y, Watanabe TK (1979) Inversion polymorphisms in Japanese natural populations of *Drosophila melanogaster*. Jpn J Genet 54:69–82

Inoue Y, Tsuno K, Watanabe TK (1981) Association of chromosome- and enzyme-polymorphism in natural populations of *Drosophila melanogaster*. Annu Rep Natl Inst Genet, Misima, Jpn 32:105–106

Johnson FM, Powell A (1974) The alcohol dehydrogenases of *Drosophila melanogaster*: frequency changes associated with heat and cold shock. Proc Natl Acad Sci USA 71:1783–1784

Johnson FM, Schaffer HE (1973) Isozyme variability in species of the genus *Drosophila*. VII Genotype-environment relationships in populations of *Drosophila melanogaster* from the Eastern United States. Biochem Genet 10:149–163

Jong De G, Scharloo W (1976) Environmental determination of selective significance or neutrality of amylase variants in *Drosophila melanogaster*. Genetics 84:77–94

Kamping A, Delden Van W (1978) The alcohol dehydrogenase polymorphism in populations of *Drosophila melanogaster*. II. Relation between ADH activity and adult mortality. Biochem Genet 16:541–551

Kerver JWM, Delden Van W (1984) Development of ethanol tolerance in *Drosophila melanogaster*. (in press)

Kimura M (1968) Evolutionary rate at the molecular level. Nature (London) 217:624–626

Kimura M (1977) Causes of evolution and polymorphism at the molecular level. In: Kimura M (ed) Molecular evolution and polymorphism. Natl Inst Genet, Misima, Jpn, pp 1–28

Kimura M, Ohta T (1971) Protein polymorphism as a phase of molecular evolution. Nature (London) 229:467–469

Knibb WR (1982) Chromosome inversion polymorphisms in *Drosophila melanogaster*. II. Geographic clines and climatic associations in Australasia, North America and Asia. Genetica 58: 213–221

Knibb WR, Oakeshott JG, Gibson JB (1981) Chromosome inversion polymorphisms in *Drosophila melanogaster*. I. Latitudinal clines and associates between inversions in Australasian populations. Genetics 98:833–847

Marks RW, Brittnacker JG, McDonald JF, Prout T, Ayala FJ (1980) Wineries, *Drosophila*, alcohol and ADH. Oecologia 47:141–144

McDonald JF, Chambers GK, David J, Ayala FJ (1977) Adaptive response due to changes in gene regulation: a study with *Drosophila*. Proc Natl Acad Sci USA 74:4562–4566

McKenzie JA, McKechnie SW (1978) Ethanol tolerance and the Adh polymorphism in a natural population of *Drosophila melanogaster*. Nature (London) 272:75–76

McKenzie JA, Parsons PA (1974) Microdifferentiation in a natural population of *Drosophila melanogaster* to alcohol in the environment. Genetics 77:385–394

Mettler LE, Voelker RA, Mukai T (1977) Inversion clines in populations of *Drosophila melanogaster*. Genetics 87:169–176

Mukai T, Voelker RA (1977) The genetic structure of natural populations of *Drosophila melanogaster*. XIII. Further studies on linkage disequilibrium. Genetics 86:175–185

Mukai T, Watanabe TK, Yamaguchi O (1974) The genetic structure of natural populations of *Drosophila melanogaster*. XII. Linkage disequilibrium in a large local population. Genetics 77: 771–793

Nevo E (1978) Genetic variation in natural populations: patterns and theory. Theor Popul Biol 13:121–177

Oakeshott JG, Gibson JB, Anderson PR, Knibb WR, Anderson DG, Chambers GK (1982) Alcohol dehydrogenase and glycerol-3-phosphate dehydrogenase clines in *Drosophila melanogaster* on different continents. Evolution 36:86–96

Parsons PA (1973) Genetics of resistance to environmental stresses in *Drosophila* populations. Annu Rev Genet 7:239–265

Perry AS, Agosin M (1974) The physiology of insecticide resistance by insects. In: Rockstein M (ed) The physiology of insecta, vol VI. Academic Press, London New York, pp 3–121

Powell JR (1975) Protein variation in natural populations of animals. Evol Biol 8:79–119
Richmond RC (1970) Non-Darwinian evolution: a critique. Nature (London) 225:1025–1028
Vigue CL, Johnson FM (1973) Isozyme variability in species of the genus *Drosophila*. VI. Frequency-property-environment relationships of allelic alcohol dehydrogenases in *Drosophila melanogaster*. Biochem Genet 9:213–227
Voelker RA, Mukai T, Johnson FM (1977) Genetic variation in populations of *Drosophila melanogaster* from the Western United States. Genetica 47:143–148
Voelker RA, Cockerham CC, Johnson RM, Schaffer HE, Mukai T, Mettler LE (1978) Inversions fail to account for allozyme clines. Genetics 88:515–527
Wills C (1973) In defense of naive panselectionism. Am Nat 107:23–34

Genetic Polymorphism and its Ecological Background in *Tephritid* Populations (Diptera: Tephritidae)

A. SEITZ and M. KOMMA[1]

Introduction

Fruit flies of the family Tephritidae have been well researched for several decades because they include pests of many cultivated plants. Examples are the Mediterranean fruit fly *Ceratitis capitata* Wied., which feeds on a broad range of fruits, the olive fruit fly *Dacus oleae* Gmel., the European cherry fruit fly *Rhagoletis cerasi* L., and the apple maggot fly *Rhagoletis pomonella* Walsh. Other Tephritids which feed on weeds are host specific and therefore potential agents in biological control. Because of their economic importance detailed research has been carried out on Tephritid physiology, ecology, genetics, and evolution.

We have initiated a series of investigations (Eschenbacher 1982, Komma 1982, Romstöck 1982, Seitz 1982) on populations of two Compositae infesting tephritid flies, namely, *Urophora cardui* L. and *Tephritis conura* Lw. The *U. cardui* is monophagous on the creeping thistle *Cirsium arvense* (L.) Scop. (Peschken and Harris 1975) and has therefore been introduced for the biological control of this weed in Canada (Peschken et al. 1982) and other countries. *Tephritis conura* infests *Cirsium oleraceum* (L.) Scop., *C. heterophyllum* (L.) All., *C. acaule* (L.) Weber, *C. erisithales* (Jacq.) Scop., and *C. spinosissimum* (L.) Scop.

Both fly species are univoltine. *Urophora cardui* oviposits into the stems of *C. arvense*. The developing larvae induce the formation of multilocular galls (i.e., galls which consist of a number of cells, each bearing a single larva) which can easily be located, counted and collected. The larvae and pupae are found inside the galls between June and May the following year and the population size and area can therefore be estimated by counting and mapping the galls.

Although the host plant *(C. arvense)* is found throughout central Europe from southern France to southern Sweden, the distribution of *U. cardui* is patchy, with distinct borders between the Rhine valley, the Danube region, and the populations in the nothern German lowlands. Based on the ecological data by Zwölfer (1982) we concluded that this patchy distribution is caused by the poor dispersal capacity of the fly and by interactions with its parasitoids.

In contrast to *U. cardui, T. conura* is common throughout central Europe. It infests the flower heads of its host plants and feeds on the developing florets and receptacle tissue.

1 Institute of Animal Ecology, University of Bayreuth, 8580 Bayreuth, FRG

Population Biology and Evolution
Ed. by K. Wöhrmann and V. Loeschcke

The two common thistles which are infested in northern Bavaria by *T. conura* are *C. heterophyllum* and *C. oleraceum*, which grow at different altitudes. *Cirsium oleraceum* is found up to 600 m and *C. heterophyllum* down to 500 m above sea level, with an overlapping region between the 500 m to 600 m altitudes. In Denmark and southern Sweden the two species are sympatric (Hultén 1971). Besides these discontinuous distributions, distinct differences also exist between the morphology of their buds which demand elaborate orientation mechanisms by the female fly during oviposition. We, therefore, expected the evolution of two genetically distinct host adapted strains of *T. conura*. These differences of the population genetic structure of *U. cardui* and *T. conura* were studied by means of electrophoretic methods.

Material and Methods

Both fly species were sampled in the larval stage. We collected galls of *U. cardui* in autumn and early spring from 28 localities in eastern France, the upper Rhine valley, southern and northern Bavaria, and Schleswig-Holstein (Fig. 1). The galls were dissected in the laboratory and the gall size, the number of larvae per gall, and the number of larve killed by parasitoids was recorded. The larvae were reared to adults in controlled environmental chambers. They were collected daily and stored in liquid

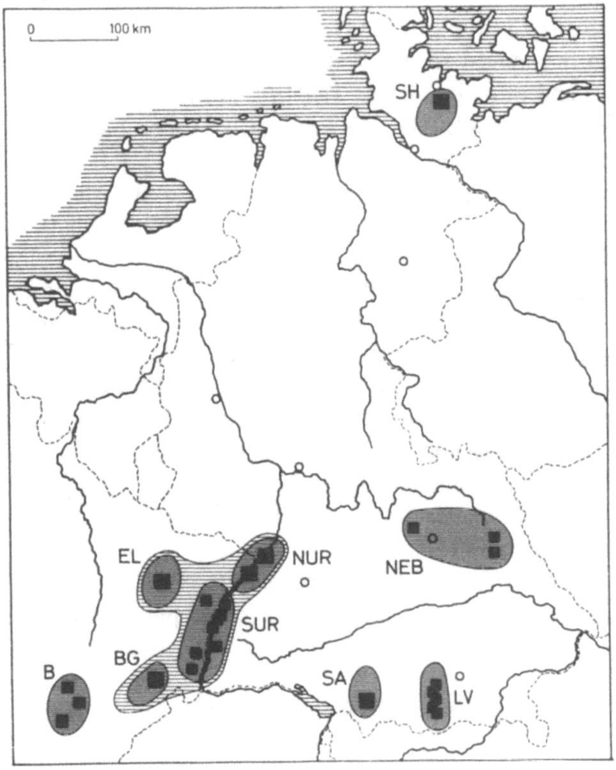

Fig. 1. Sampling stations of *Urophora cardui*. The *shaded areas* represent the pooled stations. The *light shaded* region covers the areas which are referred to in the text as Rhine region. *SH* Schleswig-Holstein; *NEB* north eastern Bavaria; *L V* Lech valley; *SA* Swabian Allgäu; *BG* Belfort Gap; *B* eastern Burgundy; *NUR* northern upper Rhine valley; *SUR* southern upper Rhine valley; *EL* eastern Lorraine

nitrogen for further analysis. The proportion of larvae which died during laboratory treatment was below 1%.

Tephritis conura were sampled in August from infested flower heads and kept in environmental chambers until all flies had emerged. These adults were stored in liquid nitrogen in the same way as *U. cardui*.

In northern Bavaria *T. conura* was sampled from 12 localities within a radius of 30 km. At five of these stations *C. oleraceum* was the only host plant species present, at three stations populations of *C. oleraceum* were mixed with *C. heterophyllum*, and at the remaining four stations pure stands of *C. heterophyllum* occurred. The flies from the mixed populations were kept separate according their host plants so that a total of 15 populations were sampled.

The electrophoresis was carried out on horizontal starch slab gels. The gels were stained using a zymogram technique based on methods in Shaw and Prasad (1970) with minor modifications. The samples from different localities were mixed on the same gel to allow direct comparisons between the relative mobilities of electromorphs from different stations. We have tested 25 enzyme systems; 13 of them could be checked routinely on *U. cardui* and 11 on *T. conura*.

The number of individuals analyzed for *U. cardui* varied from 10 to 100 per sample, which approximately reflects the spectrum size of the subpopulations. We usually analyzed 40 to 70 individuals per sample on *T. conura*, with the exception of the station Eimersmühle, where 260 flies were submitted to electrophoresis for more detailed statistics.

The resulting allele frequencies were used to test differences between sampling stations based on single loci by means of the minimum discrimination statistics 2I (Gabriel 1966) and also to calculate Nei's genetic distances (Nei 1972). These were used to construct phenograms using the unweighted pair group with arithmetic means (UPGMA) clustering technique (Sneath and Sokal 1973).

Because of the small sample sizes of *U. cardui* populations, the data were insufficient to reliably test single local populations for deviation from Hardy-Weinberg equilibrium. Therefore, the data from the 28 sampling stations were pooled into 9 regions: the northern and the southern Rhine valley, the Belfort Gap, eastern Burgundy, eastern Lorraine, northeastern Bavaria, Allgäu region, Lech valley, and Schleswig-Holstein (Fig. 1). The pooled data are not suitable to study the genotypic structure of the populations. Under the conditions of the special population structure of *U. cardui*, discussed below, pooled allele frequencies can be better used to compare more extended areas.

Results

The *Urophora cardui* System

On *U. cardui* eight loci were monomorphic: *me, hk, idh1, idh2, lap1, lap2, pgm,* and *gdh.* Three loci showed three electromorphs *acon2, got2,* and *mdh1.* The anodal *acon1* showed six electromorphs and the also anodal *got1* 14. The enzymes coded by the second locus of *idh* and *lap* migrate faster than the one of the first. *Acon2, got2,* and *mdh2* migrate cathodally. The latter produced no sharp bands and could, therefore, not be scored. The number of electromorphs of *got1* is high, but is in the range which is reported for other enzymes of tephritid populations (Tsakas and Zouros 1980, Berlocher and Bush 1982).

Based on the 29 alleles of the 5 scorable polymorphic loci, the 9 regions could be distinguished from each other at the $P \ll 0.001$ level. Their genetic distance values (Table 1) ranged from 0.005 between the populations in the Belfort Gap and the northern Rhine Valley, to 0.072 between the Schleswig-Holstein population and north-eastern Bavaria.

Table 1. Matrix of genetic distances between sampling regions of *Urophora cardui*

	(2)	(3)	(4)	(5)	(6)	(7)	(8)	(9)
(1) Northern upper Rhine	0.010	0.005	0.035	0.024	0.050	0.018	0.024	0.020
(2) Southern upper Rhine	–	0.011	0.036	0.020	0.061	0.006	0.021	0.028
(3) Belfort Gap		–	0.035	0.023	0.052	0.014	0.018	0.022
(4) Burgundy			–	0.069	0.071	0.048	0.071	0.037
(5) North eastern Bavaria				–	0.072	0.019	0.011	0.038
(6) Schleswig-Holstein					–	0.063	0.057	0.071
(7) Lorraine						–	0.015	0.043
(8) Allgäu							–	0.040
(9) Lech valley								–

We have constructed a phenogram from the matrix of genetic distances using the UPGMA method (Fig. 2). This shows a close relationship between the populations of the Rhine valley. The remaining populations show a rather random pattern which suggests a considerable amount of isolation between the areas. Consequently, we have once more condensed the data and pooled the material from the Rhine area and the Belfort Gap. Ghis finally resulted in six genetically distinct areas: Rhine region, Burgundy, Lech valley, Allgäu, northern Bavaria, and Schleswig-Holstein (cf. Fig. 1).

Major differences between these large areas are found at the *got1* and *acon1* loci: the number of alleles and the dominating allele differ from region to region (Table 2).

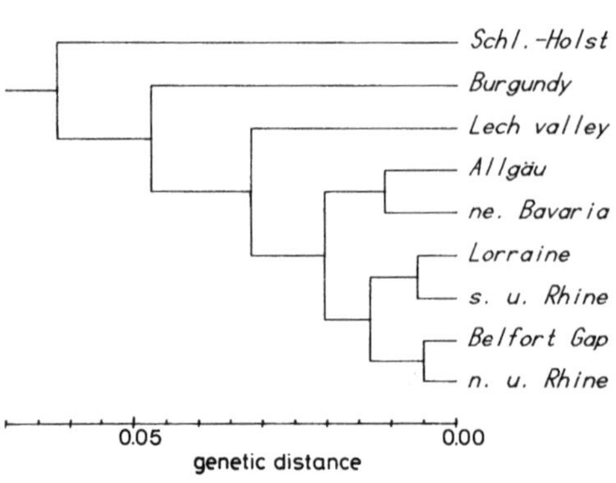

Fig. 2. Phenogram constructed by an UPGMA clustering of genetic distances between sampling regions of *U. cardui*

Table 2. Allele frequencies of *got1* and *acon1* of *U. cardui* [a]

Relative mobility	Burgundy	Rhine region	Lech valley	Allgäu	North eastern Bavaria	Schleswig-Holstein
			GOT1			
0.00	0.02	0.05	0.06	0.00	0.00	0.00
0.43	0.00	0.04	0.00	0.00	0.00	0.00
0.71	0.11	0.01	0.00	0.00	0.00	0.52
1.00	0.27	0.27	0.32	0.20	0.28	0.00
1.14	0.03	0.00	0.00	0.00	0.00	0.00
1.46	0.29	0.00	0.13	0.00	0.00	0.00
2.00	0.07	0.17	0.32	0.20	0.27	0.11
2.20	0.03	0.00	0.00	0.00	0.00	0.00
2.64	0.04	0.06	0.00	0.05	0.00	0.07
3.00	0.13	0.39	0.10	0.55	0.45	0.30
3.14	0.01	0.00	0.00	0.00	0.00	0.00
3.40	0.00	0.01	0.08	0.00	0.00	0.00
			ACON1			
0.80	0.14	0.04	0.03	0.08	0.09	0.00
0.90	0.08	0.03	0.00	0.00	0.00	0.00
1.00	0.73	0.67	0.60	0.53	0.50	0.66
1.10	0.05	0.25	0.31	0.39	0.39	0.34
1.20	0.00	0.01	0.04	0.00	0.01	0.00
1.26	0.00	0.00	0.03	0.00	0.00	0.00
Number of individuals	75	340	99	20	30	28
Number of localities	3	12	8	1	3	1

[a] Only alleles with a frequency exceeding 1% are listed

The highest degree of polymorphism is found in Burgundy, the least in Schleswig-Holstein. The total number of electromorphs of the *acon1* and *got1* loci with a frequency of more than 1%, amounts to 14 for the Burgundy and only 6 for Schleswig-Holstein. The 1% threshold level was used to prevent a distortion of the data due to unequal sample sizes (we found a total of 14 alleles of *got1* in the Burgundy sample).

The study areas were ranked according to their presumable sequence of recolonization after the last glaciation (Fig. 3) and an almost perfect correlation between age and number of alleles was found (Fig. 4). To test whether this correlation is only an effect of different sample sizes or numbers of sampling stations within an area, correlations between these factors and number of alleles were calculated. None of them showed a significant correlation with the number of alleles. The pattern of the genetic structure of *U. cardui* which we find today is, therefore, only a transient image which basically results from a predominantly unidirectional gene flow and a colonization process which is not yet completed. This hypothesis is corroborated by the distribution pattern of the fly and its host plant (Fig. 5): large areas within the distribution of the host plant are not yet colonized by the fly in spite of the fact that artificial releases have shown that these areas can successfully be colonized (Zwölfer 1979).

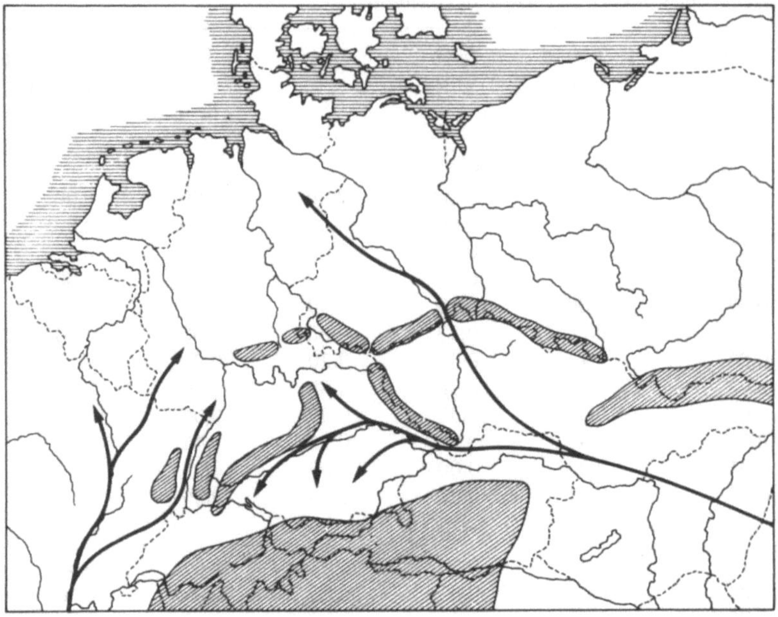

Fig. 3. Main routes of immigration of plants from the Mediterranean and pontic area after the Würm glacial period (modified after Litzelmann 1938). *Shaded areas* show the alps and extended hilly regions

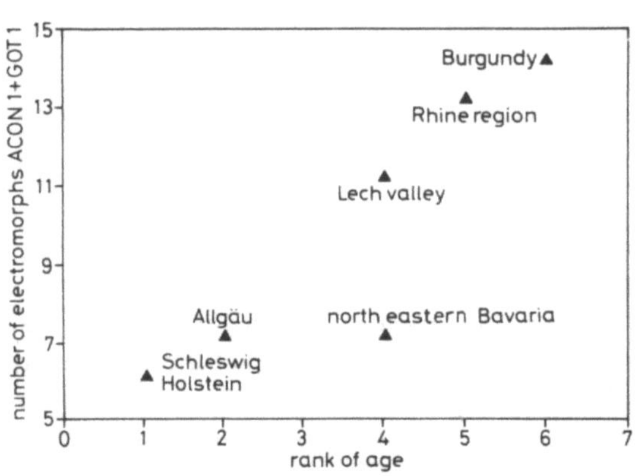

Fig. 4. Relation between estimated age of populations of *U. cardui* in the pooled areas and the total of electromorphs of *acon1* and *got1* exceeding 1% frequency. The Spearman rank correlation is highly significant ($R_{s,b} = 0.956, P < 0.01$)

To study the reasons for the slow dispersal of *U. cardui*, we have analyzed small subpopulations in the upper Rhine valley and in the Lech valley in more detail. A comparison of the allele frequencies of *acon1* and *got1* for closely adjacent, but isolated sampling stations (Fig. 6) by a 2Î-test for homogeneity showed that the populations of *U. cardui* are divided into genetically distinct subpopulations (for both loci $P < 0.001$). At the *got1* locus half of the single comparisons showed significant differences between pairs of stations ($P < 0.001$). For *acon1* all tested subpopulations differed from each other ($P < 0.001$).

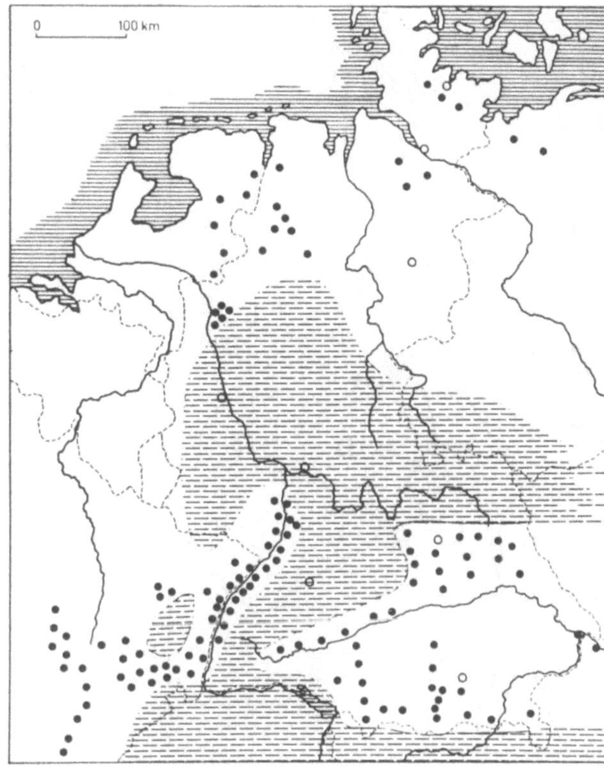

Fig. 5. Distribution of *U. cardui* in central Europe. The *black circles* refer to localities where *U. cardui* was sampled or from available literature records. Within the *shaded areas U. cardui* is not known to occur in spite of the presence of *C. arvense*

The *Tephritis conura* System

On *T. conura* three loci were monomorphic: *gdh, g3pd,* and *me*. The locus *hk2* showed two electromorphs (the slower *hk1* produced no sharp bands and could, therefore, not be scored). Four electromorphs could be identified from *idh, pgi,* and *pgm*, six from *acon* and *mdh*, and seven from *got* and *pep*.

As for *U. cardui* we have used the allele frequencies to calculate genetic distances as shown in Table 3. The values range from 0.002 to 0.102. If we group the stations according to their host plants, we find the following pattern: the genetic distances between populations from *C. oleraceum* range between 0.002 to 0.018 (mean 0.011; SE 0.001). Genetic distances between *C. heterophyllum* were similar and ranged between 0.002 to 0.01 (mean 0.005; SE 0.0005). However, the distances between populations from different host plants are significantly higher and range from 0.042 to 0.102 (mean 0.077; SE 0.001). This is true regardless of the geographical origin of the sampling station. More significant, however, is that this also holds true for mixed stations. If we compare average genetic distances between populations from sampling stations with only one host plant species and populations from localities with both host plant species, we find higher genetic distances between populations from the same station but different host plant species than between populations from different stations but the same host plant species (Table 4).

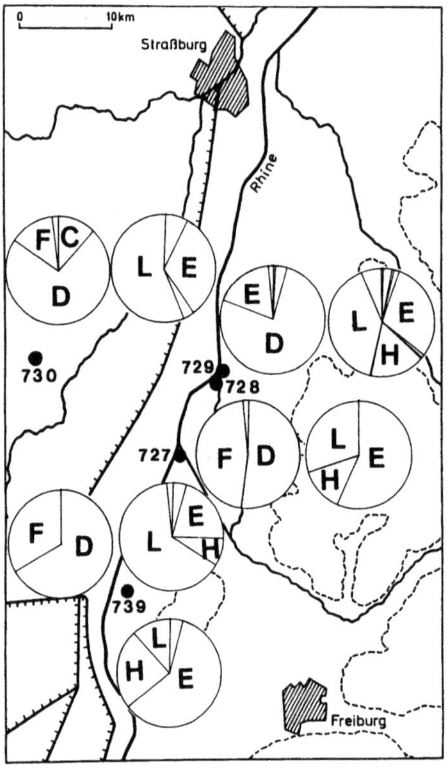

Fig. 6. Frequencies of alleles of *acon1* and *got1* for five sampling stations of *U. cardui* in the upper Rhine valley. *Left circles* show *acon1*, *right circles got1*. The *letters* designate representative electromorphs. Data for *acon1* are not available for station 739

This relationship between host plants and fly populations becomes clearer in the phenogram constructed by an UPGMA clustering (Fig. 7): the populations can be classified into two distinct groups, one consisting of populations sampled from *C. heterophyllum* and the other sampled from *C. oleraceum*. The populations from *C. heterophyllum* are more homogenous than those from *C. oleraceum* which was expected from the data in Table 4.

For more insight into the reasons for this differentiation, the single loci were investigated in more detail. Two of these, namely, *hk2* and *got* showed significant differences. With respect to these two loci, deviation from Hardy-Weinberg expectations could only be proven for *hk2* at the station Eimersmühle, from which more data were available. For *T. conura* reared from *C. oleraceum* there is a significant lack of heterozygotes. Since its A allele is almost fixed in populations from *C. heterophyllum*, data are insufficient to prove this for populations of *T. conura* from this host as well.

At the *got* locus a decrease in the frequency of the D allele of *T. conura* populations reared from *C. heterophyllum* from mixed sampling stations was found compared to those from pure *C. heterophyllum* stations, pure *C. oleraceum* stations, and from *C. oleraceum* at mixed stations (see Fig. 9a). These differences were highly significant.

The differences in the allele frequencies of these groups are obviously correlated with the flowering period of the host plants and, as a consequence, with the periods

Table 3. Matrix of genetic distances between sampling stations of *Tephritis conura*

	(2)	(3)	(4)	(5)	(6)	(7)	(8)	(9)	(10)	(11)	(12)	(13)	(14)	(15)
(1) 0[a] Eimersmühle	0.002	0.011	0.004	0.005	0.011	0.007	0.018	0.078	0.089	0.089	0.076	0.065	0.079	0.073
(2) 0 Hohenknoden	–	0.011	0.004	0.004	0.009	0.008	0.012	0.066	0.075	0.073	0.061	0.049	0.063	0.062
(3) 0 Koppel		–	0.008	0.012	0.016	0.014	0.010	0.087	0.100	0.098	0.085	0.071	0.081	0.084
(4) 0 Schönfeld			–	0.002	0.017	0.010	0.018	0.089	0.102	0.101	0.086	0.073	0.087	0.084
(5) 0 Goldkronach				–	0.019	0.010	0.019	0.089	0.102	0.100	0.084	0.069	0.087	0.083
(6) 0 Frankenhammer					–	0.015	0.014	0.047	0.056	0.058	0.048	0.042	0.047	0.046
(7) 0 Röhrenhof						–	0.014	0.090	0.102	0.099	0.082	0.072	0.085	0.086
(8) 0 Pfeifersmühle							–	0.073	0.083	0.079	0.064	0.052	0.064	0.072
(9) H[b] Standort 1								–	0.002	0.004	0.005	0.010	0.005	0.004
(10) H Bischofsgrün									–	0.002	0.004	0.010	0.005	0.005
(11) H Forst										–	0.002	0.007	0.004	0.008
(12) H Warmensteinach											–	0.004	0.002	0.007
(13) H Pfeifersmühle												–	0.005	0.008
(14) H Frankenhammer													–	0.006
(15) H Röhrenhof														–

[a] 0 = Populations reared from *C. oleraceum*
[b] H = Populations reared from *C. heterophyllum*

Table 4. Genetic distances between populations of *Tephritis conura* from mixed and nonmixed localities

Sampling station	D_o[a] ± SE	D_m[b]	D_h[c] ± SE
Frankenhammer	0.014 ± 0.0013	0.047	0.0045 ± 0.0006
Röhrenhof	0.011 ± 0.0012	0.086	0.0063 ± 0.0006
Pfeifersmühle	0.015 ± 0.0013	0.052	0.0073 ± 0.0025
Average	0.013 ± 0.0007	0.062	0.0060 ± 0.0004

[a] D_o = average genetic distance between populations from mixed and from pure stands of *C. oleraceum*
[b] D_m = genetic distance between populations from different hosts at the same locality
[c] D_h = average genetic distance between populations from mixed and pure stands of *C. heterophyllum*

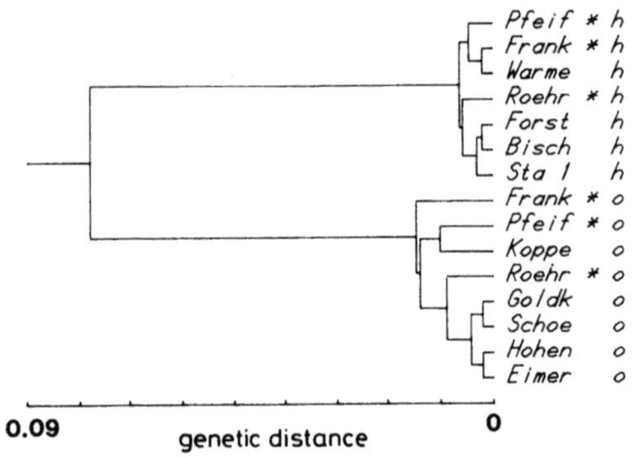

Fig. 7

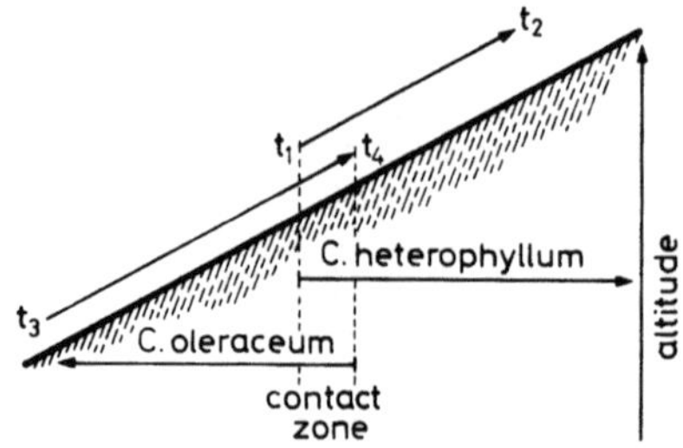

Fig. 8

Fig. 7. Phenogram of sampling stations of *T. conura* constructed from genetic distances by an UPGMA clustering. Populations marked by an *asterisk* are from mixed stations; *h* and *o* refer to populations sampled from *C. heterophyllum* and *C. oleraceum*, respectively

Fig. 8. Seasonal appearance of *T. conura* in the study area. t_1 to t_4 refer to successive flowering periods of the host plants and oviposition of the flies

at which the female flies lay their eggs (Fig. 8). The first plants accepted for oviposition by *T. conura* is *C. heterophyllum* at medium altitudes (mixed stations), then progressively on plants of this species growing at higher altitudes (pure stations). The next host selected for oviposition is *C. oleraceum* growing at low altitudes (pure stations) followed by infestation at medium altitudes (mixed stations). Maximum time differences are recorded from mixed localities which reflect similar differences in the *got* D allele frequencies. There is obviously a high selection pressure to synchronize oviposition with certain stages of host plant development (Eschenbacher 1982, Romstöck 1982). Successful oviposition is possible if the closed flower heads of the host are neither too small nor too large.

The effect of this divergent selection for oviposition periods is expressed in the *hk2* locus, the second one studied in detail. The frequency of its A allele from *C. he-*

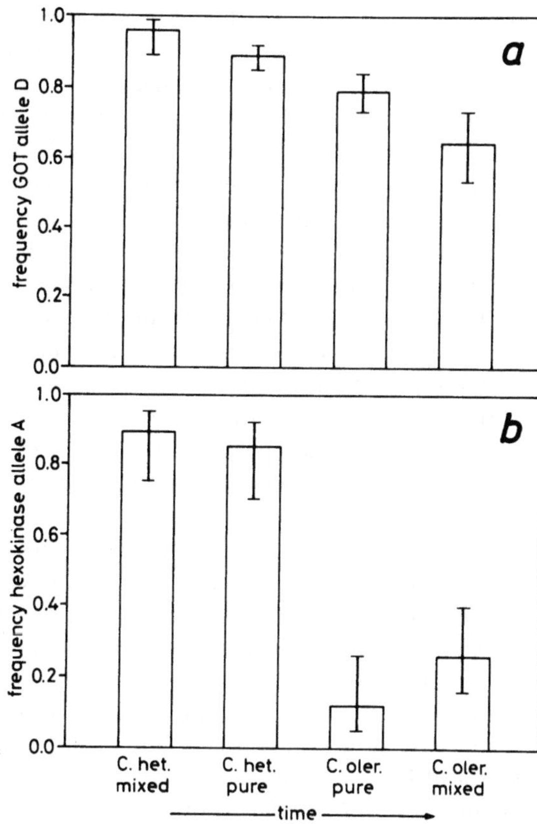

Fig. 9a, b. Frequencies of *got* allele D (**a**) and HK2 allele A (**b**) of *T. conura*. The data are averages from all mixed and pure sampling stations from both host plant species. The 1% confidence limits are shown

terophyllum populations is about 87%, regardless whether they are from mixed or pure sampling stations. In populations reared from *C. oleraceum* the frequency of the A allele ranges from 12% to 23%. Whereas differences between populations from different host plant species were highly significant, differences between populations from the same host plant species cannot be detected (Fig. 9b). From these results we conclude that the populations of *T. conura* from different host plant species are reproductively isolated. This is corroborated by Wahlund's principle: the populations of *T. conura* at the mixed stations are in Hardy-Weinberg equilibrium with respect to *hk2* if tested separately ($P > 0.3$ and $P > 0.5$). The pooled data, however, show a lack of heterozygotes. The deviation from Hardy-Weinberg expectations is highly significant ($P \ll 0.001$).

Discussion

In spite of the similar life history and ecology of *T. conura* and *U. cardui*, we have found marked differences in the genetic structure of their populations. These differences are probably caused by their different host ranges and different dispersal capacities. As shown in Fig. 5, *U. cardui* shows sharp gaps in its distribution in central

Europe which are not caused by the patchy distribution of its host plant. The northern boundary of its distribution is now in southern Sweden (Hultén 1971). In contrast, *T. conura* has a wider host range and can be found on *C. heterophyllum* up to the northern polar regions (Romstöck, pers. comm.) and also on *C. erisithales* in the Alps up to 2,000 m altitude. In central Europe both insect-plant systems are younger than 10,000 years, which is the time span after the retreat of the ice cover during the Würm glaciation. Due to its preference for moist lowlands, *C. arvense* first recolonized central Europe along the Rhone, Rhine, Danube, and Elbe valleys (cf. Fig. 3). *U. cardui* apparently could not follow the post-Pleistocene expansion of *C. arvense* (Zwölfer 1982) which was caused mainly by the introduction of agriculture in central Europe. Reasons for the slow dispersal rate of *U. cardui* might be explained by the structure of the host-parasitoid relationships between *U. cardui* and *Eurytoma* species (Zwölfer 1979). Long-term investigations (Zwölfer pers. comm.) showed tremendous density fluctuations of *U. cardui* populations down to extinction followed by recolonization. It seems that local subpopulations go extinct due to parasitoid pressure and are recolonized later.

The strong genetic differences between neighboring populations of *U. cardui* is probably the result of repeated extinction and recolonization of populations by only a few animals. The impact of genetic drift by repeated bottle necks (Nei et al. 1975) is strengthened by the generally small population sizes (ten to some thousands per local population). The low population densities together with limited flight activities contribute to the fly's poor general dispersal. Zwölfer (pers. commun.) recorded a dispersal rate of 100 m per generation and Peschken et al. (1982) have observed a dispersal of 4 km in 5 years at one of their release sites, although in others the dispersal was much lower. *U. cardui* is, therefore, able to form metapopulations, i.e., populations consisting of a number of semi-isolated subpopulations. In this respect the metapopulations of *U. cardui*, with approximately equal sized subpopulations, agree with the model described by Levins (1970) and differ from the one proposed by Boorman and Levitt (1973), in which extinguished subpopulations are recolonized from a large central population. The population structure formed by *U. cardui* should, therefore, permit a larger number of competing genotypes for the following reasons: (1) due to the subdivision into compartments, the probability of direct interaction between genotypes from different subpopulations decreases because of reduced migration; (2) as a consequence, unfavorable genotypes can exist in different subunits of the population in absence of competition. This also is in accordance with the arguments of Wright (1931) that subdivided populations with limited migration promote rapid evolution.

The distance between the Pleistocene refugia and the northern distribution border of *U. cardui* is 1,700 km. Therefore, the time span necessary to cover this distance at a pace of 100 to 800 m per year amounts to approximately 10,000 years, i.e., since the last ice cover.

Obviously colonization by *U. cardui* was not constant in all areas. Since *U. cardui* predominately colonized the riverine lowlands, the dispersal from one river system to the next must have required considerable time. This probably accounts for the pronounced genetic differences between populations from the Rhine area and the adjacent Rhone region. Naglyaki (1976) has studied the effect of variable migration rates in a diffusion model which was applied to situations with geographical barriers in the habitat and his results agree with our interpretation of the *U. cardui* data.

The progressive loss of genetic information from the southern Pleistocene refugia to the north agrees with a one-dimensional stepping stone model of dispersal and with the maintainance of this population structure for a long period of time (Kimura and Weiss 1964). A discontinuous but predictable distribution of the host plant is, therefore, the basic factor of the population structure of *U. cardui*. On the one hand, it allows the establishment of small isolated subpopulations which enable a process of hide-and-seek between *U. cardui* and its parasitoids and prevents the system from going extinct. This may be the main reason why *U. cardui* is largely limited to the riverine lowlands where *U. cardui* is the only host for these parasitoids. On the other hand, this population structure reduces the flow of genetic information which is reflected in the marked differences of the genetic constitution of subpopulations.

In contrast to *U. cardui, T. conura* occurs as contiguous populations in all areas in central Europe where host plants are available. We find, therefore, no significant differences between local subpopulations reared from the *same* host plant species. However, the differences found between populations from *different* host plant species warrant a more detailed discussion. *Tephritis conura* can clearly be divided into two groups: those reared form *C. oleraceum* and those form *C. heterophyllum*. These results can bc interpreted in three ways:

1. there exists only one species of *T. conura* which does not distinguish between different host species. In this case the observed differences are the result of differential survival of larval genotypes;
2. there exists one interbreeding species of *T. conura*, but with different genotypes which select different host plant species for eviposition;
3. the morphospecies *T. conura* consists of two sibling species or genetically distinct host races with different reproductive periods synchronized with their respective host plant species.

At present, we favor the third interpretation for the following reasons:

Case 1 can be discarded since it postulates that up to 50% of the larvae would die because of their maladaptation to one or the other host plant. Such high larval mortalities of *T. conura* which are not caused by insect parasitoids have never been observed (Eschenbacher 1982, Romstöck 1982). This type of system would rapidly evolve to a system described under 2 or 3. Case 2 is unstable since it is rather dangerous for flies to be exposed to predators while waiting for weeks for their right host. It would favor animals with a tendency to oviposit as early as possible.

Ecological studies of Eschenbacher (1982) and Romstöck (1982) have shown that females of *T. conura* accept only a limited size spectrum of buds of their host plants for oviposition. Therefore, an exact synchronization between oviposition and flowering periods is necessary. Exact synchronization with two host plant species which flower at different periods results in a divergent selection for different oviposition periods. Indeed *T. conura* is closely attuned to its host's phenology. At localities with both host plant species present, *T. conura* appears first during a period of approximately 2 weeks on *C. heterophyllum*, when this species is suitable for oviposion. No flies can be found during this time on *C. oleraceum*. Then *T. conura* disappears for 1 or

2 weeks. It then can again be found ovipositing exclusively on *C. oleraceum*. Unlike *C. heterophyllum* individual *C. oleraceum* plants produce a series of successive buds. Therefore, the timing of oviposition on this plant species is not as crucial as with *C. heterophyllum* which has a shorter flowering period. These phenological differences correlate with the higher degree of genetic polymorphism found in *T. conura* populations from *C. oleraceum* compared with those from *C. heterophyllum*. Those observations suggest that *T. conura* is capable of synchronization with and recognition of its respective host plant species. This model of host race formation and speciation is different from the types discussed by Bush and Diehl (1982). It requires no strong host discrimination and no changes in genes responsible for the survival of a parasite on its host, at least during the initial stages of the host shift. The disruptive force is the gap in the flowering periods of the host plants and the advantage to synchronize with them.

In *Tephritids*, host recognition is not only linked to oviposition, but also plays a dominant role in mating behavior. As in many phytophagous and entomophagous parasites, the host plant serves as a rendezvous place for males and females (Askew 1968, Bush 1969a,b, 1974, Zwölfer 1974) which establishes premating barriers between strains of populations with altered host specifity. As shown in two closely related species of the gall forming *Procecidochares*, the change of one single locus can alter host recognition and mate selection (Huettel and Bush 1972). If we assume that *hk2* in *T. conura* is either related to host selection in some way or is linked to the locus which determines host selection, for example, via timing, then the lack of heterozygotes proved at the station Eimersmühle is expected for this case of divergent selection.

At the moment we do not know whether the speciation process is completed. Since, under the conditions discussed above, the selection for premating reproductive isolation in the absence of the host plant is not strong (Boller and Bush 1974), shifts of the flowering periods of the host plants may, for the case that the assortment is only a matter of timing, reunite the gene pool of *T. conura* from *C. heterophyllum* and *C. oleraceum*. This may have occurred repeatedly in the past so that *T. conura* is kept in a constant state of speciation.

In view of the above discussion, the question arises why *U. cardui* shows no sign of a speciation process in spite of all the necessary prerequisites, i.e., marked population dynamics, extension of its area, repeated genetic bottle necks, small and isolated populations, etc. The answer may be found in the high degree of specialization and adaptation of *U. cardui* to its single host. The complexity of the physiology of gall formation (Lalonde 1983) is probably an evolutionary dead end which has permanently coupled host plant and parasite.

Acknowledgements. We thank Prof. H. Zwölfer for long and fruitful discussions and Prof. G.L. Bush for critical comments on the manuscript. Our special thanks are also due to Mrs. A. Servant-Miosga for lab assistance and to Dr. H. Zimmermann who checked the English. The work was partially financed by a grant of the Deutsche Forschungsgemeinschaft to the first author and by the Sonderforschungsbereich 137.

References

Askew RR (1968) Consideration on speciation in Chalcidoidea (Hymenoptera). Evolution 22: 642–645

Berlocher SH, Bush GL (1982) An electrophoretic analysis of *Rhagoletis* (Diptera: Tephritidae) phylogeny. Syst Zool 31:136–155

Boller EF, Bush GL (1974) Evidence for genetic variation in populations of the European cherry fruit fly, *Rhagoletis cerasi* (Diptera: Tephritidae) based on physiological parameters and hybridization experiments. Entomol Exp Appl 17:279–293

Boorman SA, Levitt PR (1973) Group selection on the boundary of a stable population. Theor Popul Biol 4:85–128

Bush GL (1969a) Sympatric host race formation and speciation in frugivorous flies of the genus *Rhagoletis* (Diptera, Tephritidae). Evolution 23:237–251

Bush GL (1969b) Mating behaviour, host specifity, and the ecological significance of sibling species in frugivorous flies of the genus *Rhagoletis* (Diptera – Tephritidae). Am Nat 103:669 to 672

Bush GL (1974) The mechanism of sympatric host race formation in the true fruit flies *(Tephritidae)*. In: White MJD (ed) Genetic mechanisms of speciation in insects. Sidney, pp 3–23

Bush GL, Diehl SR (1982) Host shifts, genetic models of sympatric speciation and the origin of insect species. Proc 5th Int Symp Insect – Plant Relationships, Wageningen, pp 297–305

Eschenbacher H (1982) Untersuchungen über den Insektenkomplex in den Blütenköpfen der Kohldistel *Cirsium oleraceum* L. (Compositae). Dipl Thes, Univ Bayreuth, p 100

Gabriel KR (1966) Simultaneous test procedures for multiple comparisons on categorical data. J Am Stat Ass 61:1081–1096

Huettel MD, Bush GL (1972) The genetics of host selection and its bearing on sympatric speciation in *Procecidochares* (Diptera: Tephritidae). Entomol Exp Appl 15:465–480

Hultén E (1971) Atlas över växternas utbredning i Norden. Generalstabens Litogr Anst, Förlag, Stockholm

Kimura M, Weiss GH (1964) The stepping stone model of population structure and the decrease of genetic correlation with distance. Genetics 49:561–576

Komma M (1982) Gelelektrophoretische Untersuchungen zur Populationsgenetik von Tephritis conura (Diptera: Tephritidae) auf Cirsium heterophyllum und Cirsium oleraceum. Dipl Thes, Univ Bayreuth, p 51

Lalonde RG (1983) Insect-host interactions in the *Urophora cardui* (L.) (Diptera: Tephritidae) *Cirsium arvense* (L.) Scop. (Compositae) Gall System. PhD Thes, Lauretanian Univ, Sudbury, Ontario, p 132

Levins R (1970) Extinction. In: Gerstenhaber M (ed) Some mathematical questions in biology. Lect Math Life Sci 2:77–107

Litzelmann E (1938) Pflanzenwanderungen im Klimawechsel der Nacheiszeit. Schr Dtsch Naturkd Neue Folge 7:48

Naglyaki T (1976) Clines with variable migration. Genetics 83:867–886

Nei M (1972) Genetic distance between populations. Am Nat 106:283–292

Nei M, Maruyama T, Chakraborty R (1975) The bottle neck effect and genetic variability in populations. Evolution 29:1–10

Peschken DP, Harris P (1975) Host specifity and biology of *Urophora cardui* (Diptera: Tephritidae). A biocontrol agent for Canada thistle *(Cirsium arvense)*. Can Entomol 107:1101–1110

Peschken DP, Finnamore DB, Watson AK (1982) Biocontrol of the weed Canada thistle *(Cirsium arvense)*: Releases and development of the gall fly *Urophora cardui* (Diptera: Tephritidae) in Canada. Can Entomol 114:349–357

Romstöck M (1982) Untersuchungen über den Insektenkomplex in den Blütenköpfen von *Cirsium heterophyllum* (Cardueae). Dipl Thes, Univ Bayreuth, p 115

Seitz A (1982) Populationsgenetische Untersuchungen zur Verbreitung der Bohrfliege *Urophora cardui* L. (Dipt.: Tephritidae) in Mitteleuropa. Verh Dtsch Zool Ges 1982:297

Shaw CR, Prasad R (1970) Starch gel electrophoresis of enzymes – a compilation of recipies. Biochem Genet 4:297–320

Sneath PHA, Sokal RR (1973) Numerical taxonomy. Freeman, San Francisco, p 573

Tsakas SC, Zouros E (1980) Genetic differences among natural and laboratory – reared populations of the olive-fruit fly *Dacus oleae* (Diptera: Tephritidae). Entomol Exp Appl 28:268 to 276

Wright S (1931) Evolution in mendelian populations. Genetics 16:97–159

Zwölfer H (1974) Das Treffpunkt-Prinzip als Kommunikationsstrategie und Isolationsmechanismus bei Bohrfliegen (Diptera: Trypetidae). Entomol Germ 1:11–20

Zwölfer H (1979) Strategies and counterstrategies in insect population systems competing for space and food in flower heads and plant galls. Fortschr Zool 25:331–353

Zwölfer H (1982) Das Verbreitungsareal der Bohrfliege *Urophora cardui* L. (Dipt.: Tephritidae) als Hinweis auf die ursprünglichen Habitate der Ackerdistel (*Cirsium arvense* (L.) Scop.). Verh Dtsch Zool Ges 1982:298

Multilocus Organization of Plant Populations

A.H.D. BROWN[1]

Introduction

The study of multilocus systems is claimed as "one the most important unresolved problems of population genetics" (Ginzburg and Braumann 1980). Multilocus organization at the population level means that certain combinations of the allelic variants at different loci are present more frequently than expected, whereas other combinations are infrequent or absent. In a simple example of two alleles at each of two loci, the extremes of organization are random association of alleles into four gametic genotypes on the one hand or their absolute association in two gametic genotypes on the other.

Wright (1965) has argued that "linkage disequilibrium" – which is the traditional yet somewhat unsatisfactory term used in this area – is a phenomenon generally only of secondary importance in evolution. His reasons were that in outbreeding populations, deviations from random association would be slight, except for tightly clustered genes. Functionally related loci were apparently scattered throughout the genome. Further, the exact mathematical treatment was complex so that results are of little use in general, However, Wright recognized that such second-order complications of general theory should be thoroughly explored and that in some organisms, increased linkage could convert these complications into first-order importance.

These views just predated an upsurge of interest in multilocus systems which was stimulated by Lewontin and his colleagues who emphasized that chromosomal segments, if not whole chromosomes, were the basic units of selection, rather than single independent loci. "Context and interaction are not simply second-order effects ... Context and interaction are of the essence" (Lewontin 1974). Much of the ensuing experimental and theoretical effort has been put into two-locus studies. It is apparent from the reviews by Hedrick et al. (1978) and Barker (1979) that it is time to assess the fruits of this enthusiasm.

The evolutionary dynamics of multilocus systems reflect the operation of several processes, namely, chromosomal rearrangements, amplification, recombination, small population size, partial isolation, hybridization, mating system and selection. When compared with the behaviour of single loci, the essential, new processes are recombination and selection determined by interlocus interaction. The other clear contrast with the study of single loci is a much increased role of stochastic effects, arising either

1 CSIRO, Division of Plant Industry, Canberra A.C.T. 2601, Australia

Population Biology and Evolution
Ed. by K. Wöhrmann and V. Loeschcke

from natural bottlenecks or from inadequate samples. Experimentalists define their task as to find examples of particular associations among certain pairs of loci and then to distinguish among the several forces to account for any multilocus structure observed. Here we examine an additional, broader approach, namely, the measurement of multilocus structure, based on a distribution of the number of heterozygous loci or sites when two random gametes or copies are compared. The problems of interpretation and the role of research on multilocus structure are briefly considered.

Measurement

The common basis for measuring associations among alleles is the linkage disequilibrium parameter D_{ik}. This is defined as the difference between the frequency of joint occurrence of A_i with B_k and its expected frequency assuming random combination of alleles A_i and B_k.

$$D_{ik} = g(A_iB_k) - p(A_i)\,p(B_k)\,.$$

This measure can be generalized to higher order disequilibria (three or more loci). Usually D_{ik} has been understood as a gametic measure. Weir (1979) has also shown how disequilibria can be defined at other levels of joint occurrence (e.g. between uniting pairs of gametes). Here we restrict the notion of structure to gametic association. The parameter D_{ik} and related measures are relatively manageable when there are few alleles at a small number of loci. However, with multiple alleles and many loci, the number of values quickly multiplies beyond comprehension. A smaller set of summary statistics would be useful to compare populations or different species.

Sometime ago Sved (1968) suggested that the variance of the number of heterozygous loci among individuals in a randomly mating population is a convenient summary of pairwise associations over many loci. More generally, the distribution f(K) of the number $\{K = 0, 1, ..., m\}$ of loci having different alleles, when two randomly chosen gametes are compared locus by locus at m loci, can be defined irrespective of the breeding system and the actual level of heterozygosity (Brown et al. 1980). In random mating populations, f(K) becomes the distribution of the number of heterozygous loci and is of direct interest in its relation with the distribution of fitness and of plasticity or homeostasis. In inbreeding populations, the distribution is a conceptual abstraction.

The most useful statistic is the variance (σ_K^2),

$$\sigma_K^2 = \sum^m h_j - \sum^m h_j^2 + 2 \sum_j^m \sum_{\ell > j}^m \sum_i \sum_k \left[2\, p_{ji} p_{\ell k} D_{ik}^{j\ell} + (D_{ik}^{j\ell})^2\right], \tag{1}$$

where

p_{ji} is the frequency of the ith allele at the jth locus,
$p_{\ell k}$ is the frequency of the kth allele at the ℓth locus,
$D_{ik}^{j\ell}$ is the disequilibrium between these alleles, and
$h_j = 1 - \sum_i p_{ji}^2$

is the gene diversity at the jth locus. The variance (σ_K^2) is independent of higher order associations. It is cumulative over locus pairs. It therefore tends to increase with "more" disequilibrium both within a pair of loci and over pairs. Higher moments of the distribution f(K) are increasingly complex functions of higher order disequilibria which may have limited application. An alternative expression for Eq. (1) is

$$\sigma_K^2 = (m - \Sigma h_j)(\Sigma h_j + 1 - m) + 2 \sum_j \sum_{\ell > j} \sum_i \sum_k (g_{ik}^{j\ell})^2 \,, \quad (1\text{ A})$$

where $g_{ik}^{j\ell}$ is the frequency of the two-locus gamete with the ith allele at the jth locus and the kth allele at the ℓth locus.

Standardization

As a basis for drawing inferences about an observed distribution, we need to specify what the distribution would look like with the same level of polymorphism at the contributing loci, but *assuming independence* among the loci on the one hand or assuming "maximum association" on the other.

In the case of independence, the complete distribution can be obtained as the convolution of the generating functions for heterozygosity at each single locus. This generating function is

$$G(X) = \prod_{j=1}^{m} [(1 - h_j) + h_j X] \,. \quad (2)$$

The function is expanded numerically and the coefficient of the term X^i is the probability that K equals i. All the moments can then be calculated. In particular,

$$\sigma_K^2 = E[K - E(K)]^2 = \Sigma h_j - \Sigma h_j^2 \,, \quad (3)$$

$$E[K - E(K)]^3 = \Sigma h_j - 3\Sigma h_j^2 + 2\Sigma h_j^3 \,, \quad (4)$$

$$E[K - E(K)]^4 = \Sigma h_j - 7\Sigma h_j^2 + 12\Sigma h_j^3 - 6\Sigma h_j^4 + 3[\Sigma h_j - \Sigma h_j^2]^2 \,, \quad (5)$$

where all summations are over $j = 1,2,\ldots,m$.

At the other extreme, the concept of "maximum association" is harder to define. Since most attention is on the variance of f(K), Brown and Burdon (1983) suggest that a suitable yardstick would be a distribution with maximum dispersion. Such a multilocus zygotic distribution can be constructed by ranking the given $\{h_j; j = 1,\ldots,m\}$ such that

$$h_1 > h_2 > h_2 \ldots > h_m \,.$$

A set of hypothetical multilocus diploid genotypes can be assembled so that $n(h_m)$ are heterozygous at all m loci, $n(h_{m-1} - h_m)$ are heterozygous at $(m - 1)$ loci, etc., and $n(1 - h_1)$ are heterozygous at no loci. The variance of this hypothetical distribution is

$$s_K^2 \max = 2\,\Sigma j h_j - \Sigma h_j - (\Sigma h_j)^2 \,. \quad (6)$$

However, it is usually impossible to define an array of multilocus gametes which on random union would generate this extreme zygotic distribution. Alternatively, expression (1 A) shows that the variance is maximized when the sum of squares of two-locus gametic frequencies is maximum. Yet, a method for finding the set of multilocus gametes with maximum variance, given the $\{h_j\}$ needs to be developed.

Using the above extreme models, one can define certain statistics for comparative purposes. Ideally, such statistics should remove the effect of sample size or of the level of polymorphism at individual loci. Brown et al. (1980) suggested a series of suitable indices of multilocus association as follows. Let $\mu(i)$ denote the values of the ith central moment computed from the observed distribution and let $\mu(i)$ denote the expected value of the ith central moment, given the single locus diversities (h_j) and the assumption of independence of loci [computed from Eqs. (3), (4), and (5) above]. Then we define:

$$X(i) = [m(i)/\mu(i)] - 1 \qquad (i = 2,3,...,)$$

as indices of multilocus structure. [For i = 1, X(1) equals zero.] For i = 2,

$$X(2) = s_k^2/[\Sigma h_j - \Sigma h_j^2] - 1 \,. \tag{7}$$

The division by $\mu(i)$ renders the X measures less dependent on the values of h_j and substracting one gives the expected values of X(i) of zero for i under independence.

However, values of X(2) can be affected by the number of loci included in any study, so that comparison among different studies is difficult. Using Eq. (6) above, we can obtain a corresponding maximum value for X(2) and finally a standardized index of intensity of association (I) relative to the maximum possible

$$\begin{aligned} I &= X(2)/X(2)\,\max \\ &= [s_K^2 - \Sigma h_j + \Sigma h_j^2]/[s_K^2\,\max - \Sigma h_j + \Sigma h_j^2] \,. \end{aligned} \tag{8}$$

Such values should be readily comparable among populations and species.

Estimation and Testing

Essentially two procedures are available for deriving the observed distribution. The first is applicable to predominant inbreeders or where the multilocus gametic frequencies are known. In this case the distribution is obtained by comparing each gamete in the sample of size n with every other and noting the number of heterozygous loci. The single locus values for the $\{h_j\}$ should be estimated by the same procedure. Thus, in the case of a very small size, the probability model is one of sampling without replacement. Although formulae such as Eq. (1) above, do not strictly apply to this situation, their essential structure holds. The terms in the expression for σ_K^2 are biased in the order of n^{-1}. However, association among the loci will usually inflate the variance above a value predicted by the $\{h_j\}$. The second procedure applicable to panmictic populations is to use the observed distribution directly. Further discussion of the problems of estimation (incomplete data, etc.) is given by Brown et al. (1980).

Let s_K^2 denote the observed value of the variance, which is an estimate of the population value (σ_K^2). The null hypothesis to be tested is

$$H_0 : \sigma_K^2 = \Sigma h_j - \Sigma h_j^2 .$$

Assuming that H_0 is true, the sampling variance of the observed variance (σ_K^2) is obtainable approximately as:

$$\text{var}\,[s_K^2 | H_0 \text{ is true}] = \{\mu(4) - [\mu(2)]^2\} /n .$$

From results Eqs. (3) and (5) above

$$\text{var}\,[s_K^2 | H_0 \text{ is true}] = \{\Sigma h_j - 7\Sigma h_j^2 + 12\Sigma h_j^3 - 6\Sigma h_j^4 + 2\,[\Sigma h_j - \Sigma h_j^2]^2\} /n , \quad (9)$$

where the parametric values of the $\{h_j\}$ are replaced by their estimates. Using this variance and assuming the sampling distribution of s_K^2 approximates normality, the upper 95% confidence limit for s_K^2 is:

$$L \cong \Sigma h_j - \Sigma h_j^2 + 2\,\{\text{var}[s_K^2 | H_0 \text{ is true}]\}^{1/2} . \quad (10)$$

Thus, if the observed s_K^2 exceeds L, the null hypothesis of independence at the level of locus pairs is rejected. The behaviour of this test in a wide range of simulated situations is yet to be investigated.

Advantages and Disadvantages of the Heterozygosity Measures

The following is a summary of the advantages of the approach outlined above (see Brown et al. 1980 for further discussion). The main interpretative properties which are helpful are: (1) the variance in heterozygosity, either actual or potential variance, is a biologically meaningful concept; (2) ready comparisons among species or populations are possible; and (3) values do not increase merely because of increasing sample size. From an analytical view, heterozygosity distributions (1) have received much theoretical attention, (2) are subject to testing procedures, and (3) can be calculated or observed in most situations. Further, this approach can be extended to the analysis of subdivided populations and used to quantify the extent to which the patterns of multilocus association are repetitive among populations (Brown and Feldman 1981).

The principal disadvantages of the variance approach are: (1) it represents a severe reduction in information compared with the many estimable D parameters; (2) the variance depends on pairwise two-locus disequilibria and ignores higher order association; (3) it assumes each locus is equivalent; (4) there is not a monotonic relationship between increasing D and the variance; and (5) it is severely distorted by certain patterns of high allele multiplicity, e.g. in the HLA incompatibility system in man. Karlin and Piazza (1981) have suggested several alternative summary statistics for this situation. It appears that the procedure works optimally for values of h_j around 0.5. For loci with extreme diversity, one option would be to lump some alleles until h_j approaches 0.5. This also remains to be investigated.

The above treatment of multilocus structure is based on whether or not two copies of a single locus are dissimilar. The treatment can readily be extended to summarizing structure over loci for other two-state, single-locus variables. For example, Brown and Burdon (1983) suggested one variable of interest would be based on whether or not the diploid genotype at each locus in each plant was the commonest homozygote for that locus. Equations 2–10 are applicable with h_j redefined as the population frequency of the commonest homozygote for the jth locus. If the observed variance for this variable significantly exceeded its expected value, the interpretation might be that genetic variation is concentrated in a few individuals, whereas much of the population tends to be a single optimal multilocus genotype.

Examples

The examples summarized in Table 1 serve to illustrate the above procedure and highlight the typical results for plant populations. The first case derives from a study

Table 1. Estimates of multilocus associations in various plant populations

Species/Source	Total Sample	P.L.[b]	$\bar{h}$	Variance statistics			
				s_X^2	X(2)	X(2)max	I × 10^2
Allozyme polymorphism[a]							
1. *Zea mays*							
10 generations	1,672	8	0.48	1.87	0.05	5.2	0.8
2. *Echium plantagineum*							
1 large population	193	13	0.36	2.84	0.15	6.9	2.2
3. *Pinus ponderosa*							
1 heterogeneous popn.	199	7	0.31	1.16	−0.08	3.5	− 2.3
4. *Hordeum vulgare*							
CC II 7	1,044	4	0.42	1.20	0.29	2.4	12.4
18	2,087	4	0.41	1.07	0.17	2.3	7.5
41	2,868	4	0.22	1.66	1.44	2.3	63.3
CC V 5	1,452	4	0.40	0.91	0.04	2.1	1.9
17	2,443	4	0.40	0.93	0.05	2.2	2.5
26	3,049	4	0.44	1.16	0.22	2.6	8.6
5. *Hordeum spontaneum*							
Shifon site	50	6	0.07	0.97	0.13	2.3	5.8
Gadot site	49	11	0.25	3.55	0.53	6.2	8.5
Bor Mashash site	52	8	0.06	3.82	2.79	4.7	59.7
Average of 26 sites	1,118	8	0.14	2.81	0.80	4.5	19.3
DNA sequences							
6. *Zea mays*							
Knob heterochromatin	12 copies	35	0.23	23.33	2.93	26.9	10.9

[a] Source of data: 1. Brown and Allard (1971) and unpublished; 2. Brown and Burdon (1983); 3. Linhart et al. (1981) and Mitton unpublished; 4. Clegg et al. (1972); 5. Brown et al. (1980); 6. Brown and Clegg (1983)

[b] Abbreviations: P.L. = Number of polymorphic loci; $\bar{h}$ = mean diversity

of ten generations of reciprocal recurrent selection in maize (Brown and Allard 1971). Each seedling was assayed for eight polymorphic allozyme loci and the observed and expected distribution of heterozygous loci compared for each generation. The observed variance (1.87) was marginally, but not significantly higher than the expected. The X(2) value indicated 5% inflation of the variance due to allele associations. Estimates of disequilibrium among pairs of loci indicated that two linked esterases were in significant disequilibrium, following the bottlenecks associated with selection. However, the X(2) value based on all eight loci did not detect this single association. In this agricultural, randomly mating and thoroughly mixed population, the degree of association for these loci approached zero.

The second example is a large population of the annual, outbreeding weed *Echium plantagineum* (Brown and Burdon 1983). This natural population was expected to exhibit the genetic effects of its weedy habit, namely, limited polymorphism and high disequilibrium. In contrast it was highly polymorphic and showed only small, nonsignificant multilocus association. The outbreeding system was evidently very effective in removing transient disequilibrium. A third outbreeding example is a population of *Pinus ponderosa* which displayed spatial and temporal heterogeneity (Linhart et al. 1981). Again the X value is close to zero.

Turning to predominantly inbreeding populations, the fourth example is based on two composite cross (C.C.) populations of barley analysed by Clegg et al. (1972). The genetic data are for four esterase loci, three of which are closely linked, the fourth being on a separate chromosome. Estimates of X showed associations present in the early generations. In both populations these increased in later generations, although not monotonically. In the F_{41} generation of CCII, the index reached over 60% of its maximum possible. These trends were confirmed by Muona (1982) with an additional locus. Some form of direct or indirect selection was presumably responsible for these evolutionary shifts.

The fifth example is natural populations of the inbreeding species, *Hordeum spontaneum*, which is the evolutionary progenitor of barley. Using allozyme variation at 20 loci, estimates of the variance in numbers of heterozygous loci in two randomly chosen gametes were made (Brown et al. 1980). For three-fourths of the populations, these estimates were larger than the values predicted by assuming no association among genes. Values of the association index ranged from 0.13 (a level almost as low as those of outbreeders) to 2.79. The most polymorphic population had X(2) = 0.53. On average the variance was inflated 80% due to multilocus associations. A comparison of the estimates of standardized index I for cultivated and wild barleys shows the range was similar, although the wild populations had a high average value.

An example of the use to which the summary statistic can be put is to plot relationship between it and other genetic or ecological variables. Figure 1 is a scatter diagram of the X(2) values for the *Hordeum spontaneum* populations against the mean genetic distance of each population to every other population. It is evident that the more genetically divergent populations are those showing highest indices of multilocus structure.

In summary, these five examples of allozyme data show the striking dichotomy between outbreeders and inbreeders. In outbreeders the levels of association among allozymes are generally low, whereas in the inbreeders so far examined they are usually

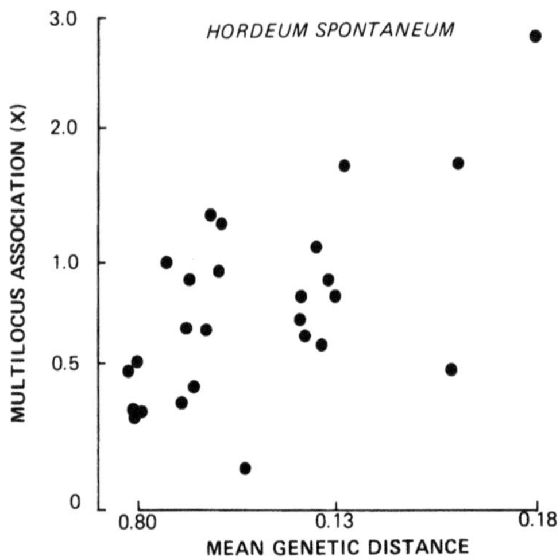

Fig. 1. Scatter diagram of the mean genetic distance of each polymorphic population of *Hordeum spontaneum* to all other sampled populations from Israel, plotted against its index of multilocus association, X(2). Rank correlation is R = 0.55 ($P <$ 0.01). [Brown et al. (1980)]

high. The deterministic theory for neutral genes in infinite populations in which some outcrossing occurs, does not predict this dichotomy.

Interpretations

To students of *Drosophila* populations, the common finding of linkage equilibrium among allozymes was somewhat unexpected. Langley (1977) suggested that one explanation was the failure of electrophoresis to resolve all allelic differences and that there might be hidden disequilibrium between hidden alleles. However, this could not be a major factor, as disequilibria were readily apparent in inbreeding plants and indeed among loci within or close to inversions in *Drosophila*.

On the other hand, the common findings of disequilibria in inbreeding plants may have turned into an embarassment of riches. To explain such disequilibria, experimentalists readily invoked selection based on interlocus interactions among the gene markers themselves or involving closely linked or functionally related loci (Allard at al. 1972). However, much recent theoretical work has led to criticism of this conclusion. The associations could arise by chance from population subdivision, hitchhiking or founder effects. For example, Golding and Strobeck (1980) found in their analysis of disequilibria in partially selfing, finite populations, that the effective recombination rate in predominant selfers is reduced by a factor of two times the outcrossing rate. In small populations of inbreeders (with say 1% outcrossing), the maximum effective recombination rate between any of their genes is about 1%. Hence all the genes of inbreeders behave as do closely linked genes in outbreeders. Significant linkage disequilibria due to random drift alone would arise when population size is less than the inverse of this effective recombination rate.

Indeed, for closely linked genes, Avery and Hill (1979) have computed that chance effects at the two-locus level can be so large as to lead to severe, persistent distortions of pattern. They argued that observations on disequilibria among linked loci are unlikely to be informative about any selective forces present. This seems unduly pessimistic.

Two approaches can be made to re-examine this problem. One is to compile estimates of the intensity of association for various kinds of species, populations or loci and look for broad trends. The variance measures may be a convenient way to do this. The comparison of related outbreeders and inbreeders is one obvious dimension. Others are annual versus perennial species, young colonizing versus old stabilized situations, marginal versus central populations, classes of loci with respect to linkage, species with few versus many chromosomes, etc.

The second approach is to look for specific kinds of association - particular multilocus genotypes - and especially for apparent independent repetitions of pattern in space. The variance measure can be extended to the analysis of subpopulation structure of associations (Brown and Feldman 1981). In this analysis, the total variance is partitioned into components such as the variance due to Wahlund effects in two loci on the one hand and the persistent and the variable disequilibrium across populations on the other. Analysis of the cultivated and wild barley data highlighted the repetitive pattern of associations between the two composite crosses and the regionally localized pattern in natural populations.

Is it Worthwhile to Study Multilocus Structure?

Perhaps the initial enthusiasm for experimental studies of disequilibrium was misplaced. However, the extreme pessimism of some current theory is equally so. Such pessimism ignores the classic evolutionary studies on heterostyly, mimicry, and inversion and translocation systems. Obviously such studies do not rely on estimates of D alone, but are integrated with other sorts of evidence, namely, functional or ecological. They have provided among the best evidence of natural selection in operation, less equivocal than much single locus work. Barker (1979) concluded his review with the point that the body of theoretical knowledge appears to be somewhat disjunct from our understanding of interlocus interactions in the real world. This needs to be rectified.

For almost 2 decades population geneticists and population ecologists have claimed that the most significant advances in their disciplines would come from the interface between them. During this period population genetics has been vigorously stimulated by the development of gel electrophoresis. In plants, the discipline of population ecology has come of age. Yet the question raised in several papers at this meeting concerns whether we have indeed seen the synthesis of these two fields. The recent prominence of molecular biology and its application to population research portends a far richer genetic component to this synthesis. Novel complexities of genome organization, of level and structures of genetic variation, and novel evolutionary mechanisms are now being characterized. The study of multilocus organization is needed to develop the genetic component of synthetic population biology. One key element will be an overview of the general features of multilocus organization, for all components of the genome.

To support this point, we refer to the last set of data in Table 1. This is an analysis of DNA sequence variation among 12 copies of a highly repeated sequence (more than 10^6 copies) associated with the knob heterochromatin in a line of maize (Dennis and Peacock 1983). The sequence was 180 base pairs in length and base substitutions were detected at 35 of these positions. Brown and Clegg (1983) used the formal analogy between site in a copy of the DNA sequence and locus in a multilocus gamete to form the distribution of the number of dissimilar sites when two randomly chosen copies are compared nucleotide by nucleotide. Figure 2 contrasts the observed distribution with that assuming independence of substitutions [from Eq. (2)] and with that assuming maximum variance. Clearly the sequences are exhibiting a marked degree of association among base substitutions. Presumably they are not independent in origin and indicate that concerted evolution or the propagation of one mutation to other copies of the sequence is a feature of their evolution. Thus, the analysis of the "multilocus structure" of these sequences gives additional evidence as to the mode of sequence divergence during amplification.

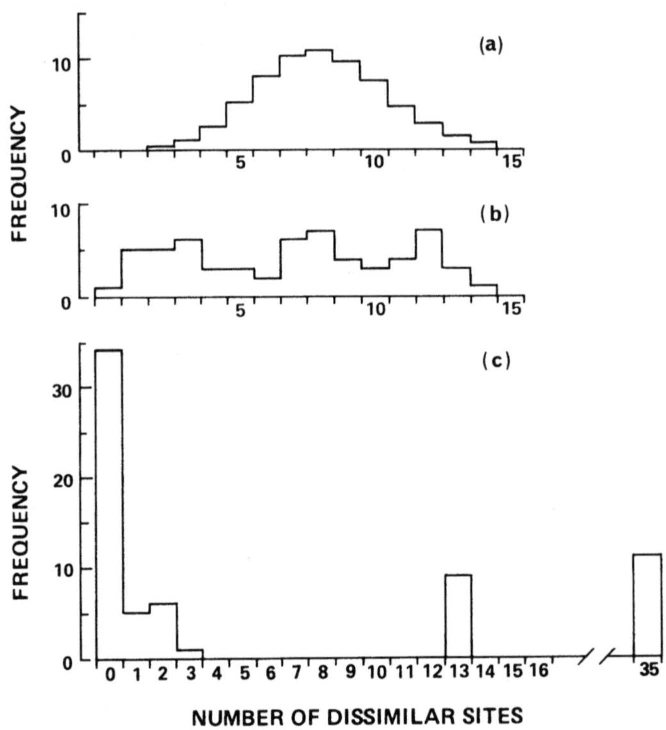

Fig. 2a–c. Variation among copies of a highly repeated DNA sequences from *Zea mays*. Histograms show the expected distribution of the number of dissimilar sites (a) assuming independence or (c) assuming maximum variance in comparison with (b) the observed distribution when two random copies are compared

References

Allard RW, Babbel GR, Clegg MT, Kahler AL (1972) Evidence for coadaptation in *Avena barbata*. Proc Natl Acad Sci USA 69:3043–3048

Avery PJ, Hill WG (1979) Distribution of linkage disequilibrium with selection and finite population size. Genet Res Camb 33:29–48

Barker JSF (1979) Inter-locus interactions: A review of experimental evidence. Theor Popul Biol 16:323–346

Brown AHD, Allard RW (1971) Effects of reciprocal recurrent selection for yield on isozyme polymorphisms in maize (*Zea mays* L.). Crop Sci 11:888–893

Brown AHD, Burdon JJ (1983) Multilocus diversity in an outbreeding weed, *Echium plantagineum* L. Aust J Biol Sci 36:503–509

Brown AHD, Clegg MT (1983) Analysis of variation in related DNA sequences. In: Weir BS (ed) Statistical analysis of DNA sequence data. Dekker, pp 107–132

Brown AHD, Feldman MW (1981) Population structure of multilocus associations. Proc Natl Acad Sci USA 78:5913–5916

Brown AHD, Feldman MW, Nevo F (1980) Multilocus structure of natural populations of *Hordeum spontaneum*. Genetics 96:523–536

Clegg MT, Allard RW, Kahler AL (1972) Is the gene the unit of selection? Evidence from two experimental plant populations. Proc Natl Acad Sci USA 69:2474–2478

Dennis ES, Peacock WJ (1983) Knob heterochromatin homology in maize and its relatives. J Mol Evol (in press)

Ginzburg LR, Braumann CA (1980) Multilocus population genetics: relative importance of selection and recombination. Theor Popul Biol 17:298–320

Golding GB, Strobeck C (1980) Linkage disequilibrium in a finite population that is partially selfing. Genetics 94:777–789

Hedrick P, Jain S, Holden L (1978) Multilocus systems in evolution. Evol Biol 11:101–184

Karlin S, Piazza A (1981) Statistical methods for assessing linkage disequilibrium at the HLA, –A, B, C loci. Ann Hum Genet 45:79–94

Langley CH (1977) Nonrandom associations between allozymes in natural populations of *Drosophila melanogaster*. In: Christiansen FB, Fenchel T (eds) Measuring selection in natural populations. Springer, Berlin Heidelberg New York, pp 265–273

Lewontin RC (1974) The genetic basis of evolutionary change. Columbia Univ Press, New York

Linhart YB, Mitton JB, Sturgeon KB, Davis ML (1981) Genetic variation in space and time in a population of ponderosa pine. Heredity 46:407–426

Muona O (1982) A multilocus study on an experimental barley population. Hereditas 96:247–254

Sved JA (1968) The stability of linked systems of loci with a small population size. Genetics 59:543–563

Weir BS (1979) Inferences about linkage disequilibrium. Biometrics 35:235–254

Wright S (1965) Factor interaction and linkage in evolution. Proc R Soc London Ser B 162:80–104

Genetic Structure and Demography in Plants

Age-Specific Fitness Components in Plants: Genotype and Phenotype *

B.A. SCHAAL[1] and W.J. LEVERICH[2]

Introduction

One area of population biology that has been of recurrent interest is the relationship between the phenotype of an organism and its genotype. That is, how much of the phenotypic variation that is observed in natural populations has an underlying genetic basis? Most natural populations of both plants and animals show enormous variation in phenotypes, specifically in life history components that affect fitness. Some individuals survive to reproductive age; others die young. Some individuals produce many offspring, while others produce none. The critical question for the evolutionary biologist is: what is the underlying genetic basis for this variation in phenotype? Natural selection can exert its evolutionary influence only on genetically based variability. Does most phenotypic variation observed in natural populations reflect underlying genetic variation? Or, does the variability reflect phenotypic plasticity? In the following we will be concerned with phenotypic and genotypic variation, specifically in higher plants.

Seed plants have a number of important characteristics which distinguish them from animals and which strongly affect their population biology. First, plants are sedentary for most of their life cycle. Movement occurs only at certain times in the life cycle, the times of seed and pollen dispersal. Hence, the site where a seed lands after dispersal is the precise location where that individual must germinate, mature, flower, and set seed. An individual plant must survive in its specific microhabitat in order to leave descendents in future generations. Plants do not have the option of simply moving away from a local, unfavorable site. This inability to move has a strong influence on the population biology of plants, since plant environments are highly heterogeneous, even within small sections of single populations.

A second major aspect of plants which differentiates them from animals is their range of phenotypic plasticity. Many plants adjust to the environmental heterogeneity of their habitats by phenotypic adaptation. Many plants are highly plastic; single genotypes can yield a wide range of phenotypes. Phenotypic variation can occur for many traits, such as physiology, secondary compounds, and overall intensity of reproduction (e.g., Clausen et al. 1940, Levin 1976, Mooney et al. 1977).

---

* This work was supported by National Science Foundation Grant DEB 8141023

1 Department of Biology, Washington University, St. Louis, MO, 63130, USA
2 Department of Biology, St. Louis University, St. Louis, MO, 63103, USA

Population Biology and Evolution
Ed. by K. Wöhrmann and V. Loeschcke

One might conjecture, a priori, that in plant populations which exhibit a great deal of phenotypic variability, much of this variability may *not* have an underlying genetic basis. Here, our purpose is to examine levels of phenotypic and genotypic variation in components of fitness in plants, with emphasis on the annual plant, *Lupinus texensis*.

Life History Variation in Natural Plant Populations

Plants show a great deal of variation in life history features. Within single populations much variability exists; for example in *Lupinus texensis*, the distribution of reproduction among individuals in field populations varies greatly. Even in this annual species, some plants fail to reproduce at all. Other individuals show very high levels of reproduction; a single plant can contribute over 6% of a population's total seed production (Schaal 1980). Such variation is typical of plant species. The number of seeds produced per individual varies among species and in most plant species, populations show a wide range in the amount of reproduction achieved. Survivorship within plant populations is, of course, also variable. Even in plant species which tend toward a Type I survivorship curve, where most mortality occurs late in the life cycle, there is variation in the individual plant time of death (e.g., Schaal and Leverich 1982).

For life history characteristics, there is often greater variation between populations than within populations. In a study of the annual plant, *Impatiens pallida*, Cid-Benevento (1979) found substantial variation in life history measures among populations. Populations differed in their survivorship curves: some populations showed a typical Type I curve, while other populations approached a Type II curve, indicating density independent mortality. The populations also differed in their reproduction. The net reproductive rate, R(o), ranged from 1.5–16.3 among five populations. In addition, reproductive value, the age specific contribution to the ancestory of future generations, varied greatly among populations; this affords the potential for between population differences in the action of natural selection (Cid-Benevento 1979). Since much of the life history variation in *Impatiens pallida* is directly related to environmental favorability, a strong phenotypic response is suggested for these summer annuals (Cid-Benevento 1979).

From these studies and numerous others, it is apparent that much variation exists in natural plant populations for life history features. How much of this variation has a genetic basis? Several studies have addressed this question. In a study of the grass, *Poa annua*, Law et al. (1977) found significant differences among sibling groups of plants grown in a common garden. Differences in both age specific survivorship and fecundity were detected. The data suggest genetic differences in life history features in natural populations of *Poa*. Likewise, Fowler and Antonvics (1981) showed differences in survivorship and reproduction among microsites in *Plantago lanceolata*, using transplant experiments. These results also suggest that some phenotypic differences in life history features have an underlying genetic basis.

Experimental Studies of Life History Variation in *Lupinus texensis*

These sorts of life history differences that persist in the greenhouse, common garden, or in transplant experiments have been examined in a series of studies of demographic variability within and among populations of *Lupinus texensis. Lupinus texensis* is a winter annual, endemic to portions of Texas, USA. It is widespread throughout its range and forms large populations, comprising many thousands of plants. The species is predominantly outcrossed and is bee pollinated. Populations are highly variable for allozymes and for morphological characters, such as leaf shape, pubescence, seed color, and floral characteristics.

Demographic variation was studed among populations of *L. texensis* by a common garden experiment (Schaal unpublished). Seeds were collected from thirty plants chosen at random from each of ten natural populations. Seed progeny from each field population were germinated in the greenhouse and followed throughout their lifespan. Survivorship and reproduction were scored for each plant at weekly intervals. Life tables were constructed for each population and the life tables compared. Since the uniform greenhouse conditions eliminate site specific environmental variation, significant differences in life history features among the populations are suggestive of genetic differences among the original plants from which the population samples were derived.

The results of this study are shown in Figs. 1 and 2. Figure 1 plots age specific survivorship for three representative populations. The two populations with the most extreme survivorship schedules and one population with an intermediate schedule are shown. Survivorship differs significantly among the populations, both at the beginning of the life cycle, after germination, and during the time of maximum mortality, after reproduction peaks. Likewise, reproduction varies among the populations. Figure 2 shows the age specific distribution of 1(x)m(x), which is the average number of offspring produced per individual during a given age class, m(x), weighted by the survivorship of that age class, 1(x). In this study, the value plotted as m(x) is the number of flowers, since to count seeds would have introduced environmental variation due to

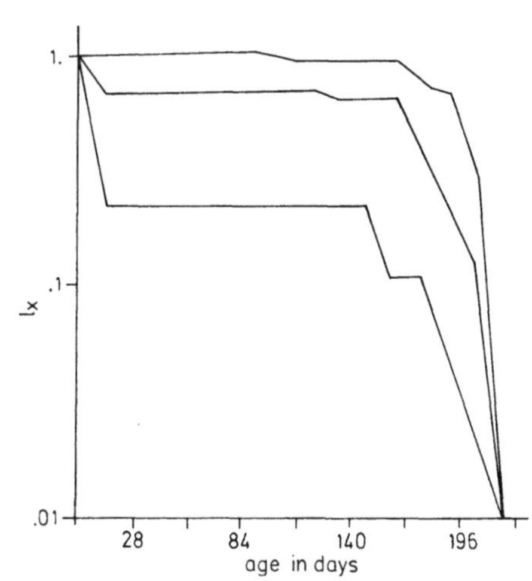

Fig. 1. Age specific survivorship for three representative populations of *Lupinus texensis*

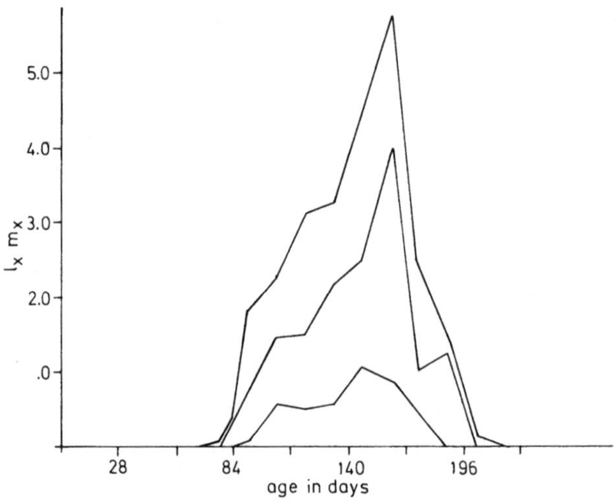

Fig. 2. Age specific reproductive curves for 3 representative populations of *Lupinus texensis*. Age specific reproduction, *m(x)* is measured in number of flowers

pollinator service. The populations again differ significantly. In this case, the differences are manifest as significant variation in the intensity and duration of reproduction. The relative reproductive performances of the entire ten populations can be compared by calculating $\Sigma 1(x)m(x)$. This value varies among the populations, from 166.3 to 861.7 flowers per individual, a fivefold difference (Table 1). These values are measures of reproductive potential analogous to R(o), the net reproductive rate (which would be based on actual production of seeds per individual). These flower values can be converted to estimates of R(o) based on typical seed set in field populations; R(o) varies among the populations from 2.7 to 14.1.

These data show differences between populations in patterns of survivorship and reproduction. They are strongly suggestive of underlying genetic differences among populations of *Lupinus texensis*. However, in this study, as in the studies cited above,

Table 1. Estimates of the net reproductive rate for populations of *Lupinus texensis*[a]

Population	R(o)
1	440.2
2	861.7
3	166.3
4	193.3
5	603.6
6	289.0
7	798.8
8	716.6
9	659.8
10	727.6
Mean	545.6

[a] Estimates based on numbers of flowers

the presence of maternal effects, the expression by offspring of phenotypic differences between mothers, may be a confounding factor.

Maternal Effects. The maternal parent, whether it be an animal or plant, often has a profound influence on the survivorship of its offspring. These maternal effects are separate from the genetic endowment a parent bestows on its offspring. In both seed plants and many animals, the maternal parent has a nurturing or nutritive role. In animals, the maternal parent may provision an egg with nutrients or directly nurture the developing embryo in the case of mammalian offspring. This nurturing role also occurs in plants where a developing seed is nurtured analogously to a developing mammalian offspring. These maternal effects can affect the fitness of offspring far beyond the time of parturition. The ability to nurture developing embryos can vary among maternal plants and this variation can be due to either genetic or environmental factors. Plants that live in favorable environments may have more resources to place in the developing seeds and, thus, the offspring may have increased fitness over seeds which have few nutrients. Likewise, some seed parents may be genetically better equipped to raise developing seeds.

In plants, maternal effects are often manifested in seed size. Plants which can do a good job of nurturing their developing embryos will produce large seeds, while plants of depauperate environments or with less fit genotypes will produce small seeds. Seed size, in turn, has a marked effect on the fitness of the offspring. Table 2 shows the relationship between seed size and fitness components for *Lupinus texensis* (from Schaal 1980). The larger the seed, the greater the probability of germination and survival. In addition, large seeds produce large seedlings which grow more rapidly. Such fitness effects of seed size are commonly observed in plants and result from the direct conversion of seed biomass to seedling biomass (Harper 1977).

Clearly, maternal effects via seed size are an important determinant of seedling fitness. Are these maternal effects predominantly genetic or environmental in origin? The heritability of seed weight can be used to assess what portion of the variance in seed weight is due to the additive genetic component. Heritability of seed weight in *Lupinus texensis* varies among populations (Schaal, unpublished), but is in general, quite small. In the population used for the above study of maternal effects, the

Table 2. Seedling survivorship and biomass as a function of seed weight

Seed size class	Number of seeds	Germination (%)	Surviving (%)	Mean seedling biomass
< 17 mg	31	41.9	84.6	8.09 ± 1.1
18–22	81	90.1	94.5	12.94 ± 0.33
23–27	120	94.2	99.1	16.52 ± 0.28
28–32	120	95.8	98.3	19.40 ± 0.32
33–37	107	95.3	100.0	21.22 ± 0.36
38–42	98	96.9	95.8	26.22 ± 0.42
43–47	50	98.0	100.0	31.37 ± 0.54
> 48 mg	52	100.0	100.0	35.48 ± 0.54
Mean		92.6	98.2	21.47 ± 0.31

heritability of seed weight is 0.09 (Schaal 1980). This low heritability implies that most of the variation in seed weight is due to environmental effects. Thus, variation in seed size and the accompanying maternal effects in *L. texensis* are predominantly the result of environmentally based phenotypic plasticity.

Given that maternal effects on the fitness of offspring occur in *L. texensis*, let us examine the intensity of the maternal influence and the length of time for which the offspring exhibit these effects. Maternal effects were again studied in a greenhouse population (Schaal unpublished). A large sample of plants was grown from seed which were collected at the same ten field sites as in the previous population study. Each greenhouse plant was repeatedly pollinated. Seeds resulting from these pollinations were collected. Populations differed from each other in mean seed weight. The seeds were weighted, germinated, and the resulting plants followed throughout their life spans in the greenhouse. Age specific survivorship and reproduction were recorded at weekly intervals and life tables were constructed. The relationship between seed size and various components of fitness was determined.

Maternal effects via seed size had a large and continuing effect on fitness. Survivorship and plant size are positively correlated with seed size in many of the populations. The duration of this positive correlation varied among populations. Some populations showed no detectable maternal effects; other populations show a significant correlation between seed size and fitness up to the seventh week after seed germination (Table 3). The time of germination was in general negatively correlated to seed size

Table 3. Correlation of plant size[a] and seed weight in ten *Lupinus texensis* populations

Weeks after germination	Number of populations with significant correlations
1	5
2	4
3	3
4	4
5	3
6	1
7	0

[a] Measured as number of leaves

(mean $r = -0.22$). Smaller seeds tend to germinate earlier. Significant differences in survivorship also occur only in the first few weeks after seed germination, although, again, the duration of the correlation varies among populations. Interestingly, reproduction also has an association with seed size. The total amount of reproduction, R(o), is not significantly correlated to mean seed weight for the ten populations [$R(o) = 0.56, P > 0.05$]. However, if reproduction is apportioned into its age specific components, a portion of the life cycle may show a significant correlation with seed weight. Table 4 illustrates one such population where weeks 14 through 20 show significant correlations between progeny seed weight and age specific reproduction.

Table 4. Age specific reproduction: correlation with seed size

Age	Correlation
8 weeks	r = 0.30
10 weeks	r = 0.16
12 weeks	r = 0.34
14 weeks	r = 0.56[a]
16 weeks	r = 0.56[a]
18 weeks	r = 0.62[a]
20 weeks	r = 0.49[a]
22 weeks	r = 0.34
24 weeks	r = 0.23
26 weeks	r = 0.10
28 weeks	r = 0.21

[a] $P < 0.05$

This relationship between seed size and reproduction is most likely due to a secondary effect. Large seed produce large seedlings. Plants which come from large seedlings, in turn, tend to be the most reproductive at this time.

These studies indicate a large, long lasting maternal effect in *L. texensis*. Maternal effects strongly influence the fitness of plants in the early stages of life, from the time of fertilization through seedling establishment. In addition, maternal effects have a secondary influence which is seen in reproductive performance.

Genotypic Effects. The above studies indicate that the environment, via maternal effects, has a major influence on the fitness of a seedling. What fitness effects does the genotype of the seedling have? In another study seed progeny were produced by outcrossing or by selfing. These two groups of progeny are genetically quite different from each other. In a highly variable, outbred species like *L. texensis*, we expect that the inbred progeny should be more homozygous, while the outbred progeny should be much more heterozygous. Thus, the two progeny groups differ in genotype and a comparison of the two progeny groups can be used to determine when genotypic effects may be important in the life cycle. These differences in phenotype should be detectable as differences in fitness, since in a species such as *L. texensis* which is predominantly outbreeding, inbreeding depression might be expected.

First consider the effects of the type of mating (outcross vs self) on seed production. Somewhat surprisingly, there is little evidence of inbreeding depression early in the life cycle. The percentage seed set, the number of seeds per fruit, the percent seed abortion, and the mean seed weight do not differ significantly between the two types of matings (Table 5). In addition, neither the time nor the percentage of seed germination appears to be affected by the type of mating which produced the seed. It can be concluded that early in the life cycle there is little evidence for inbreeding depression and the variance in fitness at this early part of the life cycle does not appear to have a strong genotypic component.

After seedling establishment, 4 weeks beyond germination, differences in fitness between inbred and outbred progeny begin to appear. Outcross progeny become

Table 5. Seed and seedling production selfed vs outcrossed progeny

	Selfed (n)		Outcrossed (n)	
% Seed set	12.15	(2,708)	11.46[a]	(1,571)
Seeds/fruit	4.57	(214)	4.56[a]	(104)
Viable seeds/fruit	2.86		3.08[a]	
Aborted seeds/fruit	1.71		1.48[a]	
Seed weight, g	38.65	(206)	39.58[a]	(223)
% Germination	91.55	(206)	91.03[a]	(223)

[a] Not significant

Table 6. Plant size: selfed vs outcrossed progeny

Age	Size measure	Selfed (n)	Outcrossed (n)	t
2 weeks	# of leaves	2.62 (167)	2.67 (192)	0.317
4 weeks	# of leaves	8.94 (154)	10.76 (176)	4.12[a]
6 weeks	# of leaves	20.08 (144)	23.44 (172)	2.14[a]
8 weeks	# of stems	5.22 (140)	5.56 (166)	1.99[a]

[a] $P < 0.05$

larger than inbred progeny after these initial 4 weeks (Table 6). Inbred and outbred plants have different mean life spans, 140 days vs 165 days, respectively. The overall pattern of mortality also differs among the two plant groups (Fig. 3). So, although fitness effects of mating type are not clear in early life, genotype strongly influences viability components of fitness beyond the fourth week, both in terms of plant size and survivorship.

Finally, we can consider the effects of inbred and outbred genotype on reproduction. Figure 4 plots the age specific reproduction, measured as the average number of

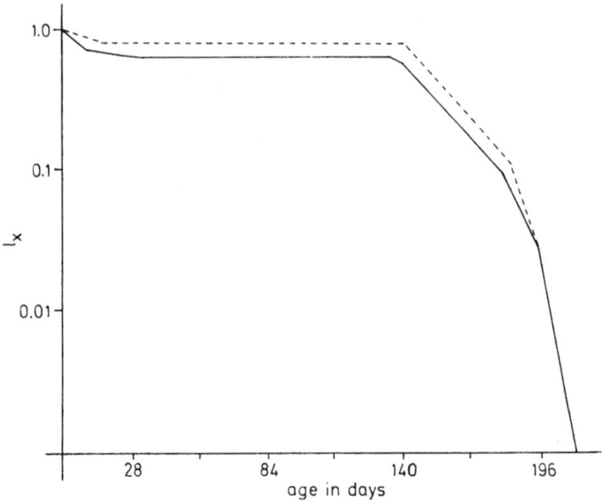

Fig. 3. Age specific survivorship curves for progeny of selfcrosses and outcrosses. *Solid line* is inbred plants; *broken line* is outcrossed plants

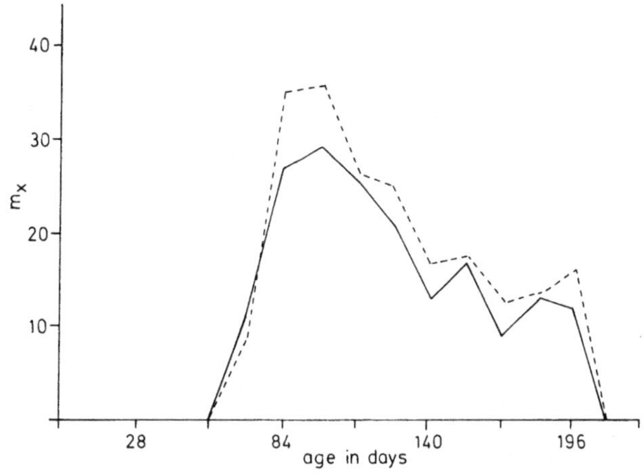

Fig. 4. Age specific reproduction for progeny of selfcrosses and outcrosses. Age specific reproduction, *m(x)* is measured in number of flowers. *Solid line* is inbred plants; *broken line* is outcrossed plants

flowers produced per individual in an age class. The outbred progeny produce many more flowers during nearly all age classes. The average number of flowers produced per individual throughout the life span reveals the relative reproductive potential of the different groups. For inbred progeny, 88.4 flowers per individual are produced on average, whereas for the progeny of outcrosses, 108.7 flowers are produced on average.

The comparison of inbred vs outbred progeny clearly demonstrates inbreeding depression and suggests that genotypic differences can contribute to the life history variation that is observed in natural populations. Inbreeding depression is clearly seen in both the fecundity and viability of *L. texensis*. Interestingly, inbreeding depression is not evident early in the life cycle.

Conclusions

Perhaps the most significant aspects of the Lupine work relate to the time that various factors are operative. Environmental influences, via maternal effects, are strong determinants of fitness early in the life cycle. Later in the plant life cycle as the influence of the seed parents phenotype diminishes, the genotype of the progeny becomes influential in establishing fitness differentials. Finally, these results underscore how complex the determination of phenotype and realized fitness of progeny must be in natural plant populations. Phenotypic and fitness differences in progeny result from the genotypic effects of both maternal plant and the progeny itself. Natural selection, thus, must act on two genotypes; the relative importance of the two genotypes changes with age. Likewise, fitness differences which result from environmentally based phenotypic plasticity must also be complex and the interaction must also change over time. At first, early in the life cycle, the environment of the seed parent has a strong influence via maternal effects. Subsequently, the environment of the progeny will induce phenotypically plastic responses to the environment. Finally, all of the genotypic and the environmental influences must interact among one another.

The timing, duration, and intensity of the interactions must vary with life cycle stage. The pattern of interaction must also vary with the environment and with the plant species. Clearly, natural selection in native plant populations does not merely operate on a single phenotype with a clearly based genotype. The interplay of environmentally and genotypically based variation is extraordinarily complex and it is a dynamic interaction of several environments and genotypes.

References

Cid-Benevente C (1979) The demography of *Impatiens pallida* Nuttal (Balsaminaceae). Ms Thes, Ohio State Univ

Clausen J, Keck D, Hiesey W (1940) Experimental studies on the nature of species. I. The effect of varied environments on western North American plants. Carnegie Inst Washington Publ 520:1–542

Fowler N, Antonovics J (1981) Small-scale variability in the demography of transplants of two herbaceous species. Ecology 62:1450–1457

Harper J (1977) Population biology of plants. Academic Press, London New York, pp 892

Law R, Bradshaw AD, Putwain PD (1977) Life-history variation in *Poa annua*. Evolution 31:233 to 247

Levin DA (1976) The chemical defenses of plants to pathogens and herbivores. Annu Rev Ecol Syst 7:121–159

Mooney HA, Bjorkman O, Collatz J (1977) Photosynthetic acclimation to temperature and water stress in the desert shrub, *Larrea divaricata*. Carnegie Inst Washington Yearb 76:328–335

Schaal B (1980) Reproductive capacity and seed size in *Lupinus texensis*. Am J Bot 67:703–709

Schaal B, Leverich WJ (1982) Survivorship patterns in an annual plant community. Oecologia 54:149–151

Influence of Population Parameters on the Genetic structure of *Capsella* Populations

H. HURKA[1]

Introduction

The genus *Capsella* is a member of the Cruciferae family and comprises several species. However, due to the enormous polymorphism within this genus, no clear cut species lines exist and the number of recognized species varies between three and ten. The Swedish botanist Almquist conducted the first comprehensive experimental survey on *Capsella* at the beginning of this century. His taxonomy was rather extreme: all individuals which produced similar progenies in cultivation, the phenotype of which remained constant over several generations of propagation in his experimental garden, were attributed to a Linnean species. He recognized 200 *Capsella* species. His classification reflects the great amount of polymorphism within *Capsella*. Furthermore, it points to the existence of a strong genetic component to the variation, an aspect which was also demonstrated by Almquist (1907, 1921, 1929).

For our purposes, we adopt a rather broad species concept and in accordance with Flora Europaea (Tutin et al. 1964) recognize three main species in Europe: *C. bursa-pastoris* (L.) Medicus, *C. grandiflora* (Fuché & Caub.) Boiss., and *C. rubella* Reuter. These species display different ploidy levels, different breeding systems, and different distribution patterns (Table 1). *Capsella* plants typically grow in weedy places. Annual forms (summer or winter annuals) predominate in all three species.

The genus *Capsella* almost certainly originated in the Old World (East Mediterranean area) and was presumably brought to Middle and Northern Europe with the

Table 1. Ploidy levels, breeding systems, and geographical distribution in the genus *Capsella*

Species	Ploidy level	Breeding systems	Geographical distribution
C. rubella	2 n	Reported to be predominantly outcrossing	Mediterranean climates
C. grandiflora	2 n	Strictly outcrossing	Only W. Greece and rarely in Northern Italy
C. bursa-pastoris	4 n	Predominantly selfing	World-wide except the hot tropics

1 Institute of Biology, Universität Osnabrück, Barbarastraße 11, 4500 Osnabrück, FRG

Population Biology and Evolution
Ed. by K. Wöhrmann and V. Loeschcke

spread of agriculture from the Middle East during the bronze and iron ages some 4 to 5 thousand years ago. From Europe it extended its range into the New World and Australasia by following European colonists. Today, the species *C. bursa-pastoris* is one of the most frequent and most widespread flowering plants on earth and is characterized by its colonizing ability and its high ecological amplitude. All this suggests that the genus *Capsella* may be very well suited to serve as a model for studying the evolutionary biology of plants.

In this paper, we shall present results of several lines of experiments which aim at understanding the underlying factors that influence variation in one of the most common plants on earth.

Experimental Procedures

Seed samples from 25 to 30 plants were randomly collected from more than 400 natural populations along a transect from northern to southern Europe. Population sizes and habitat data were recorded in the field. For the purpose of estimating variability within populations, different characters were studied and, dependent upon the specific characters, different experiments were designed as outlined below.

1. Morphological and reproductive characters such as plant height, leaf morphology, flowering time, number of fruits, and seeds were analyzed from progeny of the field collected mother plants. Progeny were grown in random block experiments in the field stations. The data were analyzed with the help of a nonparametric analysis of variance (H-test of Kruskal and Wallis).
2. Seed sizes and seed weight have been measured. Cluster analyses were carried out and the fraction of phenotypic variation that is due to genetic differences was investigated by estimating the heritability of seed size.
3. Allozyme variability was also analyzed. Progenies of the field collected seeds were grown under controlled environmental conditions in a greenhouse. Rosette leaves of about 10- to 12-weeks-old plants were analyzed for a variety of enzyme systems by means of polyacrylamid gel electrophoresis. Prior experiments had shown that leaves of that age always were in the same developmental stage, at least as far as the isozyme patterns under consideration are concerned. Using crossing experiments and progeny tests, the genetics of many isozyme bands could be established.

Patterns of Variability Within and Between Populations

Up to now all the traits studied on a sufficient number of plants show both continuous and discontinuous patterns of phenotypic or genetic variation. Different patterns of variation may superimpose, creating a complex overall pattern.

Macrogeographical Variation (Continuous Variation)

Macrogeographical variation patterns often appear to be continuous and can sometimes be in connection with environmental factors. Some examples follow. Within *C. bursa-pastoris*, four basic leaf types are observed which are named heteris, rhomboidea, tenuis, and simplex (Shull 1909, 1911). When samples from different localities are grown under uniform conditions (block experiments in a field station), a highly significant correlation between the frequency of the heteris leaf type within the populations and increasing annual temperatures at the collecting site can be observed (29 populations, Spearman's rank correlation coefficient: $0.3 < r_S \leqslant 0.5, 0.001 < 2\alpha \leqslant 0.01$). As the penetrance of the genes, which determine the leaf type, depends to a certain degree on environmental factors, it remains to be clarified whether these genes follow the same pattern. The higher the annual amount of rainfall, the later is the onset of flowering (29 populations, $0.5 < r_S \leqslant 0.7, 0.001 < 2\alpha \leqslant 0.01$). The flowering time within populations varies the more, the greater the number of days with average temperatures of above 10 °C (coefficient of variation $V_{\bar{x}} = s/\bar{x}$ increases; $0.7 < r_S \leqslant 1.0$, $0.001 < 2\alpha \leqslant 0.01$).

Besides these clinal patterns, one often observes large-scale geographic variation. With regard to seed weights, for instance, evidence is provided that *Capsella* plants, which were sampled from sites with severe climates, such as high mountain areas, produce larger seeds than those from more favorable climatic conditions. Seedlings emerging from the bigger seeds had remarkable longer hypocotyls compared to those of the smaller seeds. The possible adaptive significance of seed size could be superior seedling performance under adverse climatic conditions (Hurka and Benneweg 1979). It is reasonable to assume that in this case environmental factors should account for the observed geographic variation pattern.

The macrogeographic distribution patterns of other characters, however, are not so easy to explain. One example follows. In *Capsella*, we find at least four loci coding for the enzyme GOT. The so-called GOT 1 locus displays four alleles (a_1 ... a_4), GOT 2 three alleles (b_1 ... b_3), and GOT 3 two (c_1, c_2). GOT 4 could not yet be analyzed thoroughly enough. GOT 2 and GOT 3 are probably the result of a gene duplication (judged by the formation of an interlocus heterodimer). This gene duplication occurs throughout Europe in all three *Capsella* species. It is most frequent in the tetraploid *C. bursa-pastoris*. One particular genotype ($a_1a_1b_1b_1c_1c_1$) is typical for Scandinavia, but often lacking outside this area (Hurka 1983). Whether this distribution pattern is the direct outcome of selection on the enzyme genotype or the consequence of historic aspects and/or effective breeding or linkage systems remains to be determined. Nonetheless, all the examples cited above point to the existence of a large-scale, more or less continuous pattern of variation.

Microgeographical Variation (Discontinuous Variation)

The macrogeographical variation pattern contrasts with a microgeographical pattern. Characters, which are viewed on a large scale and appear to be continuously distributed, may often vary discontinuously when analyzed at a microgeographical scale. For instance, high seed weights which are typical in regions with severe climates, can also be found in more favorable climates. Despite large-scale geographic variation of allozyme allele frequencies, one often finds distinct and clear-cut changes of those allele frequencies within a few kilometers or even meters. The same holds true for flavonoid patterns, plant height, flowering time, germination, number of hairs per leaf area, and leaf morphology. The juxtaposition of continuous with discontinuous variation patterns becomes even more confounding when the variation within single populations is considered. Populations with varying levels of polymorphism are found. Sometimes, populations which are polymorph or monomorph for a particular character show a patchy distribution pattern. In other instances, a continuous pattern is observed.

Whereas the macrogeographical variation of at least some characters could be correlated with environmental factors, such correlations are hard to demonstrate for microgeographic variation. In many instances, it would appear that characters vary randomly. However, it is possible to relate some variation patterns with the state of the biotope (Bosbach et al. 1982). The sites which were sampled can be separated into "highly disturbed" and "less disturbed" biotopes. Highly disturbed refers to freshly filled up or broken soils, whereas less disturbed means places which have not been cultivated lately. In highly disturbed sites populations are often composed of more genotypes than in less disturbed places (Table 2).

Table 2. Number of different enzyme genotypes of *Capsella bursa-pastoris* in relation to habitat conditions. (From Bosbach et al. 1982)

Number of different genotypes		1	2	3	4	5	6	7	8	9	10
Number of	Less disturbed sites	12	5	1	–	–	–	–	–	–	–
populations	Highly disturbed sites	–	4	3	2	4	2	1	1	–	1

Analysis of Population Parameters

Relationship between environmental differences and population variability has been demonstrated predominantly by statistical inference. But the question as to why these relationships exist has still to be answered in most cases. It can be hoped that a more precise and detailed knowledge of the environment on one hand and of the biology of the investigated species on the other hand, may help in understanding the processes which regulate genetic variability. Therefore, we analyzed (1) production, dispersal, and ecology of seeds; (2) the soil seed bank; (3) the breeding system; and (4) population dynamics (demography of a *Capsella* population) (Hurka and Haase 1982, Bosbach et al. 1982).

Production, Dispersal, and Ecology of Seeds

Seed Output. The number of seeds per fruit varies between 10 to 40. These differences are partly due to different genotypes, but also to phenotypic plasticity. The estimations of seed production of several hundred plants grown in random block experiments gave minima of 5,000 seeds per plant and maxima of 90,000. Most individuals grown in these experiments produced between 30,000 and 60,000 seeds.

Seed Dispersal and Ecology. Seed dispersal of single plants grown in pots in a greenhouse was studied and the distances of seed dispersal recorded. Usually, the ripe fruits remain on the plant and split along their septa. The seeds may still remain attached to the replum for a while. The mature fruits of the greenhouse plants opened only slightly if at all. Seed scattering was thus prolonged over weeks or even delayed completely. Two events can force seed shedding: vibration and a change of atmospheric humidity. Low humidity results in a long delay, whereas high humidity causes earlier fruit opening and seed rain. High diurnal variation in air humidity has the most effective influence on seed scattering. Under natural conditions, vibration and contact caused by man, animals, wind, raindrops, etc. on the one hand, and fluctuation of humidity on the other hand, apparently favour fruit opening and seed scattering.

The seeds fall to ground very close to the mother plant. All test plants scattered their seeds into an area of 15-20 cm diameter. In the field, the distances travelled by seeds are often in the same order of magnitude which can be deduced from the following observations; distribution patterns of *Capsella bursa-pastoris* on an experimental field station of 30 m^2 were accurately registered over a period of 4 years. After this time, the initial distribution pattern was still recognizable, in spite of removing competing species and tilling the ground. Taking into account the scale of the experimental field (30 m^2) and the heavy seed output, it seems reasonable to conclude that under natural conditions dispersal distances of scattered seeds must be low.

The seed epidermis of *Capsella* produces a mucilage which, when moistened, becomes sticky and facilitates the attachment of the seeds to men, animals, transport vehicles, etc. (myxospermy). Repeated soaking and drying of the seed mucilage cover does not influence germinating capacity. Thus, despite the fact that most seeds of *Capsella* display only short-distance transport, this plant is also provided with very effective long-distance transport mechanisms.

The Soil Seed Bank

Estimation of a Buried Seed Population. Soil samples from different depths were taken from an experimental field station. The initial planting of 600 *Capsella* plants was done in 1975. The population was left undisturbed for 4 years. 100 samples of 3 cm^3 for each depth were regularly taken over the field. Samples were combined to produce one representative sample of 300 cm^3 per soil layer. *Capsella* seeds were separated by washing, flotation, and sieving procedures and counted under a stereomicroscope.

From the soil top layer down to a depth of about 10 cm, the seeds are nearly equally distributed (Table 3). Further down, a continuous decline in frequencies is observed. No seeds could be detected deeper than 40 cm, which obviously can be attributed to the soil of this particular field station: from 40 to 50 cm downward the soil profile is characterized by a sandy and stony stratum saturated with clay. Based on these findings, we estimated 3,000 *Capsella* seeds per m^2 on top of the soil and 32,000 buried

Table 3. Distribution of *Capsella* seeds in different soil depths. (From Hurka and Haase 1982)

Sampling depth (cm)	0+[a]	0–1	1–2	2–3	3–4	4–5	6–7	9–10	13–14	20–21	29–30	39–40	49–50
Number of *Capsella* seeds per 300 cm^3	89	107	140	117	96	116	138	92	66	29	2	0	0

[a] + Denotes surface layer of 100 cm^2

seeds per m^2. Estimations given in the literature are in good accordance with our own findings.

Movement of Seeds in the Soil. Seeds deposited on the soil surface can undoubtedly be passed down the soil profile with rain water in loose textured soils. In arable soils ploughing will be another important factor. In the case of *Capsella*, passage through earthworms' intestinal tract plays an important role, too. In several experiments we have shown that (1) of the *Capsella* seeds ingested by earthworms, some are digested completely, some are digested partly (removal of the testa), and some pass through seemingly undamaged. (2) The rate of germination of those seeds which did pass through the earthworms' intestinal tracts is reduced compared to that of normal seeds, but still is rather high (88%). (3) The activity of earthworms can transport seeds vertically through the soil profile.

The influence of the earthworms on the buried seed population depends on the size of the earthworm population and on their activity. Earthworms burrow through the soil and bring subsoil to the surface. Darwin (1881) estimated that the annual production of worm casts in English pastures was 18.7–40.3 tonnes per ha, which is equivalent to a soil layer 5 mm deep being deposited yearly. Similar estimates for other parts of Europe are reported in the literature. The amount of soil actually turned over by worms may even be greater because some species void their casts underground. Burrows of some species go down to a depth of 150–240 cm (cf. Edwards and Lofty 1977). There seems to be no doubt that under natural conditions earthworms can highly influence the buried seed population of *Capsella* and that they constitute an important part in the formation and dynamics of the soil seed bank.

Breeding Systems

C. bursa-pastoris is predominantly self-pollinated. The selfing rate, however, depends on environmental factors. High atmospheric humidity at temperatures over 15 °C and low light intensities, i.e., cloudy and rainy weather, lead to almost exclusive self-pollination, whereas dry and sunny weather favors outcrossing. At low temperatures (about 4 °C–10 °C) anthesis is prolonged up to fivefold, but allogamy is reduced (Hurka et al. 1976).

Despite the potential influence of environmental factors on the genetic variability of populations, one would nevertheless expect in general that the within population variability in the self-incompatible *C. grandiflora* is higher than in *C. bursa-pastoris*. However, in an analysis of an array of natural populations of *C. bursa-pastoris* and *C. rubella*, we found all levels of variation, from homomorphic to highly heteromorphic populations. Some populations were as variable as those of the strictly outcrossing *C. grandiflora*. However, there are more or less clear differences between species in the organization of the genetic variability. Individuals homozygous for allozyme loci are typical for *C. bursa-pastoris* and heterozygous individuals are typical for *C. grandiflora*, although one can find highly heterozygous *C. bursa-pastoris* and homozygous *C. grandiflora* plants. It should be noted that such conclusions are based on enzyme patterns. The morphological characters are not always in accordance with the enzyme data. An unexpectedly high segregation rate was found in an analysis of morphological characters in progeny tests of *C. bursa-pastoris* (Hurka and Wöhrmann 1977). These data indicate that qualitatively different characters can lead to different, even contradictory conclusions about genetic variability within populations.

Population Dynamics

The demography of a *C. bursa-pastoris* population has been monitored from 1976 to 1980. In 1975, a garden experiment with 600 *Capsella* plants occupying an area of 30 m^2 was begun. After harvesting the plants in late summer 1975 (cutting the stalks, but leaving roots and ground leaves), the experimental site was left alone. In 1976, this area of 30 m^2 was subdivided into 120 plots, each plot measuring 0.5 × 0.5 m. The grid was left unchanged during the following years and allowed for a precise recording of vegetation pattern. Several times a year, numbers of *Capsella* plants were censured (Fig. 1). The removal of other species and tilling the soil caused remarkable alterations in population size. The decline of the number of *Capsella* plants in consecutive years was correlated with an increase of other species, clearly demonstrating the rapid successional displacement of *Capsella*. The experiment also demonstrates the ability of *Capsella* to recruit itself very quickly from autochthonous seed material in bare soils, thus, stressing the importance of the buried seed population.

Factors Influencing the Patterns of Variability

We have seen that in *Capsella*, characters sometimes follow a continuous, even clinal variation pattern and sometimes seem to vary randomly from population to population, thus, creating a discontinuous variation pattern. Some populations are monomorph for certain characters, others are polymorph. In a second set of experiments debated above, we analyzed parameters thought to be important for the biology of the species. Below I will relate both lines of experiments in an attempt to explain the observed variation patterns.

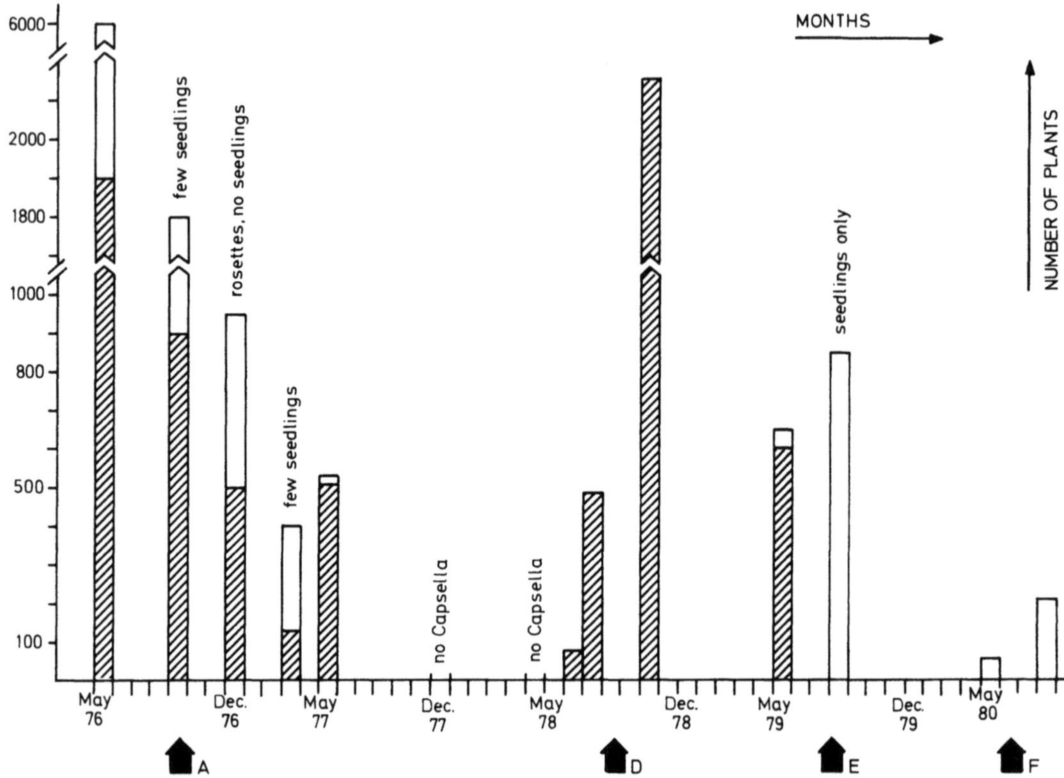

Fig. 1. Number of registered *Capsella* plants on an experimental field station of 30 m^2. *Hatched areas*: proportion of flowering plants. For the period April 1980 to July 1980 only total numbers of plants are given. *Black arrows: A:* removing competing species (22.09.1976); *D:* removing all plants and breaking the ground (08.08.1978); *E* and *F*: same procedures as in *D* (17.08.1979 and 08.05.1980)

The establishment of the clinal variation of *Capsella* can be related to seed dispersal events. Most seeds fall to ground near the parent plant and are incorporated into the soil seed bank. Therefore, populations in already occupied sites are mostly recruited from autochthonous seed material. Thus, selection can operate over many years on the same gene pool and a clinal variation pattern may eventually evolve. It seems important to stress that populations of colonizing species are not always true "colonizing" populations, but may rather be composed of gene complexes already adapted to a given locality. It is interesting to note that *Capsella* seeds buried in the soil are reported to be viable for at least 30 years.

Several factors may strongly influence the establishment of the discontinuous mosaic pattern: low gene flow between and within populations, properties of the soil seed bank, and founder effects. Long-distance transport in *Capsella*, which is facilitated by the mucilaginous seed coat, may result in natural populations arising through the accidental transport of a small number of seeds to hitherto unoccupied sites. Therefore, such populations are often genetically rather uniform (cf. populations monomorph for some characters). If different populations were established by small groups of founders drawn from a polymorphic parental population, then variation between these founder populations is to be expected. It can be assumed that the existence of genetic divergence between *Capsella* populations is partly due to such a mechanism.

The existence of a seed bank implies that the actual plant population at a given site represents only part of the total gene pool present there. Furthermore, any variation in the genotypic composition of *Capsella* populations over the years can principally be expected to be "recorded" in their respective soil seed banks. The seed bank can therefore serve as a permanent source for the introduction of "new" (or, rather "old") genotypes into plant populations, depending upon how well it is shaken up. This introduction of genetic variation will be counteracted, however, by the competitive elimination of seedlings and plants under conditions of high successional displacement stress. Therefore, one would expect less variable populations in places where succession has already started. There is still another process which enhances the differences between populations of highly disturbed and less disturbed soils. Although earthworms constitute an important factor in the formation and dynamics of the soil seed bank, they cannot compete with the rapid and drastic reorganization of the soil caused by man. It is to be expected, therefore, that in freshly filled or newly broken soils more seeds are brought up to the surface as compared with bare, but otherwise nondisturbed soils. This disturbance increases again the probability of different genotypes in highly disturbed soils.

In the case of *Capsella*, detailed investigations on dispersal mechanisms and on the role of the buried seed population opens the way for a deeper insight into the factors controlling the variation pattern of this common weed. Finally, it should not be forgotten that historical factors are also involved.

Acknowledgements. Financial support by the Deutsche Forschungsgemeinschaft (DFG) is greatly appreciated.

References

Almquist E (1907) Studien über die *Capsella bursa pastoris* (L.). Acta Horti Bergiani 4/6:1–91

Almquist E (1921) Studien über *Capsella bursa pastoris* (L.). II. Acta Horti Bergiani 7/2:41–95

Almquist E (1929) Zur Artbildung in der freien Natur. Acta Horti Bergiani 9/2:37–75

Bosbach K, Hurka H, Haase R (1982) The soil seed bank of *Capsella bursa-pastoris* (Cruciferae): its influence on population variability. Flora 172:47–56

Darwin Ch (1881) The formation of vegetable mould through the action of worms, with observations of their habitats. John Murray, London

Edwards CA, Lofty JR (1977) Biology of earthworms. Chapman & Hall, London

Hurka H (1983) Enzyme profiles in the genus *Capsella*. In: Jensen U, Fairbrothers DE (eds) Protein and nucleic acids in plant systematics. Springer, Berlin Heidelberg New York, pp 222–237

Hurka H, Benneweg M (1979) Patterns of seed size variation in populations of the common weed *Capsella bursa-pastoris* (Brassicaceae). Biol Zentralbl 98:699–709

Hurka H, Haase R (1982) Seed ecology of *Capsella bursa-pastoris* (Cruciferae): dispersal mechanisms and soil seed bank. Flora 172:35–46

Hurka H, Wöhrmann K (1977) Analyse der genetischen Variabilität natürlicher Populationen von *Capsella bursa-pastoris* (Brassicaceae). Bot Jahrb Syst 98:120–132

Hurka H, Krauss R, Reiner T, Wöhrmann K (1976) Das Blühverhalten von *Capsella bursa-pastoris* (Brassicaceae). Plant Syst Evol 125:87–95
Shull GH (1909) *Bursa bursa-pastoris* and *Bursa Heegeri*: biotypes and hybrids. Carnegie Inst Washington Publ 112
Shull GH (1911) Defective inheritance – ratios in Bursa hybrids. Verh Naturhist Vereins Brünn 49:156–168
Tutin TG, Heywood VH, Burges NA, Valentine DH, Walters SM, Webb DA (eds) (1964) Flora Europea, vol I. Univ Press, Cambridge

Population Differentiation and Asexual Reproduction

Demographic Implications of Genetic Variation in Zooplankton Populations

P.D.N. HEBERT[1]

Introduction

A study of population demography involves an investigation of birth, death, and developmental rates. Demographic studies on natural populations of a single species are difficult and an attempt to assess the effects of genotypic diversity on population demography is enormously more complicated. Thus, the lack of studies which comprehensively address this issue is unsurprising. However, over the past decade a rather extensive data base has accumulated on the freshwater zooplankton, which can be pieced together to provide an indication of the importance of genetic variation to the demography of these organisms.

In keeping with the data, this paper is itself fragmented into three sections. The first deals with taxonomic difficulties and has a simple theme – species complexes provide fertile ground for those who aim to document intraspecific variation in life history characteristics. The second section of the paper reviews information on genetic variation in zooplankton populations, concentrating on a discussion of factors influencing the extent of intrapopulational divergence. The data support the conclusion that ecological differences among genotypes are likely to be considerably more marked in populations reproducing by obligate or continued parthenogenesis than in those which reproduce sexually. The final section of the paper deals with the nature of genetic variation in three traits of demographic importance – birth rates, susceptibility to predation and tolerance to physical factors. This section reveals that variation in natality is largely environmentally induced and that genotypic differences are small. By contrast, tolerance to physical factors and predation is often under simpler genetic control, and individual variation may have significant consequences on both local abundance and regional distribution patterns.

Taxonomic Considerations

Striking examples of polymorphism and associated life history differences have been reported for a number of zooplankton species (Green 1967, O'Brien and Vinyard 1978). Zaret (1972a,b) noted a three-to-six fold difference in the rate of increase of

1 Department of Biology, University of Windsor, Windsor, Ontario, N9B 3P4, Canada

Population Biology and Evolution
Ed. by K. Wöhrmann and V. Loeschcke

two forms of *Ceriodaphnia cornuta* from Lake Gatun. The forms showed differences in fecundity, lifespan, viability and developmental rate. Similarly Kerfoot's (1975, 1977) intensive studies on *Bosmina longirostris* revealed the presence of two forms with marked differences in their rates of increase and susceptibility to predation. There was, however, no proof in any of these cases that the forms were conspecific, as the authors failed to show either that the forms could be interbred or that a female could produce progeny of both forms. The failure to carry out breeding studies is significant in light of the apparent prevalence of undescribed species in the major groups (copepods, cladocerans, rotifers) of zooplankton. Thus, Price's (1958) breeding studies showed that *Cyclops vernalis* could be divided into seven reproductively isolated groups. Similarly, Frey (1982) has suggested that many of the cladoceran species recognized by classical systematists are in fact species complexes. As a case in point, it can be noted that *Daphnia carinata* was regarded as a single variable species until allozymic and morphological study indicated that it included seven species in eastern Australia (Hebert 1977). Taxonomic work on the rotifers lags behind that on cladocerans and copepods and problems with unrecognized species are probably even worse in this group. Certainly attempts to cross "conspecific" individuals from different populations have frequently revealed reproductive isolation (King 1977).

The taxonomy of rotifers and cladocerans is further complicated by their parthenogenetic mode of reproduction. Although most species in these groups are cyclic parthenogens, a significant proportion (perhaps 10%) are obligate parthenogens. The experiences of plant systematists make it clear that such apomictic groups offer a fine opportunity to reveal the subjectivity of the taxonomic process.

The hybridization of closely related species is a regular if low frequency event in many animal groups. However, there is a complication when hybridization occurs between two cyclic parthenogens. This was already apparent some 30 years ago. In his monograph of the North American *Daphnia* fauna Brooks (1957) noted a number of populations which consisted largely or entirely of individuals intermediate in morphology between two recognized species. Brooks assumed that such cases were a consequence of hybridization, but he failed to explain the abundance of hybrid individuals. Laboratory studies failed to provide support for Brooks' hypothesis because the few cladoceran hybrids have been sterile or nearly so (Agar 1920, Shan and Frey 1968, Shan 1971). However, an instance of natural hybridization has recently been detected in the Cladocera involving two Australian taxa - *D. carinata* and *D. cephalata*. Hybrids between these two species were absent from some areas where their ranges overlapped, but common in other regions (Hebert, in preparation). In habitats where the hybrids were present, they often made up more than 90% of the population. Allozyme data indicated that only F1 hybrids were present (Table 1); there was no evidence of introgression with either parent species. Their ability to reproduce parthenogenetically was undoubtedly responsible for the unusual abundance of hybrid individuals. One might anticipate that interspecific hybrids would be commonest in lake habitats, where the absence of any requirement for diapause egg formation ensures that hybrids could perpetuate themselves indefinitely. This may explain at least in part, the particularly confused taxonomy of lake cladocerans and rotifers (Pejler 1956).

Table 1. Allozyme phenotypes of *Daphnia* from a pond near Canberra, Australia[a]

	D. carinata	Hybrids	*D. cephalata*
PGM	33	13	11
IDH	22	12	11
6PGDH	11	12	22
PEP	33	13	11
GOT	22	23	33
PGI	11	12	22

[a] All taxa had similar phenotypes at G6PDH, AO, LDH, TO, HEX, G3PDH, ME, MDH-1, MDH-2, TPI, MPI, FUM

The preceding discussion makes clear the likely sources and nature of taxonomic confusion in the major groups of freshwater plankton. It emphasizes the need for proper taxonomic validation of specimens used in ecological studies, including the deposition of voucher specimens so that identifications can be reassigned upon future taxonomic revision. In the case of rotifers and cladocerans, there is an additional need to determine whether the clones under study are members of a taxon reproducing by cyclic or obligate parthenogenesis or perhaps of hybrid origin. It is worth stressing that these distinctions cannot be made in the absence of genetic studies. Some authors have concluded that clones which produce males must be capable of sexual reproduction (Modlin 1982). Yet many obligately asexual clones retain the ability to produce males (Hebert 1982, Hebert and Crease 1983).

The value of such genetic studies have been shown in a number of cases. Although the *Ceriodaphnia* in Lake Gatun have not been reinvestigated, genetic studies carried out by Manning et al. (1978) have shown that the forms of *B. longirostris* in Lake Washington are reproductively isolated. Similarly, Schwartz (1981) and Lynch (1983) found clear life history differences between clones of *Daphnia pulex* from North America, but subsequent study has shown that their collections included both *D. obtusa* and *D. pulex* (Schwartz et al. in preparation). Snell (1977) also documented more than a twofold differences in the rate of increase of two clones of the rotifer *Asplancha girodi*, but laboratory mating experiments and field allozyme studies indicated the clones were unable to interbreed. The results of these genetic studies make it clear that reports of intraspecific variation in life history characteristics should be questioned unless the reproductive compatibility of the forms is documented.

Patterns of Genotypic Diversification in Zooplankton Populations

A comprehensive understanding of the genetic structure of a species requires knowledge of the diversity within local populations and of the extent of divergence among local and distant populations. This hierarchical distinction is somewhat artificial, but less so in the zooplankton than in many other taxa. Local zooplankton populations are discrete and areas with an abundance of pond habitats are often separated by some distance from other similar regions. In the present analysis the extent of genotypic variation within local populations is of primary concern because demographic interactions occur at this level.

Breeding systems have important effects on the genotypic characteristics of populations. Indeed aberrant breeding systems have often ben recognized as a consequence of these effects. Variation in the reproductive behaviour of the zooplankton is considerable. Nearly all copepods reproduce sexually, but the cladocerans and rotifers show a range in reproductive behaviour from quasisexual to completely asexual.

Studies have shown that the genotypic characteristics of populations reproducing by cyclic parthenogenesis are affected by environmental conditions (Hebert 1978). In intermittent habitats where populations are reestablished from sexual eggs on a regular basis, the number of coexisting genotypes is large. Genotypic frequencies are normally in Hardy-Weinberg equilibrium indicating that populations are refounded each year from a large number of individuals and that mating within populations is random. Fitness differences among clones appear small insofar as genotypic frequencies at allozyme loci remain stable, despite the opportunity which several generations of parthenogenetic reproduction affords for the amplification of fitness differences. By contrast, in permanent populations, where sex is infrequent, genotypic diversity decays and there are probably often fewer than a hundred clones present. In fact, in many habitats the population seems to be dominated by just a few clones. Genotypic frequencies in these populations often deviate markedly from Hardy-Weinberg equilibrium, ordinarily due to a heterozygote excess. Clonal frequencies are unstable – dramatic shifts in frequency often occur within a period of several months. Changes of this magnitude imply fitness differentials among clones of 10%–20%.

Populations reproducing by obligate parthenogenesis contain fewer clones than populations of cyclic parthenogens. Thus, populations of the two best studied obligate parthenogens, *D. pulex* and *D. middendorffiana*, contained an average of only three clones per habitat, with a range from 1 to 10 (Hebert 1982). Clonal diversity in *D. pulex* showed no clear latitudinal trend – clonal richness was similar in temperate and arctic habitats. There was also no difference in diversity between regions which had been glaciated during the Pleistocene and glacial refuges indicating that regional diversity reaches an asymptote fairly rapidly (Loaring 1982). Aside from their reduced clonal diversity, the population characteristics of obligate parthenogens differ in two important ways from those of cyclic parthenogens. Firstly, the clonal composition of populations reproducing by obligate parthenogenesis is conservative, while in populations reproducing by cyclic parthenogenesis, the synthesis of new clones by sex initiates clonal turnover. Secondly, co-occurring clones of obligate parthenogens have genetic similarities as low as 0.60 (Hebert and Crease 1983), while co-occurring clones of cyclic parthenogens have similarity values of 0.95 or greater. Thus, local genotypic divergence in obligate parthenogens rivals that which occurs over large geographical distances in species capable of sex. This can be attributed to the fact that clones of obligate parthenogens may invade a local region without losing their distinctiveness. Presumably as a consequence of these large genetic differences, major shifts in clone frequencies often occur in populations of obligate parthenogens in a period of a few

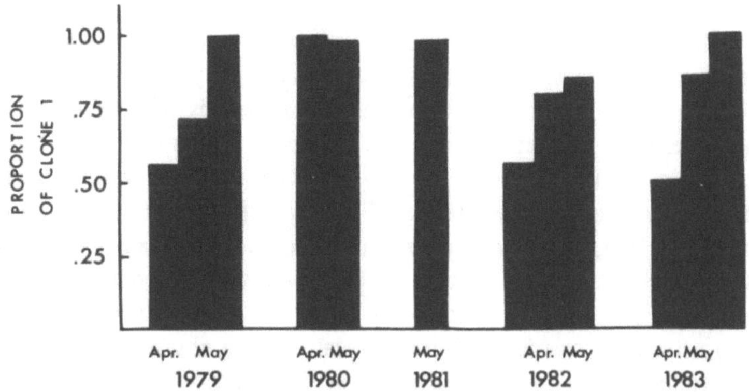

Fig. 1. Proportions of individuals of *Daphnia pulex* belonging to clone 1 in the Windsor 1 habitat

weeks (Hebert and Crease 1983). In some habitats these frequency shifts show a clear seasonal pattern (Fig. 1).

As several of the studies considered in this paper deal with genotypic variation between rather than within populations of a particular species, it is worth considering the extent of genetic divergence among populations. All three major zooplankton groups rely on passive dispersal and there is reason to assume that gene flow among populations is infrequent. Patterns of genotypic variation have been examined in local populations of two species reproducing by cyclic parthenogenesis – *D. magna* and *D. carinata*. Both studies revealed large gene frequency differences between populations only a few km apart. Inbreeding coefficients due to local population subdivision ranged from 0.25 to 0.75 (Hebert and Moran 1980). These values are much higher than those observed in actively dispersing taxa, such as *Drosophila*, and indicate that gene flow between zooplankton populations is low even on a microgeographical scale. Comparison of *D. magna* populations in Canada and the U.K. revealed even more pronounced genetic divergence (Loaring 1982), supporting the contention that local genetic differentiation is a consequence of founder effects and limited gene flow rather than natural selection.

The genetic data provide a strong basis for believing that divergence in ecological characteristics should be more pronounced in populations reproducing by obligate than cyclic parthenogenesis, so long as the former are polyclonal. In taxa reproducing by cyclic parthenogenesis a distinction needs to be drawn between those populations which reproduce by brief parthenogenesis and those which reproduce by continued parthenogenesis. Clonal diversification is likely to be considerably more pronounced in the latter populations.

Genotypic Differences in Natality

Birth rate is determined by the time to reach maturity, by the size of individual broods and by the interval between them. There is abundant evidence of intraspecific variation in each of these characteristics (Hutchinson 1967). However, it is clear that much of this variation is nongenetic in origin. Brood size in copepods and cladocerans is strongly correlated with food availability. Many rotifers lay only a single egg at a time, but the proportion of reproductive females is affected by food levels. Inasmuch as developmental rates and brood interval are largely controlled by temperature (Hutchinson 1967), much of the variation in these traits is also environmentally mediated. The likely existence of intraspecific genetic variation in components of natality is suggested by comparison of closely related taxa. Thus, in the genus *Daphnia* there is considerable variation among species in regard to brood size. Some species have a maximum brood size of 20 eggs, while others produce as many as 300 eggs. A portion of this variation can be explained on a physical basis – large species generally produce larger broods. However, similar sized species do produce divergent brood sizes. Females of *D. middendorffiana* with a core body size of 2.5 mm rarely produce more than 25 eggs, while females of *D. cephalata* with the same size produce up to 300. Variation in developmental rate and brood intervals among species seems more constrained than variation in brood size. For example, brood intervals show a similar decline with increasing temperature in a number of *Daphnia* species (Hebert 1978). The existence of these major environmental effects on egg production and developmental rates coupled with their concordant effects on different species suggest that genotypic contributions to phenotypic variation will be a minor component in comparison to environmental effects.

Only a few studies have aimed to determine the extent of intraspecific genetic variation in egg production and developmental rates. These studies report data on the rates of increase (r) of zooplankton clones under laboratory conditions. These studies exclude mortality due to extrinsic factors, and thus, documentation of significant heterogeneity in r is equivalent to the documentation of clonal differences in natality. Marked differences in r have been noted in three rotifer species. King (1972) isolated 13 clones of *Euchlanis dilatata* from a single habitat and found that their rates of increase varied from 1.64 to 2.68 at 27 °C and from 0.49 to 0.83 at 19 °C. Clones collected in the spring had the highest rates of increase in the low temperature environments, while summer clones did best at high temperature. Snell (1979) documented similar large differences in r between two clones of *Asplanchna brightwelli* from different geographic localities. No attempt was made in either of these cases to ascertain if the clones were capable of interbreeding or to establish if the clones were cyclic or obligate parthenogens. There was evidence in the case of *Euchlanis* that the populations were regularly reestablished from resting eggs as the habitat was drained each year. In a later study Snell (1980) studied clonal variation in a population of *Asplanchna girodi* known to reproduce by cyclic parthenogenesis. The three clones studied showed differences in r from 0.29 to 0.69 when cultured on *Paramecium*. Addition of the blue green algae *Anabaena* or *Lyngbya* depressed their rates of increase, but the three clones showed a differential response. The existence of clonal variation in r has also been documented in two cladoceran species. The first of these studies involved *Daphnia magna*, a species known to reproduce by cyclic parthenogenesis; and included clones isolated from both local and geographically distant sites. Woodrich (1980) found no significant differences in r between two clones from Churchill, Manitoba and between two other clones from a pond near

Cambridge, England. However, significant differences in r were apparent between clones from the two areas. Despite their marked genetic divergence, the clonal differences in r were small; in all test environments the differences rarely exceeded 10% of the mean. The similar influences of environmental factors such as temperature upon egg production and developmental rates of the clones is striking (Fig. 2). Analysis of variance confirms that the vast majority of variation in rates of increase is attributable to environmental influences, while the genetic component is minor (Table 2). Similar results were obtained in studies on obligately parthenogenetic clones of *D. pulex* from the same local habitat (Loaring and Hebert 1981). Not surprisingly, Schwartz (1981) has reported significant differences in r among clones of this same species from geographically distant localities. In both instances the divergence in r

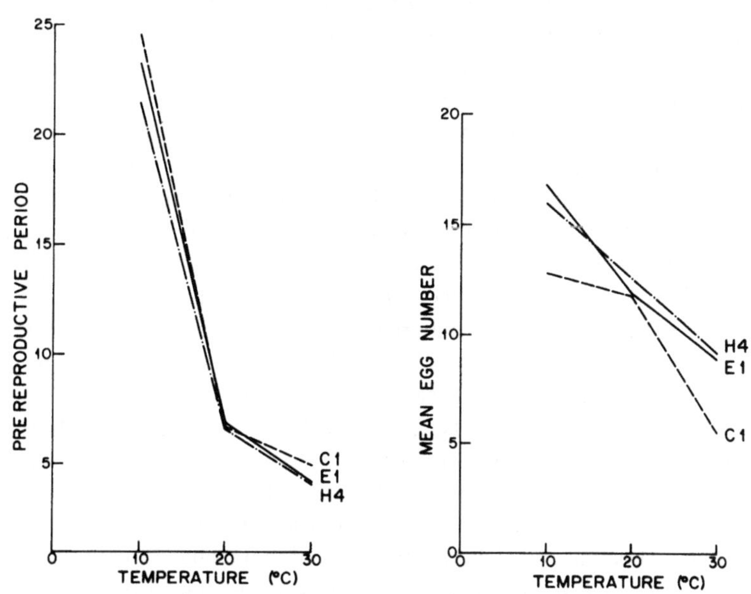

Fig. 2. Shifts in two life history characteristics of three clones of *Daphnia magna* in response to temperature change. Animals were grown individually and fed unicellular food

Table 2. Three-way analysis of variance of temperature, clone and experiment effects on r value for ten clones of *Daphnia magna*[a]

Source	df	F	Prob. F
Temperature	2	2,058.9	0.0001
Clone	9	12.6	0.0001
Experiment (temp.)	13	33.0	0.0001
Temp. × Clone	18	6.1	0.0001
Clone × Expt (temp.)	98	5.1	0.0001
Error	810		

[a] Clones were grown at 10°, 20°, and 30 °C

among genotypes in any test environment was small compared with environmental effects.

The experimental results on both rotifers and cladocerans show the major impact of environmental conditions on rates of increase and confirm that different genotypes invariably show a qualitatively similar response to shifting food levels, temperatures, etc. The studies also suggest that genotypic differences in rates of increase of about 10% are common in cladoceran populations reproducing by continued or obligate parthenogenesis, while differences as large as 50% exist in rotifer populations. These estimates of the magnitude of genotypic differences are subject to criticism. Firstly, current experimental protocols make it impossible to distinguish between differences in r that originate from maternal effects and those which arise as a consequence of genetic differences. Confidence limits on clonal rates of increase have ordinarily been derived by measuring the rates of increase for 10–20 individuals of a clone. In most cases, all clones are analyzed on a single occasion to avoid difficulties in maintaining uniform environmental conditions for an extended period. This procedure sets the stage for the intrusion of maternal effects, as a consequence of physiological differences among the mother(s) of each group of neonates. In the Cladocera, the size of a neonate is affected by its mother's age and nutritional status (Green 1954). Aside from variation in size, neonates receive a variable donation of lipids from their mother; again depending upon the nutritional status of the female (Goulden and Horning 1980). Such differences in neonate size and lipid deposits alter juvenile growth rates and, thus, have a direct effect on r. Until experimental protocols are modified to minimize the possibility of maternal effects, it would be unwise to assume that clonal differences are solely genetic in origin. Laboratory studies suffer from a second weakness. They aim to assess the magnitude of genotypic differences which exist in nature, yet laboratory environments are atypical. Temperatures and photoperiods are held constant and the food supply is simple. Exposure to these unusual conditions may reveal variation which does not exist in nature, in the same way that temperature shock can expose otherwise cryptic variation in the visible phenotype. In a study on *D. pulex*, Loaring and Hebert (1981) found that clonal differences in r were constrained in some environments, but pronounced in others. In this case it was evident that the large differences in r occurred when environmental conditions deviated from those normally encountered in nature. Clearly, there is a need to ascertain if the differences in r noted in the laboratory mirror those in nature.

Natural population studies on genotypic variation in factors affecting natality have been more restricted in scope than the laboratory studies. Work has been restricted to the cladoceran genus *Daphnia* and while considerable evidence of genetic variation in brood sizes has been obtained, developmental rates have not been studied because of practical difficulties. In populations reproducing by continued parthenogenesis, genotypic differences in egg production are not infrequent and in some cases it has proved possible to predict shifts in clone frequencies from the data on relative egg production (Hebert 1974, Young 1979). As noted in the laboratory studies, different clones showed largely concordant trends in absolute egg production so that clonal differences in egg production were small in relation to shifts resulting from food availability. Nonetheless, it does seem that natality differences bring about major shifts in clonal composition. A similar instability of clone frequencies is seen in cladoceran populations reproducing by obligate parthenogenesis (Hebert and Crease 1983). In the single survey carried out on egg production in such populations, differences among clones were apparent. The results of laboratory and field investigations are in general agreement. They indicate that most variation in natality is caused by environmental factors and that genotypic differences are small.

Genotypic Variation in Susceptibility to Predation

Studies over the past 20 years have made it clear that predation is an important source of mortality to many zooplankton populations (Zaret 1980). Death rates attributable to predation often nearly match birth rates and on some occasions exceed them. Studies have shown that the minor morphological differences which exist between closely related species, often have marked effects on the extent of predation. Thus, any reduction in body size or pigmentation reduces susceptibility to vertebrate predators, while increases in body size and cuticular outgrowths, such as spines and crests, reduce susceptibility to invertebrate predators.

There is good evidence of intraspecific variation in morphological features which significantly alter susceptibility to invertebrate predation. It seems certain that a major portion of the phenotypic shifts (cyclomorphosis) especially evident in cladoceran and rotifer populations are involved in reducing predation (Dodson 1974). Indeed, in a number of cases it has been shown that the predators themselves have important effects on their prey's phenotype - their metabolic by-products induce crest or spine formation in the prey. For instance, a compound released by the predaceous rotifer *Asplanchna* causes spine growth in its prey *Brachionus* and *Filinia* (Beauchamp 1952, Gilbert 1966, 1967, Pourriot 1964, 1974). Individuals with such spines are much less sensitive to predation than individuals lacking them. A similar phenomenon has been noted in the Cladocera. Several species in the *Daphnia carinata* group produce a prominent dorsal crest when grown in the presence of notonectids (Grant and Bayley 1981). The crest acts to reduce susceptibility to predation. Similarly, it has been shown that neonates of *D. pulex* produce a group of spikes of the anteriodorsal surface of their carapace when grown in the presence of *Chaoborus* (Kreuger and Dodson 1981). In each of the cases the compound inducing morphological change appears to be a low molecular weight excretory product. There is evidence of genotypic variation in respect to spine formation in response to *Asplanchna* factor. Beauchamp (1952) found that some clones of *Brachionus calyciflorus* produced long spines, while others produced none when grown with *Asplanchna*. Halbach and Jacobs (1971) found that a natural population of *Brachionus* exposed to *Asplanchna* predation showed an increasing predominance of clones capable of induction. Using a *Chaoborus* extract, Schwartz and Hebert (in preparation) found that some clones of *D. pulex* failed to respond, while others produced a crest with multiple spikes. They also showed that the distribution of clones capable of crest induction correlated well with *Chaoborus* abundances. The results of this study and that of Halbach and Jacobs (1971) make it clear that predators can affect the clonal composition of zooplankton populations. To assess the demographic effects of such genotypic diversity one needs information on the fitness differentials involved. It seems likely that genotypic differences with regard to predator susceptibility are greater than those which exist in natality. Certainly five- to tenfold differences in susceptibility have often been noted among closely related species (Hebert and Loaring 1980).

Genetic Adaptation to Heterogeneity in the Physical Environment

Physical variables play an important role in limiting the distribution and abundances of aquatic organisms for they experience variation in the concentration of dissolved ions and gases as well as temperature and light. Yet few studies have attempted to delimit the extent or nature of intraspecific variation with regard to physical factors. Work on the temperature tolerance of the copepod *Eurytemora affinis* stands out as a striking exception. In a series of detailed studies Bradley and his co-workers have determined the genetic and environmental contributions to the variation evident among individuals of this species to acute temperature stress (Bradley 1978, 1982, Bradley and Ketzner 1982, Ketzner and Bradley 1982). Their results provide clear evidence of both physiological adaptation and sexual differences. However, when comparison was restricted to individuals of one sex that had been kept under uniform conditions, residual variation in survival time was apparent. Among males this variation was largely genetic in origin. There was, however, no good evidence that the genetically determined differences in temperature tolerance were of adaptive significance. Surveys failed to provide any evidence of a seasonal shift in the temperature tolerance of males. Indeed, as the acute test involved exposure to temperatures beyond those ordinarily experienced in the environment, it can be suggested that the differences observed among males might ordinarily be selectively neutral. As in the case of natality, it seems that resistance to temperature stress in *Eurytemora* is a trait with strong environmental effects on the phenotype. The phenotypic variation is continuous, differences among genotypes are small, and the probable demographic significance of the variation is limited.

It is worth cautioning against using the results of this study of temperature tolerance as a model of the likely importance of genotypic variation with respect to other environmental variables. There is good evidence that the genetic basis of tolerance may often be simpler. Indeed there are two apparent reports of discontinuous variation with respect to temperature tolerance in the Cladocera. By inbreeding a clone taken from nature, Banta and Wood (1928) produced a clone of *D. longispina* (= *laevis*) with an acute temperature tolerance 5 °C higher than that of its parent. Johnson (1952) identified a clone of *D. atkinsonii* with a temperature tolerance 5 °C below that normal for the species. Such instances of discontinuous variation may not be infrequent. Genetic adaptations to the deleterious effects of solar radiation are, for example, also likely to have a simple genetic basis. The zooplankton in arctic and alpine ponds are heavily pigmented; the pigment functions to prevent photodamage from near UV and blue radiation (Hairston 1976, 1978, 1979, Siebeck 1978). In all copepods and some cladocerans the pigments are carotenoids, derived by a simple pathway from precursors present in the algal food. In other cladocerans melanin synthesis occurs. In either case a single gene change might well result in the suppression of pigment formation. Indeed in several copepod species the intensity of pigmentation varies markedly among populations; and in some habitats pigmented and unpigmented forms co-occur (Ringelberg 1980). The distribution of the two colour forms is apparently determined by a complex interaction between both visual and tactile predators and light levels (Hairston 1980). Unfortunately there is no good evidence that the differences in pigment levels of these copepods are genetically caused. There is, however, clear evidence of genetically determined differences in pigmentation of two cladoceran species. For many years it was assumed that the presence of melanin in the cuticle was a monophyletic trait diagnostic of *Daphnia middendorffiana* (Brooks 1957). However, allozyme studies have indicated that the trait is polyphyletic - in northern Canada both *D. pulex* and *D. middendorffiana* produce pigmented cuticles (Hebert and McWalter 1983). In *D. pulex* there is a clear dimorphism - some clones are heavily pigmented, while others lack the pigment. In *D. middendorffiana* the variation is more continuous, with clones ranging from pale to dark. At low arctic sites where both clone types of *D. pulex* occur; clone distributions are not random (Emery 1984). The pigmented clones occur in shallow ponds with clear water, while the nonpigmented clones occur in deeper ponds with coloured water. The water colour results from dissolved humic materials, which absorb UV light very effectively. Over large areas of the high arctic only pigmented clones of *D. pulex* are present. Water clarity in these regions is high as a consequence of the sparsity of terrestrial vegetation. These regions may well have lacked *Daphnia* until melanin synthesis evolved.

Conclusion

Despite the problems introduced by taxonomic difficulties, it is evident that genetic variation in important life history parameters is abundant. Differences in natality of 10% or more are, for instance, common between genotypes isolated from asexually reproducing populations of zooplankton. However, differences of this magnitude must be placed in perspective; changes in natality of more than an order of magnitude occur as a consequence of environmental variation. The genetic contribution to the phenotypic variance in this character is small. Recognizing the difficulties in accurately censusing zooplankton populations and the errors involved in birth rate estimates of natural populations, it is evident that population ecologists can neglect the natality differences among genotypes. However, genetic factors account for most of the phenotypic variation in other traits, of which resistance to photodamage is perhaps the best example. In such cases it is apparent that ecological work carried out without reference to genetics is at best incomplete. Clearly the need to include information on the genetic composition of populations depends upon two factors – the extent of genetic variation in the trait and the precision with which relevant ecological parameters can be measured. Thus, it is evident that as zooplankton ecology develops a more precise approach, the study of genotypic variation will become increasingly necessary.

References

Agar WE (1920) The genetics of a *Daphnia* hybrid during parthenogenesis. J Genet 10:303–330

Banta AM, Wood TR (1928) A thermal race of Cladocera originating by mutation. Verh V Int Kongr Vererbg (Berlin 1927) 1:397–398

Beauchamp De P (1952) Un facteur de la variabilité chez les rotifères du genre *Brachionus*. CR Hebd Sce Acad Sci 234:573–575

Bradley BP (1978) Genetic and physiological adaptation of the copepod *Eurytemora affinis* to seasonal temperatures. Genetics 90:193–205

Bradley BP (1982) Models for physiological and genetic adaptation to variable environments. In: Dingle H, Hegmann JP (ed) Evolution and genetics of life histories. Springer, Berlin Heidelberg New York

Bradley BP, Ketzner PA (1982) Genetic and non-genetic variability in temperature tolerance of the copepod *Eurytemora affinis* in five temprature regimes. Biol Bull 162:233–245

Brooks JL (1957) The systematics of North American *Daphnia*. Mem Conn Acad Arts Sci 13:5 to 180

Dodson SI (1974) Adaptive changes in plankton morphology in response to size-selective predation: a new hypothesis of cyclomorphosis. Limnol Oceanogr 19:721–729

Emery CJ (1984) Ecological significance of pigmentation in *Daphnia*. MSc Thes, Univ Windsor

Frey DG (1982) Questions concerning cosmopolitanism in Cladocera. Arch Hydrobiol 93:484 to 502

Gilbert JJ (1966) Rotifer ecology and embryological induction. Science 151:1234–1237

Gilbert JJ (1967) *Asplanchna* and posterolateral spine production in *Brachionus calyciflorus*. Arch Hydrobiol 64:1–62

Goulden CE, Hornig LL (1980) Population oscillations and energy reserves in planktonic Cladocera and their consequences to competition. Proc Natl Acad Sci USA 77:1716–1720

Grant JWG, Bayley IAE (1981) Predation induction of crests in morphs of the *Daphnia carinata* King complex. Limnol Oceanogr 26:201–218

Green J (1954) Size and reproduction in *Daphnia magna*. Proc Zool Soc London 124:535–545

Green J (1967) The distribution and variation of *Daphnia lumholtzi* (Crustacea, Cladocera) in relation to fish predation in Lake Albert, East Africa. J Zool 151:181–197

Hairston NG Jr (1976) Photoprotection by carotenoid pigments in the copepod *Diaptomus nevadensis*. Proc Natl Acad Sci USA 73:971–974

Hairston NG Jr (1978) Carotenoid photoprotection in *Diaptomus kenai*. Int Ver Theor Agnew Limnol Verh 20:2541–2545

Hairston NG Jr (1979) The adaptive significance of colour polymorphism in two species of *Diaptomus* (Copepoda). Limnol Oceanogr 24:38–44

Hairston NG Jr (1980) The vertical distribution of diaptomid copepods in relation to body pigmentation. Am Soc Limnol Oceanogr Spec Symp 3:98–110

Halbach U, Jacobs J (1971) Seasonal selection as a factor in rotifer cyclomorphosis. Naturwissenschaften 57:1–2

Hebert PDN (1974) Ecological differences between genotypes in a natural population of *Daphnia magna*. Heredity 33:327–337

Hebert PDN (1977) A revision of the taxonomy of the genus *Daphnia* (Crustacea, Daphniidae) in southeastern Australia. Aust J Zool 25:371–398

Hebert PDN (1978) The population biology of *Daphnia* (Crustacea, Daphniidae). Biol Rev 53: 387–426

Hebert PDN (1982) Competition in zooplankton communities. Ann Zool Fennici 19:349–356

Hebert PDN, Crease TJ (1983) Clonal diversity in populations of *Daphnia pulex* reproducing by obligate parthenogenesis. Heredity 51:353–369

Hebert PDN, Loaring JM (1980) Selective predation and the species composition of arctic ponds. Can J Zool 58:422–426

Hebert PDN, McWalter DB (1983) Cuticular pigmentation in arctic *Daphnia*: adaptive diversification of asexual lineages? Am Nat 122:286–291

Hebert PDN, Moran C (1980) Enzyme variability in natural populations of *Daphnia carinata* King. Heredity 45:313–321
Hutchinson GE (1967) A treatise on limnology, vol II. Introduction to lake biology and the limnoplankton. John Wiley & Sons, New York
Johnson DS (1952) A thermal race of *Daphnia atkinsonii* Baird, and its distributional significance. J Anim Ecol 21:118–119
Kerfoot WC (1975) The divergence of adjacent populations. Ecology 56:1298–1313
Kerfoot WC (1977) Competition in cladoceran communities: the cost of evolving defenses against copepod predation. Ecology 58:303–313
Ketzner PA, Bradley BP (1982) Rates of environmental change and adaptation in the copepod *Eurytemora affinis*. Evolution 36:298–306
King CE (1972) Adaptation of rotifers to seasonal variation. Ecology 53:408–418
King CE (1977) Genetics of reproduction, variation and adaptation in rotifers. Arch Hydrobiol Ergeb Limnol 8:187–201
Kreuger DA, Dodson SI (1981) Embryological induction and predation ecology in *Daphnia pulex*. Limnol Oceanogr 16:219–223
Loaring JM (1982) The ecological genetics of *Daphnia* species. MSc Thes, Univ Windsor
Loaring JM, Hebert PDN (1981) Ecological differences among clones of *Daphnia pulex* Leydig. Oecologia (Berlin) 51:162–168
Lynch M (1983) Ecological genetics of *Daphnia pulex*. Evolution 37:358–374
Manning J, Kerfoot WC, Berger EM (1978) Phenotypes and genotypes in cladoceran populations. Evolution 32:365–374
Modlin RF (1982) Successional changes, variation in population densities and reproductive strategies of Cladocera in two temporary ponds in north Alabama. J Freshwater Biol 1:589 to 597
O'Brien WJ, Vinyard GL (1978) Polymorphism and predation: the effect of invertebrate predation on the distribution of two varieties of *Daphnia carinata* in South India ponds. Limnol Oceanogr 23:452–460
Pejler B (1956) Introgression in planktonic rotaria with some points of view on its causes and conceivable results. Evolution 10:246–261
Pourriot R (1964) Etude experimentale de variations morphologiques chex certaines espèces de rotifères. Bull Soc Zool Fre 89:555–561
Pourriot R (1974) Relations prédateur-proie chex les rotifères: influence du prédateur *(Asplanchna brightwelli)* sur la morphologie de la proie *(Brachionus bidentata)*. Ann Hydrobiol 5: 43–55
Price JL (1958) Cryptic speciation in the *vernalis* group of Cyclopidae. Can J Zool 36:285–303
Ringelberg J (1980) Aspects of red pigmentation in zooplankton, especially copepods. Am Soc Limnol Oceanogr Spec Symp 3:91–97
Schwartz SS (1981) Variations in the life history strategy of *Daphnia pulex* along a latitudinal gradient. PhD Thes, Univ Nebraska, Lincoln
Shan RK (1971) Genetic stability in the Cladocera. Trans Am Microsc Soc: 116
Shan RK, Frey DG (1968) Induced interbreeding between two stocks of a chydorid cladoceran. Bio Sci 18:203–205
Siebeck O (1978) UV-Toleranz und Photoreaktivierung bei Daphnien aus Biotopen verschiedener Höhenregionen. Naturwissenschaften 65:390–391
Snell TW (1977) Clonal selection: competition among clones. Arch Hydrobiol Ergebn Limnol 8: 202–204
Snell TW (1979) Intraspecific competition and population structure in rotifers. Ecology 60:494 to 502
Snell TW (1980) Blue-green algae and selection in rotifer populations. Oecologia (Berlin) 46:343 to 346
Woodrich DC (1980) Ecological differences among clones of *Daphnia*. MSc Thes, Univ Windsor
Young JPW (1979) Enzyme polymorphism and cyclic parthenogenesis in *Daphnia magna*. I Selection and clonal diversity. Genetics 92:953–970
Zaret TM (1972a) Predator-prey interaction in a tropical lacustrine system. Ecology 53:48–57
Zaret TM (1972b) Predators, invisible prey and the nature of polymorphism in the Cladocera (Class Crustacea). Limnol Oceanogr 17:171–184
Zaret TM (1980) Predation and freshwater communities. Yale Univ Press, New Haven, pp 187

Population Biology of the Rose Aphid, *Macrosiphum rosae*

K. WÖHRMANN[1]

Introduction

Most aphid species reproduce by cyclic parthenogenesis, i.e., the asexual phase is succeeded by a phase of sexual reproduction (Fig. 1). Both phases can take place either on the same host or the sexual and the asexual phases can occur on different host plant species. Some aphid species omit the sexual phase and only reproduce parthogenetically. Under mild climatic conditions, holocyclic species might also reproduce asexually only. Thus, aphids seem well suited for investigations on the influence of different modes of reproduction on the pattern of genetic variation.

For this investigation *Macrosiphum rosae*, a holocyclic species, has been chosen. It uses the rose as a primary host and lays its overwintering eggs on this host. From May onward the parthenogenetic phase starts. The population peak is reached by June/July precisely at the flowering time of the roses. Thereafter, the population size decreases considerably and a part of the aphid population leaves the rose and colonizes secondary hosts which include *Dipsacum, Valeriana, Knautia,* and *Scabiosa*. In the autumn the aphids return to the roses as sexuals. The shift of hosts has a major effect on population density.

From *M. rosae*, so far, mainly population density and genetic variation data on the asexual phase are available. The population density of the sexual animals as well as the egg densities has not been determined yet. The density of populations on the

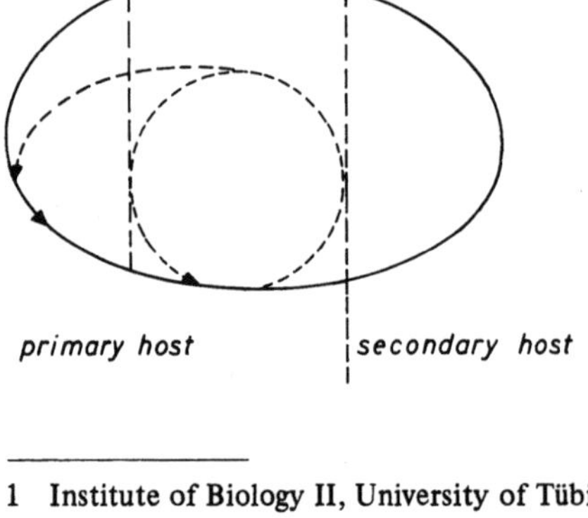

Fig. 1. Life cycles of aphids

1 Institute of Biology II, University of Tübingen, 7400 Tübingen, FRG

Population Biology and Evolution
Ed. by K. Wöhrmann and V. Loeschcke

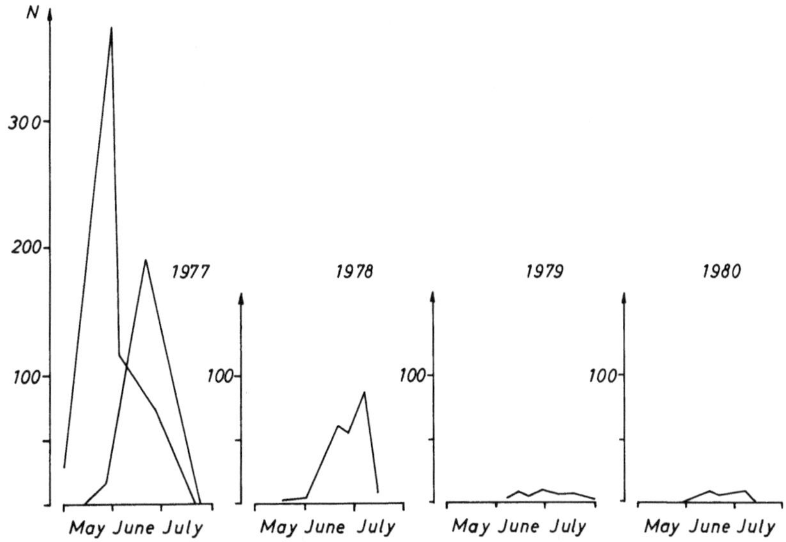

Fig. 2. The number of aphids per primary host plant (N) for two populations in 1977 and for one each in 1978, 1979, and 1980 during the growth phases

secondary hosts, however, is never as large as that on the primary host. The population size of the asexual population varies remarkably among years (Fig. 2).

Variation of Density and Genotypic Frequencies Within Populations

Various factors are known that influence population dynamics and genotypic variation in population of aphids.

Aphid populations can reach considerable densities within a colony or on the plant as a whole (Fig. 2), making them suitable for an investigation of the association between density and the reproductive rate of different genotypes (Fig. 3). The MDH

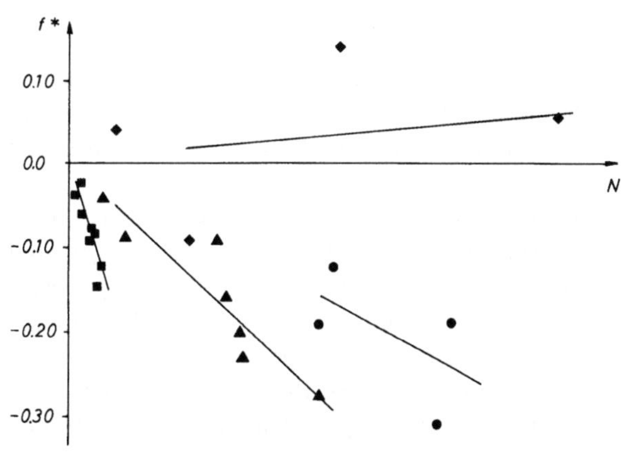

Fig. 3. Regression lines (y = a + bx) of the density (aphids per plant: N) on the frequency of the f/f genotype of MDH in the populations from Tübingen of *Macrosiphum rosae* in 1977 I (●), 1977 II (♦), 1978 (▲), and 1979 (■). The regression lines are transformed to the origin ($f^* = y - a = bx$). The correlation coefficients are 1977 (I): − 0.59; 1977 (II): 0.33; 1978: − 0.96; 1979: −0.94

locus of *M. rosae* is polymorphic with two alleles. During the years 1977, 1978, 1979 the proportion of the two homozygotes (s/s and f/f) in relation to the heterozygotes (s/f) was constant (Table 1). On the contrary, the frequencies of the two homozygotes fluctuate both within and between years.

Table 1. Numbers of MDH genotypes in samples collected in Tübingen in 1977, 1978, and 1979 at different times

Year	s/s + f/f	s/f	d.f.	χ^2 [a]
1977				
25.5.	14	14		
4.6.	39	57		
18.6.	22	16	3	3.52
4.7.	18	19		
1978				
7.6.	147	151		
15.6.	216	159		
20.6.	165	181		
27.6.	220	199	6	15.09 [b]
6.7.	57	60	5	10.00 [c]
12.7.	86	114		
17.7.	49	39		
1979				
6.6.	43	32		
13.6.	77	84		
20.6.	97	90		
27.6.	97	94	7	9.96
7.7.	89	97		
17.7.	71	55		
24.7.	115	76		
6.8.	62	47		

[a] χ^2 are calculated from R × 2 tables
[b] $P < 0.05$
[c] d.f. and χ^2 without the date of 15 th June 1978

During the phase of main growth, the frequency of the f/f genotype decreased, while the population density increased. When the density decreased the frequency of the f/f genotype increased (Fig. 3). The changes of the s/s genotype were in the opposite direction.

Similar observations were made with respect to the PGM locus, which is tri-allelic. Of the six possible genotypes, the m/m and f/m are most common in the population of Tübingen (Table 2). In this case the f/m genotype increased in frequency during the growth phase and decreased during the population crash, whereas the opposite was true for the m/m genotype (Fig. 4).

When the frequency of the f/f homozygotes at the MDH locus is plotted against the number of aphids per plant, no correlation between density and genotype fre-

Table 2. Frequencies of the MDH and PGM genotypes in 74 populations of *Macrosyphum rosae* from 15 localities in Europe[a]

Locality/year	Number of popula-tions	n	MDH			PGM					
			f/s	s/s	f/f	f/m	f/s	m/s	f/f	m/m	s/s
1. Lago Maggiore, 1981	4	180	0.97	0.00	0.03	0.14	0.00	0.00	0.00	0.86	0.00
2. Tettnang, 1981	3	77	0.30	0.47	0.23	0.33	0.01	0.00	0.10	0.56	0.00
3. Tübingen, 1979	1	1,122	0.47	0.30	0.23	0.43	0.01	0.02	0.06	0.48	0.00
4. Hamburg, 1980	5	81	0.47	0.14	0.39	0.49	0.00	0.00	0.09	0.42	0.00
5. Aalestrup	1	76	0.43	0.20	0.37	0.40	0.02	0.02	0.04	0.53	0.00
6. Kristiansand, 1981	2	184	0.43	0.29	0.28	0.24	0.17	0.01	0.42	0.11	0.05
7. Hadanger Fj., 1981	7	96	0.32	0.26	0.42	0.17	0.09	0.04	0.00	0.57	0.13
8. Gjövik, 1981	4	94	0.10	0.00	0.90	0.34	0.15	0.01	0.34	0.11	0.05
9. Hamar, 1981	3	188	0.09	0.00	0.91	0.24	0.21	0.06	0.34	0.10	0.06
10. Lillehammer, 1981	3	158	0.17	0.00	0.83	0.40	0.15	0.02	0.42	0.05	0.09
11. S. England, 1979	3	505	0.45	0.42	0.13	0.00	0.00	0.10	0.00	0.90	0.00
12. Wales, 1979	6	191	0.49	0.36	0.15	0.01	0.00	0.24	0.00	0.74	0.01
13. Ireland, 1979	11	245	0.52	0.20	0.28	–	–	–	–	–	–
14. Dublin, 1979	12	225	0.61	0.38	0.01	–	–	–	–	–	–
15. Scotland, 1979	9	212	0.46	0.37	0.27	0.12	0.02	0.00	0.00	0.86	0.00

[a] Population 1: Switzerland; populations 2–4: West Germany; population 5: Denmark; populations 6–10: Norway; populations 11–15: Great Britain; populations 13, 14: Ireland

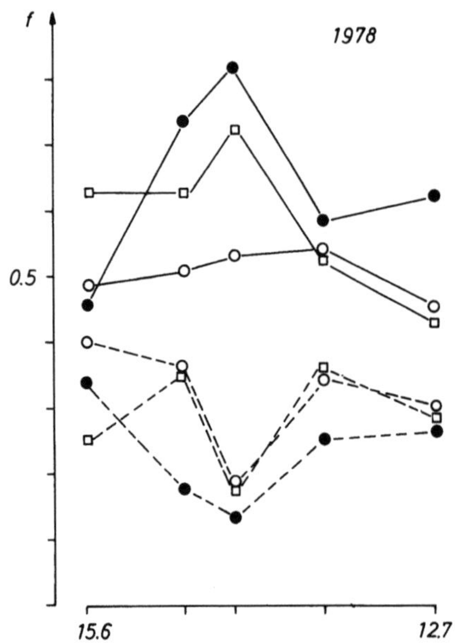

Fig. 4. The change in frequency of the two most frequent genotypes of the PGM locus (m/m = *solid lines*; m/f = *broken lines*) in three populations from Tübingen of *Macrosiphum rosae* in 1978

quency is expected as long as growth rates are unaffected by population densities. A correlation suggests an increase, respectively decrease, of genotype frequencies with density, that is the reproduction rates differ between genotypes. This is true for all populations observed in 1977 (I, II), 1978, and 1979 (Fig. 3). The slope of the regression line is determined by the selection advantage or disadvantage. The frequency change is the more significant the steeper the slope. As seen in Fig. 3 the selection disadvantage of the f/f genotypes is greatest when populations are less dense (Tomiuk and Wöhrmann 1981).

Various factors can influence population dynamics (Tomiuk and Wöhrmann 1981, 1982). The genotypic reproduction rate is modified by the climate. In 1977 and 1978 it has been shown that during the growth phase, the growth rate is correlated with temperature and humidity (Table 3). High temperatures and low atmospheric

Table 3. Correlation coefficients between climatic factors and rate of increase ($\Delta 1 n N_t$) in populations of *Macrosiphum rosae* in 1977 and 1978 (M = number of samples)

Year	M	T10[a]	T15	T20	T25	H60	H70	H90
Increasing phase[b]								
1978	6	−0.69#	−0.93#	0.93#	0.85#	0.72	0.74	−0.66
1977/78	10	−0.64#	−0.77#	0.71#	0.56	0.73#	0.69#	−0.75#
Decreasing Phase								
1978	4	−0.93	−0.58	0.58	0.46	−0.96#	−0.97#	0.96#
1977/78	5	−0.46	−0.11	0.28	0.37	−0.71	−0.76	0.89#

[a] *T10* = h per day with temperatures unter 10 °C; *T15* = h per day with temperatures under 15 °C; *T20* = h per day with temperatures above 20 °C; *T25* = h per day with temperatures above 25 °C. *H60* = h per day with humidity under 60%; *H70* = h per day with humidity under 70%; *H90* = h per day with humidity above 90%

[b] # Significant at the 5% level

humidity seem to promote reproduction. After the density peak, low atmospheric humidity delays the decrease of population size.

During the growth of the roses from sprouting until flowering time, we presume that the physiology of the plants is changing. The peak of the population size is reached at the beginning of flowering time. This coincidence could be demonstrated in 1977, when populations of two different rose beds were examined that sprouted at different points of time (Fig. 2).

Aphids may form winged morphs. A density dependent appearance of winged aphids has been reported by Harrewijn (1977), Maelzer (1977), and Shaw (1970). The winged aphids are able to disperse to new habitats and colonize new host plants. Such density dependent formation of winged morphs seems not to occur in *M. rosae*, since the percentage of winged morphs declines with increasing population density (Fig. 5).

The following taxa are known as natural enemies of aphids: *Coccinellidae, Chrysopidae, Syrphidae, Aphidius,* and *Ephedrus* species. In the case of *M. rosae* in the past, only the wasp *Aphidius rosae* and another unidentified *Ephedrus* species was included in the investigations. Parasites can be recognized later in their development, when the parasitized aphid becomes mummified. However, the presence of a parasite can be detected much earlier by the enzyme pattern of the parasite (Wool et al. 1978, Tomiuk and Wöhrmann 1981). The two parasites could be distinguished by characteristic bands at the MDII locus. During the 4 years of study, the proportion of parasitized aphids varied considerably (Fig. 6). In 1978 about 70% of the aphids were

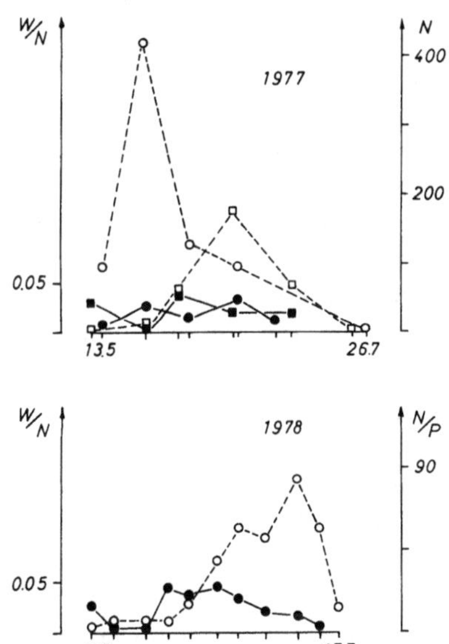

Fig. 5.

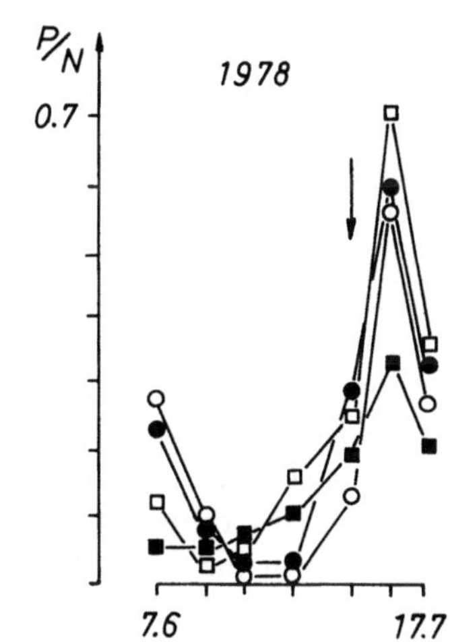

Fig. 6

Fig. 5. The number of aphids per plant (N, *broken lines*) during the growing season for two populations in 1977 (*above*) and one in 1978 (*below*) and the percentage of winged morphs (W/N, *solid lines*) in 1977 (*above*) and 1978 (*below*)

Fig. 6. The frequency of parasitized aphids (P/N) in four populations of *Macrosiphum rosae* in 1978. The *arrow* marks the population peak of the aphid population

parasitized at the time of the highest population density, while in 1977 and 1979 a maximum of 25% were parasitized. The results show the potential importance of parasites on population growth and population density.

Variation in Genotypic Frequencies Between Populations

The genotypic composition of the populations investigated in Switzerland (population 1), West Germany (populations 2, 3, 4), Denmark (population 5), Norway (populations 6, 7, 8, 9, 10), Great Britain (populations 11, 12, 15), and Ireland (populations 13, 14) is demonstrated by Table 2. From the great number of data some particularities should be mentioned in more detail. Almost 100% heterozygotes at the MDH locus are observed in the population from Lago Maggiore (Switzerland). No s/s homozygote has been found, although 34.2% should be expected when assuming Hardy-Weinberg proportions. A similar situation is found in the PGM system. In this case the population exists almost of only two genotypes (f/m, m/m) out of six possibles, one of which reaches 86%. These genotype proportions were found in four populations collected from different villages at the Lago Maggiore.

The MDH s/s genotype is also missing in the population from eastern Norway (populations 8, 9, 10). The most frequent genotype is here the f/f homozygote (about 90%). At this locality no s/s homozygote is to be expected in a sample of 100 to 200 individuals because of the low frequency of the s allele in the populations.

As already mentioned the Lago Maggiore population consists mostly of one PGM genotype (m/m). The same is true for the British populations. However, two frequent genotypes (f/m, m/m) have been found in the continental populations (populations 2, 3, 4, 5) and the Hardanger Fjord (population 7) in Norway. The populations from eastern Norway (populations 8, 9, 10) deviate from these by having three frequent genotypes (f/m, f/s, f/f).

To describe and to compare genotypic variation in the investigated populations, the genotypic heterogeneity (Hedrick 1971) is used as a measure of variation within each population. The value of 1 indicates the lack of heterogeneity. The population of the Lago Maggiore show the highest value, followed by those from Great Britain and from the continent.

The following hypothesis is proposed to explain the high amount of heterogeneity in Norway (populations 8, 9, 10) and the lower ones in Great Britain and of the Lago Maggiore. At localities with harsh winters the aphids are holocyclic and only able to overwinter in the egg stage. In mild climates, like the Lago Maggiore or the British Isles, however, the aphids may survive as adults and even propagate parthenogenetically. Therefore, selection against some genotypes may be more severe in mild climates because of the absence of sexual reproduction.

The genetic identity (Hedrick 1971) is used to compare different populations with respect to their genotypic composition. From Table 4 it becomes obvious that despite differences between populations there exist similarities within the group of populations from Great Britain, within the group of eastern Norway and within that from West Germany and Denmark, whereas the populations of eastern Norway are not like

Table 4. The intrapopulation variation (diagonal) and the genetic identity between populations of *Macrosiphum rosae* in Europe

Populations[a]	1	2	3	4	5	6	7	8	9	10	11	12	15
1. Lago Maggiore	0.70	0.28	0.50	0.41	0.62	0.09	0.38	0.03	0.03	0.04	0.63	0.79	0.68
2. Tettnang		0.17	0.78	0.66	0.69	0.29	0.83	0.25	0.21	0.21	0.72	0.46	0.77
3. Tübingen			0.16	0.90	0.87	0.39	0.82	0.39	0.31	0.41	0.66	0.58	0.76
4. Hamburg				0.15	0.82	0.67	0.61	0.49	0.37	0.50	0.44	0.45	0.55
5. Aalestrup					0.18	0.39	0.68	0.37	0.30	0.39	0.68	0.71	0.78
6. Kristiansand						0.17	0.24	0.64	0.57	0.64	0.11	0.12	0.14
7. Hadanger Fjord							0.16	0.27	0.26	0.28	0.79	0.57	0.82
8. Gjövik								0.22	0.97	0.97	0.06	0.06	0.10
9. Hamar									0.21	0.94	0.06	0.06	0.09
10. Lillehammer										0.20	0.04	0.04	0.09
11. S. England											0.31	0.81	0.98
12. Wales												0.33	0.82
15. Scottland													0.29

[a] Population 1: Switzerland; populations 2–4: West Germany; population 5: Denmark; populations 6–10: Norway; populations 11–15: Great Britain; populations 13, 14: Ireland

those of Great Britain. The large identity between populations within areas may be explained by the high migration rate which is observed in aphids. The winged morphs are mostly passively distributed over wide areas. Aphids have been caught up to a height of 1.5 km (Taylor and Palmer 1972).

The extension of the investigation to more enzyme loci showed that *M. rosae* has only a low degree of polymorphic loci. This species is polymorphic at 4 enzyme loci (MDH, PGM, SDH, EST) out of a total of 29 investigated loci. This is in concordance with the proportion of polymorphic loci in 17 other aphid species investigated which is only 0.033 on the average (Tomiuk and Wöhrmann 1983). The average degree of heterozygosity pooled over 16 aphid species is H = 0.035 which is much lower than in other invertebrates where H is estimated to be about 0.12 (Ayala 1982). *M. rosae* itself shows an average degree of heterozygosity of 0.032 considering 16 loci.

Acknowledgements. I thank Dr. V. Loeschcke and Dr. P. Hebert for reading the manuscript and for useful comments.

References

Ayala FJ (1982) Genetic variation in natural populations: The problem of electrophoretically cryptic alleles. Proc Natl Acad Sci USA 79:550–554

Harrewijn P (1977) Nutritional aspects of development and wing dimorphism in the aphid *Myzus persicae*. Acad Proefschr, Wageningen, Ponsen En Looijen

Hedrick PW (1971) A new approach to measuring genetic similarity. Evolution 25:276–280

Maelzer DA (1977) The biology and main causes of changes in numbers of the rose aphid *Macrosiphum rosae* (L) on cultivated roses in South Australia. Austr J Zool 25:269–284

Shaw MJP (1970) Effects of population density on allienicolae of *Aphis fabae* Scop. III. The expression of migration of migratory urge among alatae in the field. Ann Appl Biol 79:1–7

Taylor LR, Palmer JMP (1972) Airial sampling. In: Emden Van HF (ed) Aphid technology. Academic Press, London New York, pp 189–234

Tomiuk J, Wöhrmann K (1981) Changes in the genotype frequencies at the MDH-locus in populations of *Macrosiphum rosae* (L) (Hemiptera, Aphididae). Biol Zentralbl 100:631–640

Tomiuk J, Wöhrmann K (1982) Effect of temperature and humidity on natural populations of *Aphis pomi* De Geer and *Macrosiphum rosae* (L) (Hemiptera, Aphididae). Z Pflanzenkrankh Pflanzenschutz 89:157–169

Tomiuk J, Wöhrmann K (1983) Enzyme polymorphism and taxonomy of aphid species. Z Zool Syst Evol

Wool D, Emden Van HF, Bunting S (1978) Electrophoretic detection of the internal parasite, *Aphidius matricariae* in *Myzus persicae*. Ann Appl Biol 90:21–26

Ecological Differentiation Among Clones: The Frozen Niche Variation Model

R.C. VRIJENHOEK[1]

Introduction

Sex and its evolutionary significance have received considerable attention from theoretical biologists in recent years (see Ghiselin 1974, Williams 1975, Maynard Smith 1978, Stanley 1979, Bell 1982, Shields 1982). Sexual reproduction is a highly coordinated and potentially costly process that results in the production of genetic diversity among the progeny of an individual. Since the rate of adaptive evolution is believed to be a function of this diversity, the recombination of genes in sexual lineages provides an obvious long-term benefit in a changing environment (Fisher 1930, Muller 1932). However, asexuality provides an immediate "twofold" numerical advantage due to all-female reproduction (Williams 1975, Maynard Smith 1978). Despite this apparent short-term benefit of asexual reproduction, sexuality clearly predominates in all groups of plants and animals. How is the "monopoly of sex" maintained in the presence of this cost? Perhaps an answer to this question requires studies of the rare exceptions, the asexual populations, and the peculiar genetic and ecological conditions under which they sometimes prosper. A surprising result of such studies recently has challenged a common assumption of most theoretical considerations of sex – that asexual populations lack genetic diversity. Natural populations of asexual organisms are often composed of several genetically (and sometimes ecologically) distinct clones (Solbrig 1971, Vrijenhoek 1978, 1979a, Mitter et al. 1979, Parker 1979, Vepsalainen and Jarvinen 1979, Hebert and Crease 1980). What is the source of this clonal variation? How is it maintained? Does it contribute to the ecological and evolutionary success of asexual populations? Most importantly, we should not forget that asexual "species" of the higher plants and animals are all derived from sexually reproducing species. What then does the distribution of genotypic and phenotypic variation within and between clones tell us about the nature of genetic variation in the sexual populations from which these asexual forms arose? In this paper, I will attempt to relate the many facets of these questions by exploring further the "frozen niche variation" model, the major features of which were outlined in an earlier publication (Vrijenhoek 1979a).

Besides the recent theoretical interest in asexual populations, the study of clones provides a useful tool for dissecting the components of phenotypic variation. The

1 Department of Biological Sciences, Rutgers University, New Brunswick, NJ 08903, USA

Population Biology and Evolution
Ed. by K. Wöhrmann and V. Loeschcke

relative contributions of genetic and environmental factors to the various phenotypic characteristics of sexually produced individuals are difficult to assess. Each sexual genotype is unique. The recombinant processes and environmental interactions that created it will also replace it in the next generation. In order to provide experimental tests of the affects of "whole genotype" combinations in evolutionary processes, one must have methods for preserving genotypes and for assessing the adaptive value of their corresponding phenotypes (Annest and Templeton 1978). Fish in the genus *Poeciliopsis* (Peociliidae) provide an ideal group of organisms for such a test. Our experimental system exploits the fact that we can study genetically diverse clones of these fish in their native habitats and that we can synthesize new clones in the laboratory.

Reproduction in *Poeciliopsis*

The sexual species of *Poeciliopsis* use conventional Mendelian processes during reproduction: segregation, assortment, crossing-over, and syngamy (Leslie and Vrijenhoek 1977, 1978, 1980, Leslie 1982). However, hybridization between several species has given rise to a series of diploid and triploid, all-female, biotypes (Table 1). The

Table 1. The unisexual forms of *Poeciliopsis*: their ploidies, sexual hosts, and reproductive modes

Unisexual biotype	Ploidy	Sexual host	Reproductive mode
P. monacha-lucida	2 n	*P. lucida*	Hybridogenetic
P. monacha-occidentalis	2 n	*P. occidentalis*	Hybridogenetic
P. monacha-latidens	2 n	*P. latidens*	Hybridogenetic
P. 2 monacha-lucida	3 n	*P. monacha*	Gynogenetic
P. monacha-2 lucida	3 n	*P. lucida*	Gynogenetic
P. monacha-viriosa-lucida	3 n	*P. viriosa*	Gynogenetic

hyphenated names of each of the biotypes reflects its hybrid composition and genomic dosages (Schultz 1969, 1977). It should be noted that *P. monacha* is the common ancestor to all the unisexual biotypes, but formal genetic analyses indicate that it is not predisposed to unisexuality (Leslie and Vrijenhoek 1977, 1978).

The three triploid biotypes are gynogenetic (Fig. 1), a strictly asexual form of reproduction. Fertilization of the eggs by sperm from a coexisting sexual species is required to activate embryogenesis, but the sperm contribute nothing genotypically or phenotypically to the offspring (Schultz 1967). In these females, the chromosomes are doubled during an endomitotic division prior to meiosis. Pairing of replicated sister chromosomes preserves the maternal genotype (Cimino 1972a). Their clonal pattern of inheritance was confirmed using biochemical genetic and immunological studies (Vrijenhoek 1972, Moore 1977a, Eisenbrey and Moore 1981).

The three diploid all-female biotypes are hybridogenetic, a semisexual reproductive mechanism (Schultz 1969). During hybridogenetic oogenesis, only the haploid

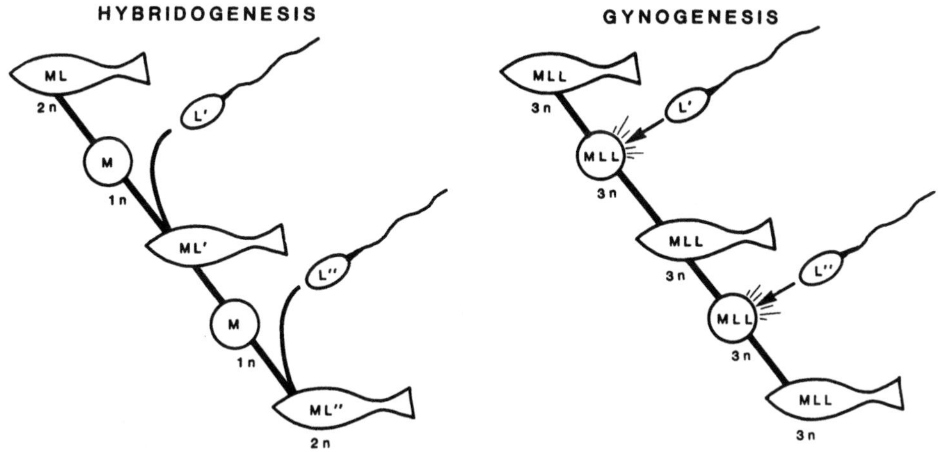

Fig. 1. Hybridogenetic and gynogenetic reproduction in *Poeciliopsis*. The letters *M* and *L* are used to represent the *monacha* and *lucida* genomes, respectively. The *primes* (′) represent different allelic markers associated with the paternal *lucida* genomes. During hybridogenesis the *lucida* genomes are substituted each generation; during gynogenesis they are not – the sperm serve only to activate embryogenesis in the triploid eggs

chromosome set derived from *P. monacha* (designated M in Fig. 1) is transmitted to the ova; the paternal set (designated L) is expelled from the oogonium, thereby precluding synapsis and potential recombination (Cimino 1972b). The haploid *monacha* ova produced by these hybrids are fertilized by sperm from a coexisting host species, thus reestablishing diploid hybrids that express morphological and electrophoretic traits encoded by both parental genomes (Schultz 1966, Vrijenhoek 1972). Hybridogenetic strains are not clones in the strict sense because their paternal genomes are substituted in each generation; thus, natural populations of hybridogens have access to all the allelic variation in the gene pool of the sexual host (Virjenhoek et al. 1977). Nevertheless, the *monacha* genome is inherited clonally, therefore, discussions of clonal variation in the hybridogens refer only to the genotypic differences among the haploid *monacha* genomes, which we have previously referred to as "hemiclones" (Vrijenhoek et al. 1977, 1978).

Various hemiclones of the hybridogenetic biotypes and clones of the gynogenetic biotypes were identified by electrophoretic phenotypes (Roman numerals) based on 22–25 gene loci and tissue grafting analyses (lower case letters; e.g., *IIa, IIb*, or *IIIa*; Vrijenhoek et al. 1977, 1978, Angus and Schultz 1979, Angus 1980). To distinguish among various clones of the various biotypes, I will assign a biotype prefix (Table 1) to each clonal designation (e.g., MO/IIa is a diploid *P. monacha-occidentalis* clone, IIa and MML/IIa is a triploid *P. 2 monoach-lucida* clóne IIa).

Ecological Models

Unlike truly parthenogenetic species, gynogenetic and hybridogenetic *Poeciliopsis* can never escape from their sexual hosts to invade new habitats, nor can they competitively exclude their hosts, for in doing so they lose their sperm source and ensure their own extinction. Unable to play the roles of "weedy" or "fugitive" species, as has been argued for parthenogenetic lizards (Wright and Lowe 1968, Maslin 1968) and cockroaches (Parker et al. 1977), unisexual *Poeciliopsis* are forced to coexist and compete with their sexual hosts, with which they share considerable genetic, and presumably ecological, overlap. Theoretically, the all-female fish have a "two fold" reproductive advantage over the sexual females which produce 50% male offspring. However, Moore and McKay (1971) identified a frequency dependent process that reduces their reproductive potential. When sexual males are rare, they mate only with conspecific sexual females, leaving the unisexuals uninseminated (McKay 1971), and, thus, the sexuals will rise in frequency relative to the unisexuals.

However, when males of the sexual species become frequent, dominance hierarchies develop and subordinate and inexperienced males are responsible for unisexual inseminations. Moore (1976) proposed three basic criteria to evaluate the relative fitnesses in the sexual/unisexual "species" complexes: (1) the probability of producing female offspring; (2) frequency dependent mating success; and (3) "primary fitness," a residual component including survivorship, fecundity, and all other factors affected by the local environment. Both computer simulations and laboratory population experiments suggested that as long as primary fitness of the unisexuals equals or exceeds that of the sexual host, the unisexuals can achieve a maximum frequency of about 84% of the female population (Moore 1975, 1976).

In nature, the unisexual females most often do not attain this theoretical upper limit (Table 2). Thus, three main hypotheses have been forwarded to explain the differential success of sexual and unisexual populations in various habitats: (1) the "ecotonal specialist" model; (2) the "heterosis" model; and (3) the "frozen niche

Table 2. The proportion of unisexual females relative to the total number of females in unisexual/bisexual complexes of *P. monacha-occidentalis* and *P. monacha-lucida*[a]

River systems[b]	Number of localities	Percentage unisexuals[c]	Range
	Monoclonal Rivers		
R. de la Concepción	4	1.5 ± 1.1	0 - 3.0
R. Sonora	10	9.6 ± 9.9	0 - 30.0
R. Matape	3	6.0 ± 4.6	0 - 11.0
R. Yaqui	16	7.5 ± 10.0	0 - 38.0
Mean:		7.2 ± 9.3	
	Multiclonal Rivers		
R. Mayo	7	82.1 ± 10.6	65.0 - 97.0
R. Fuerte	4	63.8 ± 20.6	33.6 - 91.0
R. Sinaloa	2	45.2 ± 5.2	40.0 - 50.4
R. Mocorito	2	36.9 ± 5.3	31.6 - 42.4
Mean:		66.3 ± 21.60	

[a] Data for the first five rivers are taken from Moore, Miller, and Schultz (1971)
[b] Rivers are listed in order from north to south. The first five contain *P. monacha-occidentalis* and the last three contain *P. monacha-lucida*
[c] The percentage of unisexual females ± its standard deviation

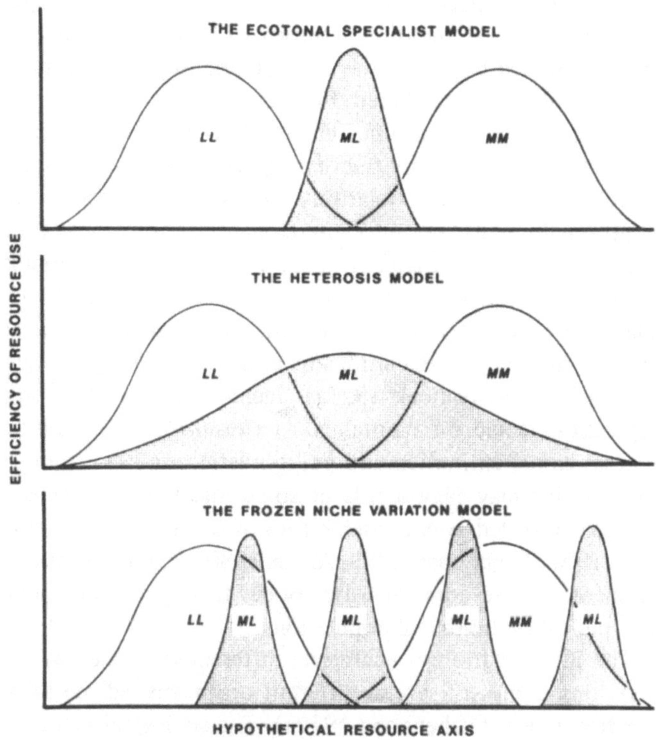

Fig. 2. Three models concerning the niche structure of unisexual hybrid (*ML*) forms of *Poeciliopsis* and their sexual ancestors, *P. monacha* (*MM*) and *P. lucida* (*LL*)

variation" model (Fig. 2). According to the ecotonal specialist model, the primary fitness of the unisexuals should be highest in habitats that are intermediate to those of the parental species (Thibault 1974, Moore 1976, 1977b). The hybrids would achieve their greatest numbers in ecotones, due to weaker competition from their sexual progenitors. This model resembles the "fugitive (weedy) species" model in its requirement that the unisexual hybrids be disadvantaged when competing with the parental species in their preferred habitats. According to the heterosis model, the unisexual hybrids have the ecological specializations of both parental species, and thus, are capable of occupying a broader niche than either of the progenitors (Schultz 1971). Bulger and Schultz (1979) found that *P. monacha-lucida* had thermal tolerances that spanned both the parental species. However, Moore (1976) identified a fundamental conflict within the heterosis model. If the unisexuals are truly heterotic, that is, having primary fitness that exceeds either of their sexual ancestors, the dynamics of his relative fitness model would predict the maximum frequency of unisexuals (about 80%) everywhere they occur. This is clearly not the case (Table 2; and Moore 1976, 1977b). Either Moore's (1976) model is incorrect in some of its assumptions or the fish are not heterotic for primary fitness, or both. Furthermore, Bulger and Schultz (1982) subsequently found that the heterotic thermal tolerances in *P. monacha-lucida* could not be generalized for other *Poeciliopsis* biotypes.

The ecotonal specialist and the heterosis models were proposed before our discovery of extensive clonal variation in unisexual populations. Since both models portray the entire unisexual population as a single phenotype (with regard to niche use), neither model could have predicted the discovery of unisexual assemblages composed of several genotypically distinct clones. It is possible that nonequilibrium conditions predominate in the highly fluctuating desert arroyos and springs of Sonora, rendering interclonal competition relatively unimportant. Such a condition should lead to monoclonal populations since interclonal selection would favor the single most broadly tolerant, generalist clone (Parker 1979). Indeed, if population sizes become very small, random processes (clonal drift) also should lead to monoclonal populations (Maynard Smith 1978). The fact that this has not occurred in many of the small, endemic headwater populations of unisexual *Peociliopsis* suggests that other factors are involved (Vrijenhoek 1979a). Jaenike et al. (1982) suggested that multiclonal populations could be maintained as a result of an equilibrium between the rates of random clonal extinction and polyphyletic origins. Although an equilibrium process such as this may play a role at some localities, our genetic studies over the past 10 years revealed stable assemblages of clones at many other localities (Vrijenhoek et al. 1977, Vrijenhoek 1978, Angus 1980). The latter discovery required the formulation of a more compehensive model to explain the coexistence of sexual species with multilocal unisexual populations.

The identification of ecological differences (niche use, life histories, etc.) among coexisting clones of several different organisms led me to propose the frozen niche variation model (Vrijenhoek 1979a). The ecological success of an asexual population relative to its sexual competitors depends largely on three factors: (1) polyphyletic origins of new clones from genetically variable sexual ancestors provide a source of genotypic variation; (2) interclonal selection acts upon this variation; and (3) niche subdivision among clones leads to an efficient exploitation of natural resources. Recombination within the sexual ancestors can generate a tremendous diversity of phenotypes and cloning should lead to the rapid fixation of epistatic gene combinations that intensify niche specialization (Roughgarden 1972, Wright 1977, Vrijenhoek 1978, 1979a). For *Poeciliopsis*, each hybridization between *P. monacha* and either *P. lucida, P. occidentalis*, or *P. latidens*, isolates another haploid *monacha* genome from the sexual gene pool. If the new hybrids prove fertile and hybridogenetic, they then "freeze" whatever combinations of morphological, physiological, behavioral, and ecological characteristics the *monacha* genomes encode (Fig. 2). Polyphyletic hybrid origins from a broad niched, genetically variable sexual ancestor should give rise to an array of new clonal genomes, each with a unique combination of life history characteristics and niche requirements (e.g., temporal, spatial, and trophic). This process may logically be extended to the origins of new triploid clones from diploid ancestors. Interclonal selection should eliminate those clones with substantial niche overlap, resulting in a highly structured unisexual population, as long as selection is not overridden by a very high rate of clonal synthesis, a point I shall return to later.

Of course, since new clones are being "frozen" from the sexual gene pool, interclonal selection also occurs within the context of potential competition from the sexual species. An additional model, termed the "tangled bank," was recently put forward by Bell (1982). Although interesting, Bell's proposal is not novel for most of its components were similarly stated in an earlier paper (Vrijenhoek 1979a). His model does extend the frozen niche variation model, however, by proposing that the twofold fecundity advantage of individual clones is likely to be offset by their more restrictive niches, and thus, prevents the unisexuals from replacing the sexuals. Sperm dependence and frequency dependent mating success in *Poeciliopsis* probably play a more important role in preventing the unisexuals from completely eliminating the particular sexual species that they rely upon as sperm sources. On the other hand, it is possible that the ecological success of hybridogenetic *Poeciliopsis* may have resulted in the exclusion of *P. monacha* from a historically broader niche. None of the hybridogens are dependent on *P. monacha* for sperm, thus, it could be displaced by a multiclonal assemblage to which it gave rise. In fact, *P. monacha* apparently has a relictual distribution, occurring in a few scattered headwater streams and springs of the Rios Mayo, Fuerte, and Sinaloa (Miller 1960). It is rarely found coexisting with the numerically dominant hybridogens in these three rivers. Unless it was displaced from a broader and more continuous distribution at some time in the past, it is hard to imagine how it arrived at these isolated headwater localities.

It should be noted that the frozen niche variation model does not preclude certain features of the heterosis or the ecotonal specialist models. The latter two models simply fail to consider the complexity of a multidimensional niche. It is conceivable that individual clones might have generalist phenotypes for some niche axes and specialist phenotypes on others (e.g., a food generalist, but habitat specialist). As opposed to Bell's model, the frozen niche variation model does not require that clones be specialized in all their niche requirements. If "generalist" vs "specialist" genotypes segregate in the sexual ancestors, these genotypes can also be "frozen" in different clones.

Ecological Studies of Sexual/Unisexual Complexes

According to the frozen niche variation model, the numerical success of unisexual populations should be a function of clonal diversity. With a greater diversity of clones, a larger portion of the available niche may be captured from the sexual competitors. In a southern group of river systems where recurrent hybridizations can give rise to multiclonal populations (Vrijenhoek 1979a), the unisexuals generally comprise from 30% to 80% of the *Poeciliopsis* females (Table 2). The northern rivers lack this opportunity for generating clonal diversity since *P. monacha* is absent. Thus, only mutation and migration can contribute to clonal diversity. These rivers have monoclonal unisexual populations that comprise less than 10% of the *Poeciliopsis* population, a result that contrasts strongly with the multiclonal rivers.

It might be argued that the northern monoclonal rivers contain a single generalist clone. However, examination of samples collected in diverse habitats in these rivers reveal that the unisexuals are only abundant in peripheral pools (Table 3). The multiclonal populations of the Rio Mayo provide a sharp contrast, for here the unisexuals occupy both current and pool habitats with equal success and outnumber the sexual form in both. Since extensive clonal diversity in the Rio Mayo was discernable only

Table 3. The frequencies of sexual and unisexual females in river systems containing the *P. occidentalis* and the diploid hybridogenetic biotype *P. monacha-occidentalis*[a]

River	Habitat	Percentage		
		Sexual	Unisexual	Total
	Monoclonal Rivers			
Concepción	Current	95.3	4.7	106
1976	pool	34.3	65.7	102
Matape	Current	89.9	10.1	69
1976	pool	69.5	30.5	95
Yaqui	Open current	91.7	8.3	157
1980	Protected current	90.9	9.1	88
	pool	81.5	18.5	157
	Multiclonal Rivers			
Mayo-1[b]	Current	17.7	82.3	96
1976	pool	19.6	80.3	214
Mayo-2[c]	Current	27.1	72.9	107
1976	pool	9.8	90.2	153

[a] These samples were taken by Vrijenhoek and Angus in 1976 and Vrijenhoek and Douglas in 1981. They should not be confused with the frequencies listed in the previous table because those samples were unspecified as to the habitat in which fish were taken. In the present samples we made a specific effort to sample current and pool habitats

[b] Downstream at Navojoa

[c] Upstream at El Tabelo

through difficult tissue grafting studies (Angus 1980), ecological studies aimed at identifying the niche requirements of individual clones could not be undertaken.

Studies of niche partitioning between identifiable clones of *Poeciliopsis* are presently underway in the laboratory and field. Much of the available data has been reviewed elsewhere (Vrijenhoek 1979, 1984). Our most intensive efforts have involved a pair of co-occurring clones of the triploid gynogenetic biotype *P. 2 monacha-lucida*. Based on the observation of distinct dentitional morphologies and habitat use, I hypothesized that coexistence among these two clones and their sexual host, *P. monacha*, is aided by resource partitioning (Vrijenhoek 1978). *P. monacha* dominates shaded, rocky habitats of low productivity; MML/II reaches its highest frequencies in open, sunny areas with high productivity; and MML/I is relatively low frequency throughout the system, regardless of the frequencies of the other two. Observations of foraging behavior in the laboratory revealed that MML/I directs a large proportion of its feeding pecks (about 30%) at hard surfaces, such as algae covered rocks, filters, and aquarium walls, rather than soft objects, such as filamentous algae or detritus. MML/II directs less than 5% of its feeding pecks at hard surfaces under these artificial conditions. More detailed analyses are presently underway by R. Schenck. He is finding that microgeographic spatial preferences appear to be the primary factors permitting differential food exploitation and niche partitioning among these forms. Both triploid clones prefer slight currents to still pools and increase the proportions of insects in their diets when captured in currents. The opposite is true for *P. monacha*. Significant trophic differences are often found between the two clones when taken together in the same habitats. During seasons when a broad range of foods are available, MML/II shows a higher degree of dietary overlap with the sexual species *P. monacha*, whereas MML/I has considerably less dietary overlap with either of these two forms. Schenck's observations on the dietary overlap within pools are consistent with the previous observations I have made on the numerical relationships between these forms between pools (e.g., that the proportions of MML/I are independent of the proportions of *P. monacha* and MML/II).

Schenck's study also examines a unisexual/sexual complex composed of *P. lucida* and two hybridogenetic hemiclones ML/VII and ML/VIII, taken from a nearby arroyo. A remarkably similar pattern is emerging. *P. lucida* and ML/VII are predominantly detritivorous, but they tend to displace one another spatially. The hemiclone ML/VIII tends to be more insectivorous and again its numbers are independent of the other two. Schenck is presently engaged in manipulative field experiments that will determine the frequency and density dependent behavioral aspects of these ecological patterns.

Laboratory studies of predation behaviors in MML/VII and ML/VIII have provided strong support for the frozen niche variation model (Vrijenhoek et al. in preparation). In this study, the two hybridogenetic clones were compared to first generation laboratory stocks of the parental species, *P. monacha* and *P. lucida*, taken from the same geographical locality. The two sexual forms showed highly significant differences in both search time and handling time when presented with benthic (chironomid larvae) and pelagic (brine shrimp) prey items. After regression techniques were used to remove the size-specific effects of the predators, we found that *P. lucida* (the deposit feeder in nature) was the least efficient predator on both kinds of prey and that *P. monacha* (the more omnivorous species in nature) was a much more effective predator. Thibault (1974) observed similar differences in cannibalistic behaviors. Clones ML/VII was intermediate between the parental species in both experiments, however, ML/VIII was the most efficient predator. The results of the predation efficiency study are consistent with Schenck's food habit studies which show that ML/VIII consistently takes the most insect larvae in nature. The extreme predation efficiency of ML/VIII provides the best evidence against the ecotonal specialist model. It appears that ML/VIII has "frozen" genetic variation from its *P. monacha* ancestors which provides nearly complete dominance in predatory ability over the *lucida* genome.

In an independent study, Bulger and Schultz (1979) found different thermal tolerances for a pair of *P. monacha-2 lucida* clones, MLL/I and MLL/II. Life history studies of various natural clones have revealed different growth rates (Schultz 1982), and behavioral studies by Keegan-Rogers (pers. comm.) suggest that *P. lucida* males can discriminate among clones, possibly leading to a frequency dependent mating success. To date, all the detailed ecological studies of *Poeciliopsis* clones have revealed differences that might contribute to the spatial and temporal partitioning of resources and to the differential survival and reproduction in this heterogeneous environment.

Synthetic Clones

A direct test of the hypothesis that cloning can freeze ecologically relevant genetic variation is presently underway in my laboratory. Schultz (1973) synthesized several new hybridogenetic strains through crosses of *P. monacha* females with *P. lucida* males, but these corsses had a low rate of success. I have synthesized over 120 hybrids of *P. monacha* × *P. occidentalis* (cross 1, below), using laboratory reared, first-generation, *P. monacha* females from a variety of localities in the Rios Mayo, Fuerte, and Sinaloa.

cross 1 MM × OO ---- > *MO*
cross 2 MO × LL ---- > *ML* .

The MO all-female hybrids exhibited very low mortality. To permit comparisons among the clonal M genomes of these hybrids, I am currently in the process of placing them in a standardized genetic background (cross 2) using an isogenic *lucida* strain (M61-31 LL; see Vrijenhoek et al. 1978). Cross 2 is less successful and, in many cases, the MO hybrids were sterile. This variation in compatibilities suggests a first and most intense level of selection acting on potentially new hemiclones. Apparently most M genomes are sufficiently compatible with O genomes to produce viable hybrids, but significantly fewer are capable of producing fertile hybrids. Nevertheless, genetic analysis demonstrated that all fertile MO hybrids were indeed hybridogenetic. Alleles diagnostic for *occidentalis* at four enzyme loci and a locus determining a dorsal fin spot that were expressed in the MO generation were all replaced by unique *lucida* alleles in the ML generation. Hybridogenesis is not an evolved mechanism; it is a spontaneous by-product of the failure of conventional oogenetic processes. Twenty-six synthetic, standardized, ML strains are presently available for testing and we are creating more. An immediate difference was observed between many of the synthetic ML strains. Using a standardized artificial insemination technique that delivers a surplus of viable sperm, we found that some strains had very high fertility and some others are barely able to produce sufficient viable offspring to permit us to continue the strain. In fact, by the second or third lab generation, we did lose five ML strains.

Phenotypic variance (V_p) in hybridogens can be partitioned into heritable and nonheritable components:

$$V_p = V_b + V_w + V_{bw} ,$$

where V_b is due to the genetic differences between *monacha* haplogenenomes and V_w is the within hemiclone variance due to nonheritable sources, such as environmental variance (V_e) and variance due to the substutable paternal genomes (V_m). V_{bw} is an interaction component between specific clonal genomes and the environmental and paternal components. Since all the fish in our experimental studies are being reared under nearly constant laboratory conditions, we anticipate that V_e will be minimal. Because all the strains are tied to the same standardized paternal genome, V_m will be zero. Thus, significant differences in morphology, behavior, and life histories between strains must be due to their different "frozen" *monacha* genomes. As a second phase of the study, we will examine effects of the paternal genome (V_m) on phenotypic variance within and between strains.

We are presently raising fish under constant laboratory conditions for the morphological studies. In the course of this rearing we are also obtaining data on growth rates and other life history characters (J. Wetherington). Preliminary analyses of these data reveal highly significant differences among several of the synthetic clones. According to the frozen niche variation model, the amount of interclonal variation should be a function of the amount of genetic variance for ecologically relevant characters in the sexual ancestor, *P. monacha*. Synthetic clones have been made from samples of *P. monacha* from two locations where electrophoretic studies revealed major differences in the levels of electrophoretic heterozygosity. Wetherington has examined strains from both locations and the preliminary data based on size of newborn fish at birth indicate that the electrophoretic variation in the *P. monacha* is predictive of the amount of interclone variation that can be "frozen" during the formation of new clones. It has not escaped my attention that our natural and synthetic clones provide an almost unique opportunity to study genotypic and phenotypic plasticity in behavioral traits. Casual observations of these fish in the laboratory have identified several strains with unusual behavioral patterns that remain to be quantified.

Summary and Conclusions

Our studies of all-female *Poeciliopsis* populations have shown that differences in trophic morphology, food habits, use of space, and life history characters all might contribute to niche diversification among clones in a multiclonal assemblage. According to the frozen niche variation model, the differences among asexual clones are fixed from the range of genetic variance in niche related characters present in the sexual ancestors. Since the asexual forms cannot play the role of "fugitive species" and escape from competition, individual clones are forced to compete with their more plastic sexual ancestors. However, the all-female forms potentially have a "twofold" reproductive advantage, thus, a diverse multiclonal assemblage should displace the sexual forms from contested portions of the available niche. The fact that unisexual *Poeciliopsis* do not completely replace the sexual forms is undoubtedly due, in part, to their sperm dependence on males of the related sexual species. Nevertheless, within the limits of the sperm dependent relationship, the numerical success of the all-female populations apparently is a function of multiclonal niche partitioning. Ironically, the success of these nonrecombinant clonal forms derives from the recombinational genetic variation in their sexual ancestors. The recruitment of "frozen" sexual genotypes into the all-female population permits the interclonal selection and niche subdivision that lead to their numerical success. In the absence of this potential for sexual recruitment, the unisexual populations have only mutation and migration as a sources of heritable variation. The inability of these processes to generate sufficient phenotypic variation for the ecological diversification of unisexual populations is apparent in the northern rivers where the sexual ancestor, *P. monacha*, is absent. Thus, for *Poeciliopsis* at least, it is impossible to escape the conclusion that sex is also good for asexual populations.

We must consider also the possibility that clonal reproductive mechanisms might occasionally break down and permit their genotypic diversity to feed back into the sexual gene pool (Vrijenhoek and Schultz 1974, Vrijenhoek 1979b). In the long run new clonal genotypes might benefit from recombinations among old clones. Such "clonal feedback" probably does not occur commonly in *Poeciliopsis*, but even if rare, it would be erroneous to consider the sexual/unisexual complexes as communities of independent evolutionary units. It is more realistic to consider the system as a diverse assemblage of *monacha* genomes: some in the diploid sexual phase as *P. monacha*; some as haploid genomes in the hybridogenetic biotypes; and some as either single or double doses in the triploid biotypes (Table 1). However, if the clonal genomes remain isolated from this recycling for sufficiently long periods of time, mutations will inevitably accumulate as a consequence of the "ratchet mechanism" postulated by Muller (1964). Genetic dissections of clonal genomes revealed significant loads of deleterious mutations including "silent" alleles for several enzymes in various hemiclonal *monacha* genomes (Leslie and Vrijenhoek 1978, 1980, Spinella and Vrijenhoek 1982). Many of these genomes had accumulated sufficient genetic changes to make them genetically incompatible with normal, sexually produced *monacha* genomes, and thus, they are permanently isolated from the "feedback" process.

Why are asexual animals apparently more common in temperate than in tropical regions, at high vs low altitudes, in xeric vs mesic environments, and in disturbed vs. undisturbed habitats (Suomalainen 1950, Ghiselin 1974). Glesener and Tilman (1978) examined this question and concluded that increased biotic diversity (e.g., in the tropics) favors genetic systems that can respond rapidly to the diverse and unpredictable challenges of competition, predation, and disease. Others have speculated that the colonization abilities of true parthenogens provide an advantage in reaching new habitats and in reproducing under regimes of low mate density (Tomlinson 1966). I would like to extend these ideas and suggest that asexual populations are most likely to be found in "ecological vacuums," that is, in species depauperate environments where interspecific competition is low relative to intraspecific competition. In such places, the frozen niche variation model will have its greatest effect. Here, the recurrent origins of new clones from a broad-niched sexual ancestor will "freeze" genotypes that can effectively exploit an unfilled multispecies niche, and thus, increase the abundance of the asexual forms relative to the sexual species that gave rise to them. Asexual populations probably arise in the tropics as often as they do in temperate zones, but in the species packed tropical environment there is simply little room for increased niche subdivision. For example, small tropical streams contain a diverse guild of fish that effectively subdivide the available food resources through the use of spatial and morphological specializations (Lowe-McConnell 1975). However, in species depauperate areas, such as the streams of the Sonoran desert, newly formed clones might have the opportunity to survive because intense interspecific competition from a diverse guild of sexual species is lacking. The ecological differences we see between natural clones of *Poeciliopsis* are reminiscent of the niche specializations observed among legitimate species of small fishes in Central American streams (Zaret and Rand 1971). Furthermore, the differences we are beginning to detect among our synthetic clones clearly demonstrate the feasibility of the frozen niche variation model in *Poeciliopsis*.

Additional tests of this model can be undertaken with a variety of organisms that have both sexual and asexual forms. Are the sexual species generally broader in their niche use, life history characters, and behaviors than the individual clones to which they give rise? What are the patterns of character covariances? Does a new clone arise with the appropriate combinations of trophic characters, feeding behaviors, and metabolic processes? It is important to remember that natural asexual populations may have been molded by intense interclonal selection, thus, the interclonal variance observed may represent only a fraction of that contained in the sexual ancestors. Are sexual individuals just compromises among all the different genetic systems affecting niche structure? Attempts to answer these important questions will depend upon the opportunities to synthesize new clones in the laboratory. However, great caution must be exercised in using synthetic clones to study the extent of variation in the sexual species. Cloning methods that enforce homozygosity (such as those with *Drosophila mercatorum* by Annest and Templeton 1978) might reveal more about the negative consequences of inbreeding depression than about "adaptive" variation affecting life-history and morphological differences in natural, outbred sexual populations. Studies based on hybrid organisms, such as ours, present the opposite problem. Simple dominance interactions between the heterospecific genomes, or even increased canalization, might mask the phenotypic expression of variation that occurs in the homospecific sexual populations. Obviously, studies of clones generated within a species, and without increased homozygosity, would be most desirable. The dandelions studied by Solbrig (1971) would provide excellent subjects in this regard since the sexual and apomictic biotypes still coexist in Europe. Also, many facultatively parthenogenetic invertebrates, such as the cladocercans, rotifers, and aphids could be useful, unless the clonal "feedback" process discussed above creates excessive linkage disequilibrium in the sexual gene pool and gives a false impression of the covariances among the characters under study.

Obviously, a major short-term benefit of asexual reproduction is the ability to rapidly produce a diverse assemblage of cohesive genotypes that can effectively exploit a multiple niche environment. The long-term success of such a reproductive strategy is questionable, however, because the overall inflexibility of clonal niches would lead to a dead end in changing environments. Although I find no need to argue that individual clones are specialized in all of their niche related characters, I agree with Bell (1982), that ultimately in a complex, multiple niche environment (the "tangled bank"), the adaptively more flexible, and generalized, sexual forms will persist in the long run, as the specialized portions of the total niche shift in and out of existence.

Acknowledgements. I wish to express my sincerest thanks to my graduate students, Russell Schenck and Jeffrey Wetherington who allowed me to cite unpublished results from their Ph.D. studies. Many of the ideas expressed here have grown from seeds that germinated while I was a graduate student with R.J. Schultz. It has been rewarding to see many of these ideas constantly challenged, expanded, and nurtured by my own students. I am grateful to M.E. Douglas, R. Schenck, and J. Wetherington for their helpful comments and careful readings of the manuscript. The research was funded by the U.S. National Science Foundation (grants DEB 79-16620 and Deb 82-12150). I am deeply indebted to the Secretaria de Pesca and Dra. Edith Polanco Jaime, La Directora General de Administracion de Pesquerias for issuing the permits (13 and 4962) that allowed us to undertake our field studies in Mexico.

References

Angus RA (1980) Geographical dispersal and clonal diversity in unisexual fish populations. Am Nat 115:531–550

Angus RA, Schultz RJ (1979) Clonal diversity in the unisexual fish *Poeciliopsis monacha-lucida*: a tissue graft analysis. Evolution 33:27–40

Annest JL, Templeton AR (1978) Genetic recombination and clonal selection in *Drosophila mercatorum*. Genetics 89:193–210

Bell G (1982) The masterpiece of nature: the evolution and genetics of sexuality. Univ Cal Press, Berkeley

Bulger AJ, Schultz RJ (1979) Heterosis and interclonal variation in thermal tolerance in unisexual fish. Evolution 33:848–859

Bulger AJ, Schultz RJ (1982) Origins of thermal adaptation in northern vs. southern populations of a unisexual hybrid fish. Evolution 36:1041–1050

Cimino MC (1972a) Meiosis in triploid all-female fish (*Poeciliopsis*, Poeciliidae). Science 175: 1484–1486

Cimino MC (1972b) Egg production, polyploidization and evolution in a diploid all-female fish of the genus *Poeciliopsis*. Evolution 26:294–306

Eisenbrey AB, Moore WS (1981) Evolution of histocompatibility diversity in an asexual vertebrate, *Poeciliopsis 2 monacha-lucida* (Pisces: Poeciliidae). Evolution 35:1180–1191

Fisher RA (1930) The genetical theory of natural selection. Clarendon Press, Oxford

Ghiselin M (1974) The economy of nature and the evolution of sex. Univ California Press, Berkeley

Glesener RR, Tilman D (1978) Sexuality and the components of environmental uncertainty: clues from geographic parthenogenesis in terrestrial animals. Am Nat 112:659–673

Hebert PDN, Crease TJ (1980) Clonal coexistence in *Daphnia pulex* (Leydig): another planktonic paradox. Science 207:1363–1365

Jaenike J, Ausubel S, Grimald DA (1982) On the evolution of clonal diversity in parthenogenetic earthworms. Pedobiologia 23:304–310

Leslie JF (1982) Linkage analysis of seventeen loci in poeciliid fish (genus: *Poeciliopsis*). J Hered 73:19–23

Leslie JF, Vrijenhoek RC (1977) Genetic analysis of natural populations of *Poeciliopsis monacha*. J Heredity 68:301–306

Leslie JF, Vrijenhoek RC (1978) Genetic dissection of clonally inherited genomes of *Poeciliopsis*: I. Linkage analysis and preliminary assessment of deleterious gene loads. Genetics 90:801–811

Leslie JF, Vrijenhoek RC (1980) Consideration of Muller's ratchet mechanism through studies of genetic linkage and genomic compatibilities in clonally reproducing *Poeciliopsis*. Evolution 34:1105–1115

Lowe-McConnell RH (1975) Fish communities in tropical freshwater: their distribution, ecology and evolution. Longman, London

Maslin TP (1968) Taxonomic problems in parthenogenetic vertebrates. Syst Zool 17:219–231

Maynard Smith J (1978) The evolution of sex. Cambridge Univ Press, London

McKay FE (1971) Behavioral aspects of population dynamics in unisexual-bisexual *Poeciliopsis* (Pisces: Poeciliidae). Ecology 52:778–790

Miller RR (1960) Four new species of viviparous fishes, genus *Poeciliopsis* from northwestern Mexico. Occ Pap Mus Zool Univ Mich 433:1–9

Mitter C, Futuyma DJ, Schneider JC, Hare JD (1979) Genetic variation and host-plant relations in a parthenogenetic moth. Evolution 34:777–790

Moore WS (1975) Stability of small unisexual-bisexual populations of *Poeciliopsis* (Pisces: Poeciliidae). Ecology 56:791–808

Moore WS (1976) Components of fitness in the unisexual fish *Poeciliopsis monacha-occidentalis*. Evolution 30:564–578

Moore WS (1977a) A histocompatibility analysis of inheritance in the unisexual fish *Poeciliopsis 2 monacha-lucida*. Copeia 1977:213–223

Moore WS (1977b) An evaluation of narrow hybrid zones in vertebrates. Q Rev Biol 52:263–277

Moore WS, McKay FE (1971) Coexistence in unisexual-bisexual complexes of *Poeciliopsis* (Pisces: Poeciliidae). Ecology 52:791–799
Muller HJ (1932) Some genetic aspects of sex. Am Nat 66:118–138
Muller HJ (1964) The relation of recombination to mutational advance. Mut Res 1:2–9
Parker ED Jr (1979) Ecological implications of clonal diversity in parthenogenetic morphospecies. Am Zool 19:753–762
Parker ED Jr, Selander RK, Hudson RO, Lester LJ (1977) Genetic diversity in colonizing parthenogenetic cockroaches. Evolution 31:836–842
Roughgarden J (1972) Evolution of niche width. Am Nat 106:683–718
Schultz RJ (1966) Hybridization experiments with an all-female fish of the genus *Poeciliopsis*. Biol Bull 130:415–429
Schultz RJ (1967) Gynogenesis and triploidy in the viviparous fish *Poeciliopsis*. Science 157: 1564–1567
Schultz RJ (1969) Hybridization, unisexuality, and polyploidy in the teleost *Poeciliopsis* (Poeciliidae) and other vertebrates. Am Nat 103:605–619
Schultz RJ (1971) Special adaptive problems associated unisexual fish. Am Zool 11:351–360
Schultz RJ (1973) Unisexual fish: laboratory synthesis of a "species." Science 179:180–181
Schultz RJ (1977) Evolution and ecology of unisexual fishes. In: Hecht MK, Steere WC, Wallace B (eds) Evolutionary biology, vol X. Plenum Press, New York, pp 277–331
Schultz RJ (1982) Competition and adaptation among diploid and polyploid clones of unisexual fish. In: Dingle H, Hegmann JP (eds) Evolution and genetics of life histories. Springer, Berlin Heidelberg New York, pp 104–119
Shields WM (1982) Philopatry, inbreeding, and the evolution of sex. State Univ, New York Press, Albany
Solbrig OT (1971) The population biology of dandelions. Am Sci 59:686–694
Spinella DG, Vrijenhoek RC (1982) Genetic dissection of clonally inherited genomes of *Poeciliopsis*: II. Investigation of a silent carboxylesterase allele. Genetics 100:279–286
Stanley SM (1979) Macroevolution: pattern and process. Freeman, San Francisco
Suomalainen E (1950) Parthenogenesis in animals. Adv Genet 3:547–558
Thibault RE (1974) The ecology of unisexual and bisexual fishes of the genus *Poeciliopsis*: a study in niche relationships. PhD Thesis, Univ Conn, Storrs
Tomlinson J (1966) The advantage of hermaphroditism and parthenogenesis. J Theor Biol 11: 54–58
Vepsalainen K, Jarvinen O (1979) Apomictic parthenogenesis and pattern of the environment. Am Zool 19:739–752
Vrijenhoek RC (1972) Genetic relationships of unisexual hybrid fishes to their progenitors using lactate dehydrogenase isozymes as gene markers (*Poeciliopsis*, Poeciliidae). Am Nat 106:754 to 766
Vrijenhoek RC (1978) Coexistence of clones in a heterogeneous environment. Science 199:549 to 552
Vrijenhoek RC (1979a) Factors affecting clonal diversity and coexistence. Am Zool 19:787–797
Vrijenhoek RC (1979b) Genetics of a sexually reproducing fish in a highly fluctuating environment. Am Nat 113:17–29
Vrijenhoek RC (1984) The evolution of clonal diversity in *Poeciliopsis*. In: Turner BJ (ed) Evolutionary genetics of fishes. Plenum Press, New York, pp. 399–429
Vrijenhoek RC, Schultz RJ (1974) Evolution of a trihybrid unisexual fish (*Poeciliopsis*; Poeciliidae). Evolution 28:306–319
Vrijenhoek RC, Angus RA, Schultz RJ (1977) Variation and heterozygosity in sexually vs. clonally reproducing populations of *Poeciliopsis*. Evolution 31:767–781
Vrijenhoek RC, Angus RA, Schultz RJ (1978) Variation and clonal structure in a unisexual fish. Am Nat 112:41–55
Williams GC (1975) Sex and evolution. Univ Press, Princeton
Wright JW, Lowe CH (1968) Weeds, polyploids, parthenogenesis, and the geographical and ecological distribution of all-female species of *Cnemidophorus*. Copeia 1968:128–138
Wright S (1977) Evolution and the genetics of populations, vol III. Univ Press, Chicago
Zaret TM, Rand AS (1971) Competition in tropical stream fishes: support for the competitive exclusion principle. Ecology 52:336–342

Theoretical Aspects of Coevolution

The Interplay Between Genetic Composition, Species Number, and Population Sizes Under Exploitative Composition

V. LOESCHCKE[1]

Introduction

Evolutionary biologists will not hesitate to accept that ecology as well as genetics are essential ingredients to an understanding of community evolution. However, evolutionary processes have been considered mainly as processes that are decomposable in changes that occur in ecological time and in evolutionary time.

In the early 1950's theoretical papers by Levene (1953) and Dempster (1955) started a discussion on the relation of environmental heterogeneity and genetic variability. These papers were followed by a large number of theoretical, field, and experimental investigations, focusing on the maintenance of polymorphism in heterogeneous environments. The contribution of ecology in these studies was not very sophisticated, as the primary questions were genetical.

In the late 1960's population genetics, for at least a decade, got involved in the study of allozyme variation. As this type of variation has in only few cases shown any correlation with the variation in (fitness relevant) characters, studies on enzyme variation have not contributed much to population biology.

In addition to the fact that our knowledge on the relation of processes on the genotypic and the phenotypic level is still rather poor, another crucial point for the failure of a combined approach has been that we have been more or less unsuccessful in detecting selection in natural populations, but instead have inferred selection as having caused the patterns observed in nature on character diversity or similarity.

I will not sketch more aspects of why the wedding between population genetics and evolutionary ecology has not worked so far, but instead concentrate on which approaches might be promising. One approach that seems obviously suited for connecting evolutionary ecology and population genetics goes via quantitative genetics, as selection undoubtedly acts on phenotypes. But we have to be aware that up to the present the quantitative genetics approach is either rather descriptive or it is more or less simple population genetics.

In the following a simple population genetics model for a quantitative character is considered in an ecological frame. On the basis of this model the purpose is to study the interplay between genetics and ecology. An exploitative competition model is used and the questions to be asked are: what constraints does ecology give to the genetics and vice versa? Are species and genotypes more or less interchangable if

1 Institute of Ecology and Genetics, University of Aarhus, 8000 Aarhus, Denmark

Population Biology and Evolution
Ed. by K. Wöhrmann and V. Loeschcke

competition is defined as passive, that is, if individuals feel the presence of other individuals just by a depletion of the resources? And last but not least, does genetics actually add to the understanding of community evolution?

The idea of the model is quite simple: a genetically controlled character is assumed to be directly related to the exploitation of renewable resources. A correlation between character expression and resource utilization is known to occur in natural populations, e.g., between head size and prey size in Anolis lizards (Roughgarden 1972) or between body size and food particle sizes in mud snails of the genus *Hydrobia* (Fenchel 1975). Another example of such a correlation is given by Price and Grant (this volume, see also Boag and Grant 1981) and concerns beak size and seed size in Darwin's finches. Bird's population dynamics, however, might not be sufficiently described by the simple model of passive competition.

The Model

We consider a guild of m genetically variable species competing for renewable resources. We consider homologeous characters in each species and assume a simple one locus with n alleles autosomal inheritance of the character in a diploid, random mating species with nonoverlapping generations. The character is directly related to the way resources are exploited. The ecological frame of the model is a discrete Lotka-Volterra model of competitive interactions. Competition is assumed to act passively so that individuals realize the presence of other individuals by a depletion of the available resources. The assumption of competition being passive makes it reasonable to assume symmetry in the competition coefficients. The purely interspecific form of this model has proven to apply well to exploitative competition between Hydrobia species (Fenchel and Kofoed 1976).

The basic model is similar to those of Loeschcke (1983, 1984) and Christiansen and Loeschcke (1980a, see also Christiansen and Fenchel 1977, Matessi and Jayakar 1981). Gaussian utilization functions $U_{\pi ij}$ with mean $D_{\pi ij}$ and variance W^2 are specified for each genotype ij and species π. We will assume additive contributions of the alleles, in that the mean $D_{\pi ij}$ of the utilization function is given by as $D_{\pi ij} = D_\pi + d_{\pi i} + d_{\pi j}$, where D_π is a scaling constant referring to the location of species π in the pure ecological model and where $d_{\pi i}$ and $d_{\pi j}$ are the contributions of alleles i and j in species π.

The competition coefficients are defined through the overlap in utilization functions. The competition coefficient between genotype ij of species π and genotype kl of species ϕ is then

$$\gamma_{\pi ij,\phi kl} = \frac{\int U_{\pi ij}(\rho)\, U_{\phi kl}(\rho)\, d\rho}{\int U_{\pi ij}(\rho)^2\, d\rho} = \exp\left[-\,(D_{\pi ij} - D_{\phi kl})^2/4\, W^2\right]. \quad (1)$$

The availability of resources is described by the spectrum $S(\rho)$, which is assumed to be Gaussian with $S \sim N(0, \sigma^2)$. The carrying capacity of genotype ij of species π is then given by

$$K_{\pi ij} = \frac{\int U_{\pi ij}(\rho)\, S(\rho)\, d\rho}{\int U_{\pi ij}(\rho)^2\, d\rho} = \exp\left[-\,(D_{\pi ij}{}^2)/(2\,\sigma^2 + 2\,\dot{W}^2)\right]. \quad (2)$$

We provide only the necessary equations to determine the evolution of the population sizes:

$$x'_\pi = x_\pi \, [1 + V (K_\pi - \sum_\phi \bar{\gamma}_{\pi,\phi} x_\phi)] \,, \qquad (3)$$

and the recurrence equations for the allele frequencies

$$p'_{\pi i} = p_{\pi i} \, [1 + V (K_{\pi i} - \Sigma \, \bar{\gamma}_{\pi i,\phi} x_\phi)]/[1 + V (K_\pi - \sum_\phi \bar{\gamma}_{\pi,\phi})] \,, \qquad (4)$$

where V is a constant describing the proportionality between the growth rate parameter $R_{\pi ij}$ and the carrying capacity parameter $K_{\pi ij}$ and where

$$K_{\pi i} = \sum_j K_{\pi ij} p_j, \; K_\pi = \sum_i K_{\pi i} p_i$$

$$\bar{\gamma}_{\pi ij,\phi} = \sum_\phi \sum_k \sum_l \gamma_{\pi ij,\phi kl} \, p_{\phi k} \, p_{\phi l} \,,$$

$$\bar{\gamma}_{\pi i,\phi} = \sum_j \bar{\gamma}_{\pi ij,\phi} \, p_{\pi j} \; \text{ and } \; \bar{\gamma}_{\pi,\phi} = \sum_i \bar{\gamma}_{\pi i,\phi} \, p_{\pi i} \,.$$

The recurrences for the genotype frequencies are

$$x_{\pi ij}' = p_{\pi i} p_{\pi j} x_\pi \, [1 + V (K_{\pi ij} - \Sigma \bar{\gamma}_{\pi ij,\phi} x_\phi)] \,, \qquad (5)$$

where $x_{\pi ij}$ $(i = j)$ is half the frequency of the heterozygotes.

The Interplay Between Genetics and Ecology

The recurrence Eqs. (3) and (4) are investigated by analytical or numerical methods to describe the interplay of gene effects, genotypic composition, population sizes, and number of species. The fitnesses in our model are frequency and density dependent, so there is some interplay between genetical and ecological parameters. We will try to specify these interactions and the constraints that they place on each other. Recall that the evolutionary process in this model is determined by the variation in two components, carrying capacities and competition coefficients. Therefore, equilibria express a balance between the evolutionary forces originating from these two opposing components.

Matessi and Jayakar (1981) gave a simple characterization of the stable coevolutionary equilibria in a mathematically equivalent model on exploitative competition in genetically variable species. They found these equilibria to be the maxima of the function

$$\Gamma = \frac{\left(\sum_\pi \sum_i \sum_j K_{\pi ij} x_\pi p_i p_j\right)^2}{\sum_\pi \sum_i \sum_j \sum_\phi \sum_k \sum_l \gamma_{\pi ij,\phi kl} \, p_k \, p_l \, x_\phi \, p_i \, p_j \, x_\pi}$$

and they could deduce local stability of the equilibria from the Lyapunow theorem.

The results of Ikeda and Siljak (1980) and others show that complex Lotka-Volterra competition models of higher dimensions may exhibit chaotic or cyclic behavior.

However, we will here concentrate on properties of the model close to evolutionary or coevolutionary equilibria. We will use the term n species interspecific coevolutionary equilibrium to describe an equilibrium where n species in the community are at a monomorphic equilibrium. Interspecific evolutionary equilibria shall describe situations where the species under consideration is at an equilibrium location due to the biotic environment, i.e., the locations of the competitors in a given resource spectrum.

What Does the Ecology Do to the Genetics?

The mode of selection is determined by the environment of the species under consideration, i.e., by the location of the species' utilization functions in relation to its "optimal" locations as determined by the resource availability, the location of the utilization functions, and the densities of the competing species. Evolution in each species should be evaluated by assuming that the external environment is constant over a generation.

The mode of evolution by selection has the simplest characterization in the pure intraspecific competition model (Christiansen and Loeschcke 1980a,b, see also Matessi and Jayakar 1976). If the species exploits resources far from the resource optimum, then directional selection will favor the genotype with the utilization mean closest to the resource optimum. The evolutionary process becomes a process of allele substitution and polymorphism is a transient state.

If the goal of this process is reached, i.e., if the resources become exploited close to the resource optimum, then the evolutionary processes become crucially dependent on the resource width, σ, relative to the utilization width, W. If the relative resource width is small, i.e., if $\sigma/W < 1$, then stabilizing selecting will result and the least deviant genotype on both sides of the resource optimum will be favored. This situation will eventually lead to monomorphism for the allele with the smallest contribution. If $\sigma/W > 1$, then disruptive selection will result. The most extreme genotypes will be favored and the variance will increase. If we consider more than one locus affecting the character, then polymorphism exhibiting high linkage disequilibrium will prevail up to medium recombination rates (Loeschcke and Christiansen 1984).

These results in general extend to communities with more species, as the critical assumption for the pure intraspecific model was that the species under consideration has come to an equilibrium concerning extrapopulational forces as intraspecific competition. As evolutionarily stable coexistence of two or more species is not possible for $\sigma < W$, we consider in the following only situations where the resource width is exceeding the utilization width. Again we expect directional selection and allele substitution to occur, if the species under consideration is exploiting the resources away from its equilibrium location as given by the location of the competitors in a given resource spectrum. The nonequilibrium location of a species might result from the invasion of a new competitor into the community, as communities consisting of only few species are even at their coevolutionary equilibria very vulnerable to invasion (Loeschcke 1983). Thus, successful invasion events can lead to a decrease in the amount of genetic variation in the resident species, as it changes the mode of selection from disruptive to directional selection.

Evolution in the second stage, where the species has come close to its equilibrium location, is again characterized through niche expansion, i.e., through variance increase by disruptive selection.

Only if the species under consideration is at a rather marginal position with respect to the resource spectrum and the whole system is not at an coevolutionary equilibrium, then stabilizing selection might also occur for $\sigma > W$ (Fig. 1).

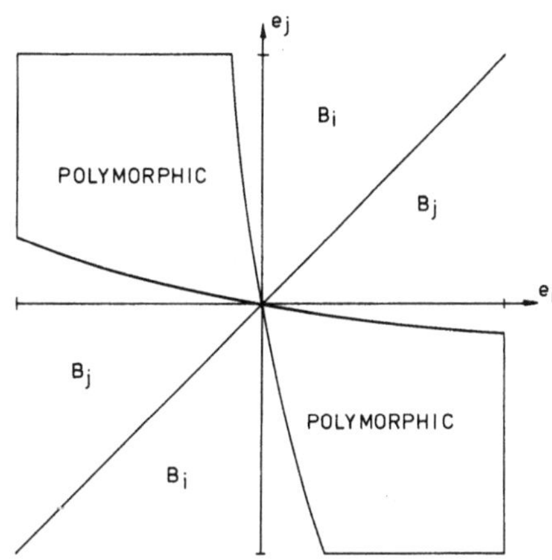

Fig. 1. Competition between two species with nonsymmetric locations of the competitors (relative resource width $\sigma/W = 1.5$). Species *1* is monomorphic with mean location $D_1 = -0.5$. The fate of genetic variants is considered in species *2* which is assumed to be close to its evolutionary equilibrium location $D_2 = 3.27$ with respect to the location of species *1*. Species *2* will be monomorphic for alleles B_i or B_j with contributions e_i and e_j or polymorphic for both alleles as indicated in the Fig. ($e_i = 0$ and $e_j = 0$ correspond to $D_2 = 3.22$). Evolution will favor the allele with the smaller contribution and, therefore, eventually lead to monomorphism, although $\sigma/W > 1$

The Build Up of Variation in a Central Species

Consider a three species system at its interspecific coevolutionary equilibrium (Fig. 2) and allow genetic variation in the central species. It can be shown that a second allele can always increase in the central species. If we assume no variation in the outer species, i.e., if their mode of resource utilization cannot react to the variance increase of the central species, then the variance in the central species will increase until an equilibrium situation with genetic variation is reached (Fig. 3). The population mean of the central species at this equilibrium with variation is at the resource optimum, while the population mean is different from, but close to, the resource optimum during the build up phase.

As the population size of the central species increases with its build up of the variance, the average competition experienced by the outer species will also increase. Thus, the optimum location of the outer species will change and be further away from the resource optimum.

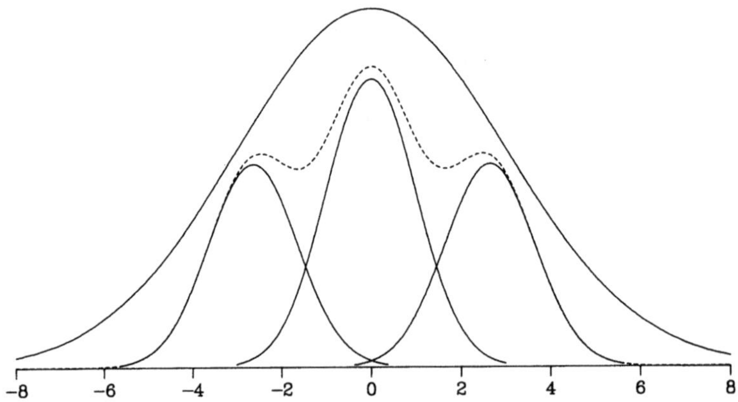

Fig. 2. The resource utilization at the three species interspecific coevolutionary equilibrium for a relative resource width $\sigma/W = 3$. The relative population sizes are shown as the heights of the utilization functions. The *dashed line* represents the sum over all weighted utilization functions, i.e., the resource utilization of the whole community. The difference to the resource availability (*straight line*) illustrates the unexploited resources

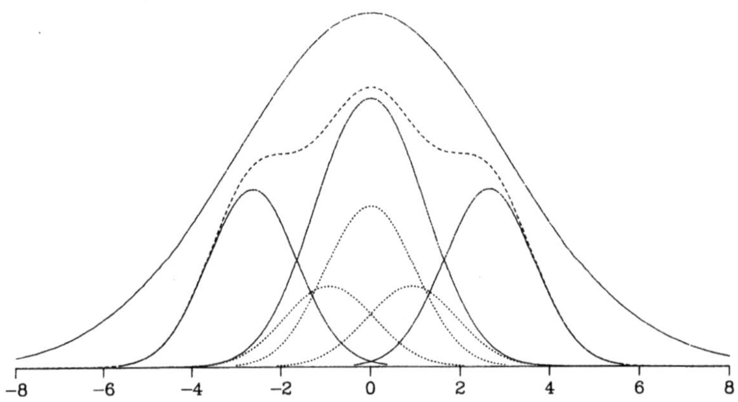

Fig. 3. The same as Fig. 2 with genetic variation in the central species. The mean locations of the genotypic utilization functions (*dotted lines*) in the central species refer to the evolutionary equilibrium after new variants have been introduced in the central species at the interspecific coevolutionary equilibrium (Fig. 2)

The Build Up of Genetic Variation in a Marginal Species

Let us consider a two species system and describe its properties by computer iterations. At the interspecific coevolutionary equilibrium as well as at the interspecific evolutionary equilibria, new alleles with contributions in the direction of the margin of the resource spectrum will increase in frequency. Alleles with more extreme contributions will be favored in the process until the variance has reached its equilibrium value.

Alleles with contributions that place their carriers closer to the competing species will only be favored if their contributions are small relative to the mean distance to the resource optimum of the species at the evolutionary equilibrium. The mean of the population will move away slightly from the competitor during this process and genotypes with utilization means closer to the utilization mean of the monomorphic species will be at high frequency (Fig. 4). As selection is weak, genotype frequencies will be close to Hardy-Weinberg proportions.

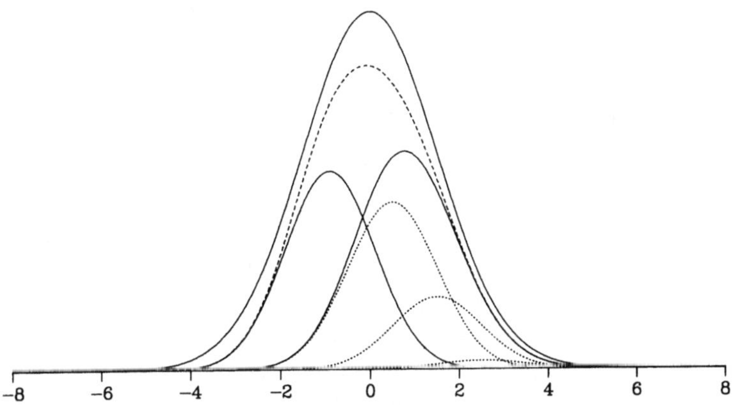

Fig. 4. Equilibrium genotype densities in one of the species in a two species competitive situation for a relative resource width of $\sigma/W = 1.5$. The equilibrium densities of genotypes are represented by the heights of the utilization functions (*dotted lines*). Species *1* is located at the two species interspecific coevolutionary equilibrium location with $D_1 = -0.9$ and relative population size $x_1 = 0.55$. In species *2* there are two alleles segregating with contributions $e_1 = -0.2$ and $e_2 = +0.8$ defined as deviations from the two species interspecific coevolutionary equilibrium location $D_2 = 0.9$. The mean location of species *2* is $D_2 = 0.96$ and the equilibrium relative population size is $x_2 = 0.67$ (*dotted line*). The utilization means of the genotypes in species *2* are $D_{211} = 0.5$, $D_{212} = 1.5$, and $D_{222} = 2.5$; and the genotype frequencies are $x_{211} = 0.67$, $x_{212} = 0.30$, and $x_{222} = 0.03$. The sum of the utilization functions of both species is the *dashed line*

What Does Genetics Do to the Ecology?

Genetic variation is a prerequisite in this type of exploitative competition model for shifting the means of the utilization functions along the resource axis. In the additive one locus model, we are justified in neglecting the genetic basis of evolutionary change in the description of the evolutionary process as long as this process is one of allele substitution (directional selection). However, if the variance within a species increases by the incorporation of a new allele (disruptive selection), the genetic variance might influence the invasion of further species and thereby the composition of the community. Similarly, the genetic basis is essential, if we drop the assumption of additivity of the allele contributions (Christiansen and Loeschcke 1980b, Christiansen 1983) or if we consider more complex genetic systems.

The genetic variation in a resident species might cause it to be pushed by competition to the margin of the resource spectrum and opened up for the invasion of a new species. If a new invasion happens to be successful, the resident species might become vulnerable to extinction due to low population size.

The matching of the resources cannot be achieved by incorporation of genetic variants at one locus alone (Fig. 3). The condition that at most two alleles can be maintained at equilibrium (Christiansen and Loeschcke 1980a), is valid up to rather large allele contributions and can be generalized to intermediate dominance (Christiansen and Loeschcke 1980b) and for multispecies systems with genetic variation.

We have assumed that the character under consideration is a quantitative character, so the allele contributions of a single locus should be of restricted magnitude. A more realistic description should be based on the assumption of many genes contributing to the character (Bulmer 1974, Slatkin 1980, Lande 1976, 1982). However, we feel that it is equally important to describe the evolutionary implications of the evolution of a character on the structure of genetic variation in simple population genetic models. The model can easily be extended to more loci ((Loeschcke and Christiansen 1984) and still keep track of the segregational process. For a set of closely linked loci, there will be a limited number of gametes segregating in the population. Thus, a perfect matching of the resources spectrum cannot be expected simply by assuming several loci with each having a small contribution to the character. A smooth matching of the resources requires first of all a certain number of species which depends on the relative resource width. If the number of species is so large that the invasion of further species is no longer possible (e.g., the situation depicted in Fig. 5), then disruptive selection within the single species will provide a further improvement of the matching of the resources.

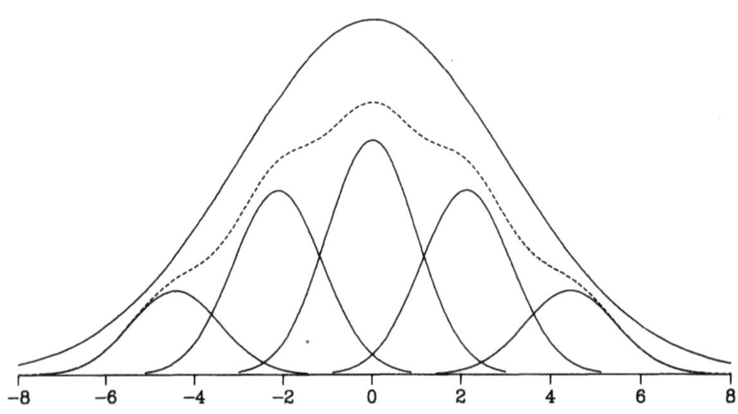

Fig. 5. Five species interspecific coevolutionary equilibrium locations and equilibrium relative population sizes (heights of utilization functions) for a relative resource width of $\sigma/W = 3$. The *slashed line* represents the sum over all utilization functions, i.e., the resource utilization of the whole community

Species Versus Genotypes

From initial increase arguments, species invasion and incorporation of new alleles can be considered as dual aspects in this exploitative competition model (Loeschcke

1983). The crucial condition that allows coexistence of two species as well as protected two allele polymorphism is that the resource width is larger than the utilization width. If the evolutionary process is considered further, then a trivial, but essential and often neglected difference between species and genotypes has to be taken into account. Species behave as independent units in the evolutionary process, whereas genotypes are the product of segregation.

The equilibrium pattern in the pure interspecific model does not resemble that of the pure intraspecific model. At equilibrium the utilization means of the genotypes differ from the utilization means of the competing species (Figs. 6 and 7). Figure 6 describes the evolutionary equilibrium situation in a three species system as a function of the relative resource width, where at the interspecific coevolutionary equilibrium, the variation in the central species has come to its equilibrium configuration. Figure 7 shows the five species interspecific coevolutionary equilibrium location as a function of the relative resource width. These equilibrium locations differ from that of the three species system, where the marginal species are assumed to be at their equilibrium location with respect to the variation in the central species. The latter situation is not illustrated by a figure, since for large relative resource widths, the allele contributions will in that case exceed the limiting value that allows a third allele to increase in frequency. Genotypes and species are, therefore, not interchangable in the process concerning the matching of the resources, i.e., further invasion of species or incorporation of new alleles are not alternatives.

Discussion

Evolutionary changes are at first approximation often decomposed into changes that occur in ecological time and into changes that occur in evolutionary time. But even if

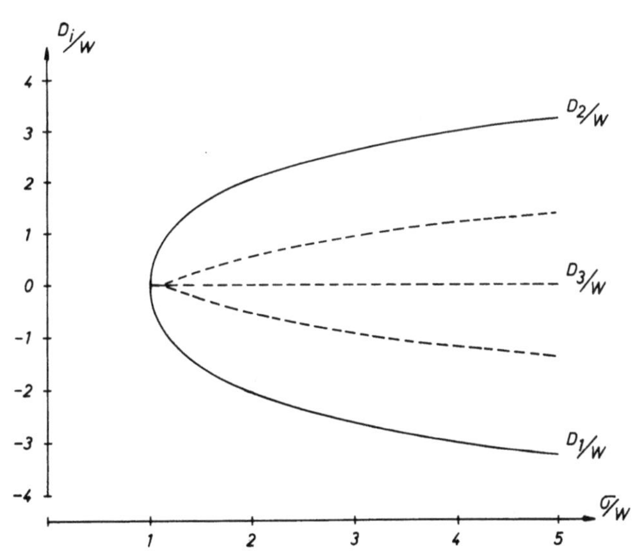

Fig. 6. Equilibrium locations of the means of the utilization functions of the three genotypes in the central species (*slashed lines*) in a three species system as a function of the relative resource width. The two outer species without variation are at the three species interspecific coevolutionary equilibrium locations (*black lines*)

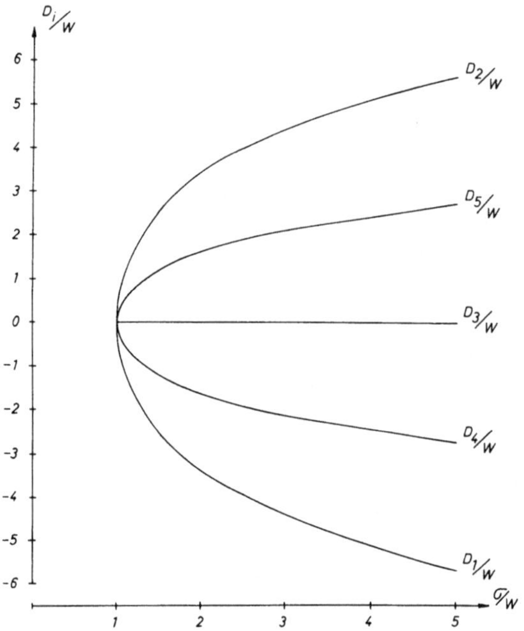

Fig. 7. Five species interspecific coevolutionary equilibrium locations, D_i/W, as functions of the relative resource width, σ/W

these differences in time scales in most cases do apply to the processes described in population genetics on the one hand and in evolutionary ecology on the other hand, some interesting interplay occurs between these processes.

We have concentrated here on the specification of some of these interactions by a simple population genetics model on the evolution of a character under exploitative competition. In our model the mode of selection is determined by two components: resource availability and competition. Equilibria are then the outcome of a balance between the evolutionary forces originating from these two opposing forces.

That opposing forces are determining evolutionary processes seems us to be a quite general phenomenon. Average selective advantage concerning one selection component will often be accompanied by a disadvantage concerning another component. Therefore, we expect our results to be structurally similar to those from other models based on evolutionary forces of opposing directions.

The basic conclusion of our investigation is that the interplay of genetics and ecology is essential in the evolutionary process and cannot be characterized sufficiently by optimizing principles. A simple characterization of only the goal of the evolutionary process does not reveal the particular interactions of genetics and ecology during this process.

The pattern of genetic variation is crucially dependent on the composition and on the history of the community. The genetic variation for the character will be subject to directional selection, if the species under consideration is not at its optimal location in relation to its competitors. The nonoptimal location might result from a recent invasion event in the community. Such invasion events are likely to occur in communities of few, not densely packed, species (Loeschcke 1983). After successful invasion, resource availability and competition pressure are modified for the resident species, thus, resulting in new optimal locations for these species. This induces directional selection. Polymorphism will be transient in this situation finally leading to monomorphism for the allele with the largest contribution in direction of the optimal location.

This result extends to multiple loci, so that we expect reduced genetic variation in those species in a community that have not yet come close to their evolutionary equilibrium location. Thus, our model predicts consequences for the pattern of genetic variation by a disturbance of a coevolutionary equilibrium situation that may persist in evolutionary time.

Resently, Jain (pers. commun.) has observed such a negative correlation between coexistence and the level of genetic variation in *Avena* and *Allocarya*. He calls this phenomenon "variation displacement."

Acknowledgements. Comments and criticism to the manuscript by Freddy Christiansen, computational help in preparing some of the figures by Hans Siegismund, and the drawing of others by Arno Jensen are gratefully acknowledged.

References

Boag PT, Grant PR (1981) Intense natural selection in a population of Darwin's finches (Geospizinae) in the Galapagos. Science 214:82–84

Bulmer MG (1974) Density-dependent selection and character displacement. Am Nat 108:45–58

Christiansen FB (1983) The definition and measurement of fitness. In: Shorrocks B (ed) Evolutionary ecology, BES Symp 23. Blackwell, Oxford, pp 65–79

Christiansen FB, Fenchel T (1977) Theories of populations in biological communities. Springer, Berlin Heidelberg New York

Christiansen FB, Loeschcke V (1980a) Evolution and intraspecific exploitative compoetition. I. One locus theory for small additive effects. Theor Popul Biol 18:297–313

Christiansen FB, Loeschcke V (1980b) Intraspecific competition and evolution. In: Barigozzi S (ed) Vito Volterra Symposium on mathematical models in biology. Lecture notes in biomathematics 39. Springer, Berlin Heidelberg New York, pp 151–170

Dempster ER (1955) Maintenance of genetic heterogeneity. Cold Spring Harbor Symp Quant Biol 20:25–32

Fenchel TM (1975) Character displacement and coexistence in mud snails (Hydrobiidae). Oecological (Berlin) 20:19–32

Fenchel TM, Koefod LH (1976) Evidence for exploitative interspecific competition in mud snails (Hydrobiidae). Oecologia (Berlin) 14:317–326

Ikeda M, Siljak DD (1980) Lotka-Volterra equations: Decomposition, stability, and structure. J Math Biol 9:65–83

Lande R (1976) The maintenance of genetic variability by mutation in a polygenic character with linked loci. Genet Res 26:221–235

Lande R (1982) A quantitative genetic theory of life history evolution. Ecology 63:607–615

Levene H (1953) Genetic equilibrium when more than one ecological niche is available. Am Nat 87:331–337

Loeschcke V (1983) Species diversity and genetic polymorphism: Dual aspects of an exploitative model. In: Jayakar S, Zonta L (eds) Evolution and the genetics of populations. Atti Ass Genet Ital XXIX (Suppl.), pp 53–66

Loeschcke V (1984) Begrenzte Ähnlichkeit und koevolutionare Gleichgewichte unter Konkurrenz: Ein Modell. Verh Ges Ökol XII (in press)

Loeschcke V, Christiansen FB (1984) Evolution and intraspecific exploitative competition. II. Two locus theory for additive gene effects. Theor Popul Biol 26 (in press)

Matessi C, Jayakar SD (1976) Models of density-frequency dependent selection for exploitation of resources. I. In: Karlin S, Nevo E (eds) Population genetics and ecology. Academic Press, London New York, pp 707–721

Matessi C, Jayakar SD (1981) Coevolution of species in competition: A theoretical study. Proc Natl Acad Sci USA 78:1081–1084

Roughgarden J (1972) Evolution of niche width. Am Nat 106:683–718

Slatkin M (1980) Ecological character displacement. Ecology 61:163–177

Coevolution at Two Trophic Levels

S.D. JAYAKAR[1]

Introduction

Most of the classical populations genetics theories have till today dealt with models regarding genetic changes in a single species as a result of its interactions with its physical environment. Models dealing with competition between species have received some attention (Roughgarden 1979, Matessi and Jayakar 1980, Christiansen and Loeschcke 1980). It is clear, however, that any serious theoretical consideration of evolutionary mechanisms must take into account the coevolution, besides its competitors, also of its prey, its predators, and its parasites. Several models of gene frequency changes of such interactions have been proposed, but they suffer from the drawback that they deal only with gene frequencies and not with population sizes or densities (Mode 1958, Leonard 1977, Lewis 1981, Jayakar 1970). Roughgarden (1979) proposed a general ecological model of species interactions and considered several particular examples. Population sizes must be an essential feature of models involving two trophic levels since the possible extinction of food sources (or hosts) must govern the choice of organisms on which to feed or parasitize.

Clearly such models introduce a higher level of complexity both in the sense that they increase the dimensionality of the system and in the sense that there are more kinds of parameters involved. A complete model of this kind would include: (1) parameters of the population dynamics of each separate type (genotype/species) at each trophic level; (2) parameters describing competition between types within each level; (3) parameters describing the interaction of each type at one level with each type at the other. It is evident then that a general analysis of such a model even if it were feasible, would make little sense in that any interpretation would be very sensitive to small changes in parameters. The most general model of this kind that can be handled is probably with two (or at most three) types at each level. Such models could on the other hand be of some use in particular situations where one wishes to examine the evolutionary predictions of a model. Such situations could involve, for example, evolutionary strategies of food resource situations or parasitization strategies.

In the present paper, we will in fact present the general model and give an application of the model to a problem of the evolution of specialists/generalists with respect to a choice of food sources.

1 Department of Genetics and Microbiology, University of Pavia, 27100 Pavia, Italy

Population Biology and Evolution
Ed. by K. Wöhrmann and V. Loeschcke

The General Model

The situation which is being modelled here concerns a set of "types" at one trophic level which feed on a set of "types" at a lower trophic level. Throughout this paper, the world "types" is used to include either species or haploid genotypes, or genotypes within a diploid species. In the most general model we will have a number of diploid species and the genotypes at one segregating locus within each species. We assume that each type at the higher trophic level is completely dependent on the set of types considered at the lower level (also called the resources) so that in the absence of this set, it would die out. Further, all interaction between the species at the higher level is due to the set of resources. The types at the lower level are assumed, in general, to be in competition with each other, but the exact nature of this competition will not be specified.

Let α_i, i = 1,...,m be the m types at the higher trophic level and β_j, j = 1,...,n the n types at the lower. We then have to define interaction parameters between each pair α_i, β_j. These parameters can be defined on biological grounds in any particular situation, but in general, let us assume that a_{ij} is the increase in fitness (decreased probability of mortality) of an individual of species α_i on contact with one of species β_j. This same pair leads to a decrease in fitness of β_j which is measured by the parameter b_{ij} (reproduction rate). The population dynamics of each β_j is assumed to be regulated by a Lotka-Volterra type of competition model. In the absence of all the resources, each α_i would decay at a characteristic rate. The model we consider is in discrete time with the usual assumptions of nonoverlapping generation and random mating. No fitness differences are assumed between types other than those induced by the above assumptions.

If A_i is the density of α_i, then in the absence of resources, the decay in this density from one generation to the next is assumed to be described by:

$$A_i' = A_i d_i, \quad 0 \leqslant d_i < 1 \tag{1}$$

d_i being the exponential decay parameter.

Further, if B_j is the density of β_j, then in the absence of all types at the higher trophic level, the dynamics of the β_j are described by

$$B_j' = B_j \left(1 + r_j - \frac{r_j}{k_j} \sum_i B_i c_{ij}\right), \tag{2}$$

where r_j, k_j are the rate of natural increase and the carrying capacity of β_j, and c_{ij} is the parameter expressing the effect of competition of β_i on β_j.

The interaction of the two trophic levels leads to a modification of Eq. (1) so that

$$A_i' = A_i \left(d_i + \sum_j B_j a_{ij}\right) \tag{3}$$

and of Eq. (2) so that

$$B_j' = B_j \left(1 + r_j - \frac{r_j}{k_j} \sum_i B_i c_{ij} - \Sigma A_i b_{ij}\right). \tag{4}$$

Equations (3) and (4) express the most general form of the model. In any given ecological situation we wish to consider in detail the model that can be modified to adapt to the particular situation. The detailed choice of population regulation of the β_j and of the survival of the α_i will not make any great difference to the analysis of the model.

A general analysis of the model can be formally outlined, but in practice is next to impossible to carry out. The equilibrium solution can be obtained by solving the set of (m + n) equations obtained by putting $A_i' = A_i, i = 1,...,m$ and $B_j' = B_j, j = 1,...,n$ in the Eqs. (3) and (4). This gives

$$\Sigma \hat{B}_j a_{ij} = 1 - d_i$$
$$\Sigma \hat{A}_j b_{ij} = r_j - \frac{r_j}{k_j} \Sigma \hat{B}_l c_{lj} \,. \qquad (5)$$

The model is applied below to investigate the evolution of resource specialization.

The Evolution of Resource Exploitation

Most models dealing with resource exploitation have considered unlimited resources distributed along a single axis. The evolution of competition between the exploiter species under these circumstances has been fairly well studied. The considerations of evolution at both trophic levels on the other hand allows us to investigate a slightly different situation.

Let us consider a number of qualitatively different resources (species) and a community of diploid exploiter species which are genotypically varied in the ability or capacity to exploit the various species of resources. Thus, different genotypes may be able to exploit either a wide range of resources or different specialized resources. Such situations are probably frequent in nature. An example is the community of aquatic isopods feeding on microfungi, though little is as yet known about this coevolutionary trophic partitioning. Some work on these aspects has been carried out by Rossi and Fano (1979) and Rossi et al. (1981) on *Asellus aquaticus* and *Proasselus coxalis*, both on intraspecific and interspecific partitioning. Their results have shown that sympatric populations of these two species have a potential trophic niche which is much broader than in allopatric populations. In general, the microfungi seem to be crucial to these exploiters in their food habits.

More specifically, within each species there seem to be genetic diversity in the capacity of individuals to utilize particular species of microfungi. Furthermore, under certain conditions, Rossi and Basset (unpublished) have found that resource generalists and specialists can coexist.

The following application of the model is addressed to the question of when resource specialists can coexist with resource generalists. The most simple situation of this type is with two species of resources and three phenotypes of a single exploiter species.

Let the two species of resources be labelled ρ_1 and ρ_2 and let their densities be represented by R_1 and R_2. Let X be the density of the exploiter species. Let us suppose that at a given locus, there are two alleles A, a (with frequencies $1 - p, p$), so that individuals of the aa genotype can only utilize the resource ρ_2 while those of the other two genotypes can utilize both ρ_1 and ρ_2. In this particular situation the Eqs. (3) and (4) reduce to:

$$R_1' = R_1 \left[1 + r_1 - \frac{r_1 (R_1 + cR_2)}{k_1} - X (1 - p)^2 b_{11} - 2 Xp (1 - p) b_{12}\right]$$

$$R_2' = R_2 \left[1 + r_2 - \frac{r_2 (R_2 + cR_1)}{k_2} - X (1 - p)^2 b_{21} - 2 Xp (1 - p) b_{22} - Xp^2 b_{23}\right] \tag{6}$$

$$X' = XF$$

$$p' = p [d + (1 - p) (R_1 a_{12} + R_2 a_{22}) + pR_2 a_{23}]/F , \tag{7}$$

where

$$F = [d + (1 - p)^2 (R_1 a_{11} + R_2 a_{21}) + 2 p (1 - p) (R_1 a_{12} + R_2 a_{22}) + p^2 R_2 a_{23}] .$$

Analysis of the Model

Rather than attempt to completely analyze the model, which is feasible but tedious, let us ask two simple questions which are perhaps those which are biologically most interesting:

(1) Under what conditions can both resources survive? (2) Under what conditions can the specialist genotype of the exploiter survive?

In order to do this, we will simply carry out boundary analyses and investigate at the relevant boundaries the conditions for their stability.

We know from competition theory that for ρ_1 and ρ_2 to coexist we must have

$$c < \min \left(\frac{k_1}{k_2}, \frac{k_2}{k_1}\right).$$

Let us consider the boundary $X = 0$ and find the conditions under which X increases when introduced at a small density, in the presence of R_1 alone, R_2 alone, or both coexistent. If only ρ_1 is present, then p must be zero, since the specialist genotype cannot utilize ρ_1. The condition for X to increase when rare is then

$$k_1 > \frac{\delta}{a_{11}}, \text{ where } \delta = 1 - d . \tag{8}$$

Similarly with ρ_2, the condition for X to increase is

$$k_2 > \min \left(\frac{\delta}{a_{21}}, \frac{\delta}{a_{23}}\right), \tag{9}$$

while if ρ_1 and ρ_2 are both present, then we must have

$$k_1 (a_{11} - ca_{21}) + k_2 (a_{21} - ca_{11}) > \delta (1 - c^2) \tag{10}$$

or

$$k_1 (a_{13} - ca_{23}) + k_2 (a_{23} - ca_{13}) > \delta (1 - c^2) . \tag{11}$$

In the former case it will be the genotype. AA which can be introduced and in the latter the genotype aa. We also want to know, however, whether in the presence of ρ_1 and the exploiter species, ρ_2 can be successfully introduced. The condition for this is

$$k_1 k_2 (\gamma_{21} - \gamma_{11}) a_{11} + k_2 \gamma_{11} \delta > k_1 \gamma_{21} c\delta , \tag{12}$$

where

$$\gamma_{ij} = \frac{r_i}{b_{ij}} .$$

The reciprocal condition in order to be able to introduce ρ_1 into a community with ρ_2 and the aa genotype of the exploiter is

$$k_1 k_2 (\gamma_{13} - \gamma_{23}) a_{23} + k_1 \gamma_{23} \delta > k_2 \gamma_{13} c\delta . \tag{13}$$

We can then set down a series of conditions for the establishment of communities including different combinations of the two resource species and the exploiter genotypes.

We turn next to the second problem, namely the one regarding the possible establishment of the specialist. This is possible in the presence of ρ_2 alone if the aa genotype can better exploit this resource, namely if $a_{22} > a_{21}$. It would be interesting, however, to see if one can also have the specialist present in a community with both ρ_1 and ρ_2 present. To do this we must consider the boundary equilibrium $p = 0$, $R_1, R_2, X \neq 0$. The condition that p increases is

$$\hat{R}_1 a_{12} + \hat{R}_2 a_{22} > \hat{R}_1 a_{11} + \hat{R}_2 a_{21} ,$$

where $\hat{R}_1$ and $\hat{R}_2$ are the equilibrium values of the above

$$\begin{aligned} \hat{R}_1 + c\hat{R}_2 &= k_1 (1 - \hat{X}\gamma_{11}) \\ c\hat{R}_1 + \hat{R}_2 &= k_2 (1 - \hat{X}\gamma_{21}) \\ a_{11}\hat{R}_1 + a_{21}\hat{R}_2 &= \delta . \end{aligned} \tag{14}$$

These solutions are in fact given by

$$\hat{R}_1 = \frac{a_{21} k_1 k_2 (\gamma_{21} - \gamma_{11}) + \delta (\gamma_{11} k_1 - c\gamma_{21} k_2)}{a_{11} (\gamma_{11} k_1 - c\gamma_{21} k_2) + a_{21} (\gamma_{21} k_2 - c\gamma_{11} k_1)} , \tag{15}$$

$$\hat{R}_2 = \frac{a_{11} k_1 k_2 (\gamma_{11} - \gamma_{21}) + \delta (\gamma_{21} k_2 - c\gamma_{11} k_1)}{a_{11} (\gamma_{11} k_1 - c\gamma_{21} k_2) + a_{21} (\gamma_{21} k_2 - c\gamma_{11} k_1)} . \tag{16}$$

To get an idea of what kind of conditions this leads to, let us consider a numerical example with the parameters given in Table 1. We can see from these numerical values that the a allele confers a certain advantage to the exploiter with respect to both re-

Table 1. Parameter values used in the numerical example

r_1 = 0.4	r_2 = 0.2	c = 0.1	d = 0.2
a_{11} = 0.0004	a_{12} = 0.0008	a_{13} = 0	
a_{21} = 0.0012	a_{22} = 0.0016	a_{23} = 0.002	
b_{11} = 0.00002	b_{12} = 0.00004	b_{13} = 0	
b_{21} = 0.00006	b_{22} = 0.00008	b_{23} = 0.0001	

source species except that it cannot utilize ρ_1. Secondly, the losses in fitness to the resources has been assumed to be proportional to the advantages gained by the exploiter genotypes which could mean that the differences are due mainly to capturing or feeding efficiency. We examine here the conditions with respect to k_1 and k_2, firstly, because these are standard parameters in competition models and, secondly, because this analysis was proposed by the experimental results of Rossi and his collaborators which suggested that the carrying capacities of the resource species were in fact crucial to the specialist.

Figure 1 shows how the conditions laid down above translate into the k_1, k_2 space. There are several interesting things to be seen in the figure. The most important fact is that in the subspace labelled (1, 2, 3, 4) all the different types are present; in other words, there is a polymorphism for feeding types of the exploiter. Secondly, in the space labelled (1, 2, 4) only the specialist genotype of the exploiter is present. Thus, depending on the values of k_1 and k_2 various types of communities can be maintained.

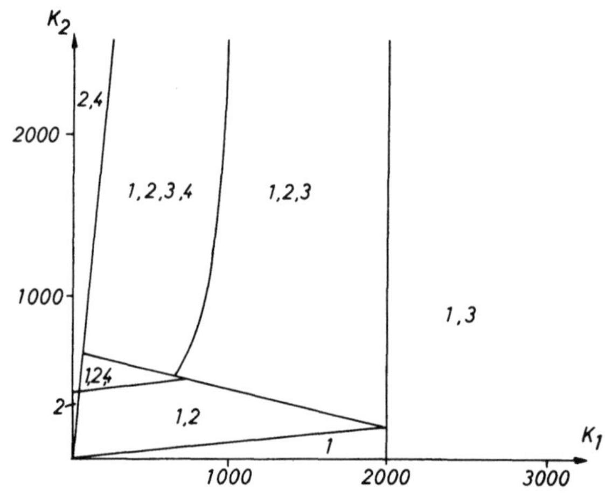

Fig. 1. Subdivision of the k_1, k_2 parameter space into regions in which the different types of resources and utilizer genotypes can coexist (*Key: 1* resource ρ_1; *2* resource ρ_2; *3* utilizer allele A; *4* utilizer allele a). Of particular interest are the areas in which the aa genotype (the specialist) can maintain itself

Discussion

Rossi and his collaborators have as yet unpublished data on two species of aquatic isopods *Asellus aquaticus* and *Proasellus coxalis* which feed on several different species of microfungi. They find that there are differences in feeding preferences within each of the species and they also have evidence for these preferences being genetic. Their most important finding is that the compositions of the species according to feeding types are different under different conditions. In each species, under conditions in which the population densities are maintained low, as in nature, by high predation by fish, specialist feeders are found in the population. When on the other hand, in the laboratory, predation is reduced and population densities increase, then the specialists are eliminated. Again in the laboratory, if predation is once more increased, the specialists can be maintained. Thus, the presence of specialized feeders depends crucially on low densities, in other words, of high availability of resources.

These findings have stimulated the study of the model discussed in this paper. From the results obtained, it seems that not only does the evolution of specialization depend on resource availability, in general, but also on the relative abundance of the different resource types. Here, of course, k_1 and k_1 are interpreted as resource availability relative to population densities of the exploiter species.

A more detailed investigation would be necessary to study more deeply the relationship between experimental data on specialized feeders and the model suggested above. What has been done here is merely to suggest a method for such a study. However, the numerical example does provide an idea of the kind of qualitative results that might be expected of such a study.

Acknowledgements. This work has been supported by the Istituto di Genetica Biochimica ed Evoluzionistica, CNR, Pavia

References

Christiansen FB, Loeschcke V (1980) Intraspecific competition and evolution. In: Barigozzi C (ed) Vito Volterra Symposium on mathematical models in biology. Lecture notes in biomathematics 39. Springer, Berlin Heidelberg New York

Jayakar SD (1970) A mathematical model for the interaction of gene frequencies in a parasite and its host. Theor Popul Biol 1:140–161

Leonard KJ (1977) Selection pressures and plant pathogens. Ann NY Acad Sci 287:201–222

Lewis JW (1981) On the coevolution of pathogen and host: I. General theory of discrete time coevolution; II. Selfing hosts and haploid pathogens. J Theor Biol 93:927–983

Matessi C, Jayakar SD (1980) Models of density-frequency dependent selection for the exploitation of resources. II. In: Barigozzi C (ed) Vito Volterra Symposium on mathematical models in biology. Lecture notes in biomathematics 39. Springer, Berlin Heidelberg New York

Mode CJ (1958) A mathematical model for the coevolution of obligate parasites and their hosts. Evolution 12:158–165

Rossi L, Fano ED (1979) Role of fungi in the trophic niche of the congeneric detritivorous *Asellus aquaticus* and *Proasellus coxalis*. Oikos 32:380–385

Rossi L, Basset A, Marchetti E (1981) Predazione ed evoluzione della nicchia trofica in cinque specie di detrivori bentonici. Boll Zool 48:97–98

Roughgarden J (1979) Theory of population genetics and evolutionary ecology: an introduction. Macmillan, New York

Comments on Models

Models in Population Genetics

S.D. JAYAKAR[1], V. LOESCHCKE[2], and G. DE JONG[3]

During the discussions at this meeting, it became evident that there was a profound misunderstanding on the part of some experimental and field workers in evolutionary ecology: firstly, as to what the conceptual bases of population genetics and evolutionary models are and, secondly, as to what kinds of conclusions can be drawn from such models. Consequently, we felt that it would be a contribution toward mutual comprehension between theoretical geneticists on the one hand and experimental and observational biologists on the other, if we could briefly set out the bases, the assumptions, the methods, and the goals of evolutionary genetics models. The following is an attempt on the part of a few of those present at Oberjoch to fill this need.

The activity of modelling in any context is to select an object of some complexity and to make a more simplified object which includes and sometimes stresses some of the features considered essential for a given purpose. A dummy in a shop window reproduces only the external anatomy of the human figure, but for the purpose of exhibiting clothes it is adequate. This means that several features of the object being modelled are being ignored either because they are considered of minor importance with respect to the purpose at hand or because they are too complex to be modelled and one simply hopes that they are not essential to the problems being addressed. So it is with population genetics models.

All models in population genetics are attempts at a mathematical description of a biological problem. They have the following essential features in common. They describe the population under consideration as absolute or relative frequencies of different genetic categories at a certain stage of the life cycle. These frequencies are then acted upon by biological processes, such as reproduction and mortality, which can be specific to each category and governed by biological phenomena, such as the laws of Mendel and Morgan, and mutation. During a single cycle, as a result of these various phenomena acting on them, the array of frequencies undergoes change. Given a numerical description of the various phenomena one can then write down the change in the frequency array in the course of a generation. The mathematical description of such a change is as follows:

$$f_i' = g_i(f_1,\ldots,f_n \mid \pi), \quad i = 1,\ldots,n,$$

1 Department of Genetics and Microbiology, University of Pavia, 27100 Pavia, Italy
2 Institute of Ecology and Genetics, University of Aarhus, 8000 Aarhus, Denmark
3 Department of Population and Evolutionary Biology, University of Utrecht, 3508 Utrecht, The Netherlands

Population Biology and Evolution
Ed. by K. Wöhrmann and V. Loeschcke

where $f_1,...,f_n$ are the initial frequencies of the n categories, $f_i',...,f_n'$ the frequencies after one generational cycle, π is the set of parameters defining the various reproductive rates, mortalities, etc., and g_i is a function which describes the effects of these parameters and of the laws of Mendel and Morgan and those of muation rates. An alternative formulation could be the description of instantaneous change, namely:

$$\frac{df_i}{dt} = g_i(f_1,...,f_n | \pi), \quad i = 1,...,n.$$

Clearly, when we write down a system we are selecting in the form of g_i particular features of the biology of the organism which we consider relevant. The choice may in any specific model be inadequate or unrealistic, but the model is valid for those choices, and so are the predictions valid under the constraints imposed by the model.

In discussing the main objectives that one wishes to achieve from the construction and analysis of mathematical models of evolution, we wish to divide the models into two types with somewhat different scopes: (1) models in which the ecological conditions are not rigidly defined, but where a certain general situation is envisaged; (2) models which apply to a very particular ecological situation of which examples have been observed in nature.

Evidently the aims for which the two types of models are studied and the questions we ask of them are very different. In the first kind of model, we are looking for rather general predictions regarding the evolutionary future of the population and for answers as to the kinds of relationships between parameters which govern the different possible evolutionary outcomes. In the second kind, we are looking for rather precise conditions for the existence or the coexistence of phenotypes and/or species. Further, it is possible in this latter kind of model to try to estimate parameters from field data. In fact, we are often in the situation where, in apparently similar field conditions, we have observed a qualitatively or a quantitatively different ecological composition of members and we wish to know why this is so – what parameters are relevant to these differences and what kind of ranges of these parameters would lead evolution to one or the other consequence, i.e., what are the domains of attraction? For both kinds of models, one of the common questions asked is – is genetic polymorphism maintained and under what conditions? This is because it is considered of some importance to a population to maintain genetic variation.

There should obviously be interaction between experimental and observational biologists on the one hand and theoretical workers on the other so as to make use of the results of models in planning field or laboratory work and vice versa to use field and laboratory data to improve models both in the more general kinds of models with respect to the assumptions themselves or in the more particular kind with respect to estimates of parameters. A comparison of observed and predicted dynamics and/or end results is the test of the goodness of the model.

In classical models of population genetics, the key parameters are "fitness coefficients" or "selection coefficients." The concept of these coefficients comes of course from Darwin's formulation of the theory of natural selection. Surprisingly, there are still serious biologists who are suspicious of these fitness coefficients. From what we gathered at Oberjoch, they criticize the fact that a fitness coefficient is a number attached to a *genotype*, whereas what happens out in nature is a complex interaction between *phenotypes* and between them and the environment. Perhaps it is unfortunate that this dichotomy of terminology (and only terminology) has led to incomprehension. But when in a model we refer to the fitness of a genotype, we do not imply in any way that (for a diploid organism) it is a property, governed in any fixed and unchangeable manner, of the two alleles which combine to make the genotype. That this terminology is indeed a short-hand, an abstraction, should be clear from the fact we even talk of the average fitness of alleles! So what is intended by the coefficient of fitness of a genotype? For simplicity let us consider viability selection alone. In a given population, in a given environment, if we consider all the individuals belonging to a given genotype, and average the contribution made by these individuals relative to that of all the individuals in the population, this is defined as the coefficient of fitness of the genotype. It is understood that there is a correlation (which can be more or less close) of the genotype with phenotypic characters, and any selective effects concern the phenotypic character or indeed more than one character. It is a simplification in that we are neglecting other factors that may influence this average, for instance, the genotypic background of which the composition will almost certainly change in time through selection. Further, we are combining various causes of selective differences into a single coefficient. The fitness is an average over the whole complex of interactions of background genotype, genetic and phenotypic correlations, pleiotropy, etc. We are neglecting such complications and hoping they do not grossly affect our predictions when we associate a fitness coefficient with a genotype. The concept is fundamentally a statistical one. The fact that the mental processes described above are not reiterated every time a model is presented does not mean that they are being ignored. Again for the above reasons, confusion and incomprehension permeate discussion regarding the unit of selection.

To defend this methodology and approach, let us cite Haldane (1964): "The existing theories of population genetics will no doubt be simplified and systematised. Many of these will have no more final importance than a good deal of nineteenth century dynamical theory. This does not mean that they have been a useless exercise of algebraical ingenuity."

Clearly, the models of population genetics have improved and will improve as more biological facts are cleared up. The information on the relationship between genotype and phenotype, and consequently, that between genotype and the selection coefficient, is still not very detailed and this is one of the limitations of population genetics models

Maximization principles independent of genetics have recently become popular in theoretical evolutionary biology. It does not seem to us an approach that is in any way an alternative to population genetics models. Even when genetics is invoked, care has to be taken when applying maximization principles, such as the Fisher Fundamental Theorem, that are based on the assumption of independent effects of loci and are, therefore, severely limited. If the "goal" of an evolutionary process is known and one constructs a model that gives a causal "explanation" of the phenomenon, it does not mean that this is the only explanation. Such maximization principles run the risk of being teleological or goal-oriented in nature.

Dawkins (1982) paraphrases population genetics in a most misleading way when he says: "To the extent that it is useful for us to think of individuals working as if to maximise fitness, we may, with precisely the same licence, think of genes as if they were striving to maximise their survival." Population geneticists do not think in this way, fortunately. Their view of selection is much more mechanistic, unoriented, and impartial. It is at the least uncritical to distort the notions of population genetics in order to justify teleological concepts, such as that of the selfish gene.

A final consideration. Perhaps this article and much discussion regarding the usefulness of population genetics could be avoided if text books on general genetics and population genetics discussed in some detail the problems, the assumptions, and the concepts involved in population genetics models before even stating the Hardy-Weinberg result, far less any results regarding effects of selection or mutation. It is essential that students and teachers who read such books know exactly what these results imply for the understanding of evolutionary theory.

References

Dawkins R (1982) The extended phenotype. Freeman, San Francisco
Haldane JBS (1964) A defense of beanbag genetics. Perspect Biol Med 7:343–359

Models in Evolutionary Ecology

S.C. STEARNS[1]

Introduction

This short chapter represents a reworking of comments made in an evening discussion and tape-recorded. Any such discussion moves from item to item, and the style differs markedly from that of a scientific paper. I have tried to preserve the tone of the remarks while making the changes required for good grammar, and I have reorganized the material to fit into a series of subject headings. I have also added a short conclusion that was not part of the discussion.

The Central Problem

The general problem facing every evolutionary biologist can be stated in a single sentence: natural selection operates on the phenotype, but only the genes record the history of selection. Population biologists have not yet been able to deal with those two questions simultaneously. The processes at each of those levels are so inherently complex that it takes the full attention and the whole life of a working scientist just to make progress on one of them. Thus, the intermediate detail, the mechanism that links them, is left out of the picture. Population biology has not yet incorporated the mechanisms of development and of physiology that link the genotype and the phenotype. I do not accept quantitative genetics as a substitute, although it is a good try, because it is statistical inference based on the phenotype and because it does not have mechanism in the sense that I understand mechanism.

Here is the current situation in population biology. On the one hand, we have had an intellectual tradition, called the Neo-Darwinian or modern synthesis, in which population genetics played a large organizing role. It showed that many kinds of information were consistent with an interpretation of evolution based on the Mendelian mechanism. It did not necessarily contain assumptions about how rapidly or slowly evolution proceeded, but it did emphasize that we could understand evolution as changes in gene frequency, and that changes in gene frequency were described by the equations of population genetics. We have had a long tradition of that, and it is an impressive and beautiful body of knowledge. However, if left out the mechanisms of

1 Institute of Zoology, University of Basel, 4051 Basel, Switzerland

Population Biology and Evolution
Ed. by K. Wöhrmann and V. Loeschcke

selection operating on whole organisms, which have been the thrust of research in evolutionary ecology. The two trains of thought are not antithetical to each other, but complementary in an important sense. The purpose of this symposium is to see how closely they can be linked up.

Genetics is Not Necessary for Ecology

Neither population nor quantitative genetics is logically necessary for any of the major predictions of evolutionary ecology that have been confirmed experimentally or in field work. They are consistent with those predictions, but not a necessary part of them. Evolutionary ecology has involved a lot of people working very hard for 10 or 15 years to test fairly new ideas. These tests have resulted in the integration of new with old observations over many taxa. This has been done without having to appeal to the apparatus of Mendelism.

This does not mean that evolutionary ecology is inconsistent with Mendelism. It means we are now faced with an exciting problem in the long run – how could evolutionary ecology get away with this trick? If in fact gene frequency change is the stuff of evolution, then there must be some important intermediate filter between the genotype and the phenotype that has made it possible for us to make successful predictions using optimality theory in evolutionary ecology. We may be getting the right answers for the wrong reasons; that is always a possibility. Nevertheless, one has to admit that there have been impressive successes. Some concern predictions of age at maturity. These essentially are predictions about the integration of the phenotype, the coevolution of some traits with others. The other successful predictions come from sex allocation theory and deal with how effort should be allocated to male vs female function. Again, these predictions can be made by optimality models, they function quite well, and they predict surprising observations that have been confirmed experimentally – for example, the work by Charnov and van den Assem on sex ratio evolution in the Hymenoptera or by Werren on blowfly parasites.

Comments on Fitness

Within the broader community of evolutionary biologists, population geneticists are distinguished by their unique conviction that they know what fitness is. To a population geneticist, fitness is a technical measure describing selection on whole organisms whose sole function is to predict changes in gene frequencies. Population geneticists usually think of fitness as viability or fecundity, sometimes as their product. So long as gene frequency change is the only problem to be solved, such a definition seems reasonable. However, other specialties within evolutionary biology have other questions on their agenda. For example, the goal function in optimality theory is a fitness definition. In this context, fitness is not being used to define changes in gene frequencies, but to understand the evolution of the equilibrium properties of the phenotype. Such a fitness definition embodies many assumptions about natural selection. It is a shorthand method of representing the notion that natural selection has certain consequences. In nature, organisms are born, they reproduce, and they die. We see the ones that are left around. That is all that goes on in nature. There is no definition of fitness in that statement. To transform the process into a model, we make an assumption about natural selection that becomes a fitness definition in the model. Because the models are criticized, and because one can ask how well different goal functions reflect real birth and death processes, this sort of argument is not teleological. There is just as much mechanism - of the demographic sort - present in life history theory as there is mechanism - of the Mendelian sort - present in population genetics. That the organism grows up, has a certain age at maturity, a certain fecundity, reproduces a certain number of times, then dies, is a mechanism as explicit and detailed as recombination, and one with direct consequences for population dynamics. Of course, it is population dynamics that drives both life history evolution and genetic change.

At this point, one could challenge evolutionary ecology and claim that it is inadequate because it contains no representation of genetic variation. To that I have two replies. First, I do not see why checking for the existence of genetic variability has logical priority over checking for the existence of phenotypic effects - they both are necessary. Second, there has recently been a surge of interest in the quantitative genetics of life history traits. The point of this work is to see whether the heritable variability exists upon which selection could act as conceived in optimality models. The answer, so far, is that it usually does, and this is an important extension of the field which, thus, is moving on its own toward questions more directly relevant to population geneticists.

Critique of Optimality and Microevolution

Within evolutionary ecology itself, there has been much debate over optimality theory. There is considerable evidence that the naive application of optimality theory must be wrong. For example, r and K selection was an optimality model that argued that gradual changes operating through microevolution and adapting populations to local environments would produce a coordinated syndrome involving five or six different phenotypic traits - age at maturity, length of life, reproductive effort, size of offspring, size of parent, and so forth. The whole syndrome was thought to be linked together and explained by microevolutionary processes.

I have been engaged for the past year in an analysis of the taxonomic levels at which the pattern arises. In fact, the pattern does exist. It is the dominant pattern found when one looks broadly across higher taxa in birds, mammals, and reptiles. By that I mean that it is the first principal component in a factor analysis done on the life history traits. However, if one subtracts the mean value of each trait for each order from the records of the species in each order, and then does a factor analysis on the residuals, one finds that the strength of the pattern called "r and K selection" is greatly reduced. If one repeats the analysis for families, and also controls for the effects of body size, one finds that the pattern of covariation called r and K selection is largely explained in the mammals by family and order effects and correlations with adult weight, in the birds by family and order effects, and in the reptiles by family and genus effects and correlations with adult length. In a separate analysis of the raw data, if one analyses each lineage separately, one finds a different correlation matrix of the life history traits in each lineage.

This means that microevolution could not have produced the pattern because it results from differences among higher taxa that presumably arose a long time ago, and has not changed as dramatically within lineages since then. This sort of correlation analysis cannot get at cause and effect, but it does refute the hypothesis that the pattern called r and K selection is caused by microevolutionary adaptation to local environments. Thus there is, within evolutionary ecology, a tradition of criticizing optimization – in this case, the r and K selection model – without invoking genetic constraints. The form of the critique raises a question: what is the nature of the evolutionary processes in each lineage that results in the observed fact that life history traits are tied together in different ways in different lineages? The answer must have something to do with genetics, and something else to do with regulatory mechanisms in development, and perhaps something else to do with functional morphology, but we do not yet know any details. I am agnostic about the causes.

Two Points of View

We have had people approaching evolution from two points of view. Population geneticists have started from the Mendelian mechanism and have looked out toward ecology to see what kinds of phenotypic phenomena Mendelism can be used to explain. They have found some cases where it can be done – the body of work called ecological genetics. One finds a gene whose function can be traced through physiology to an ecologically significant trait, then interprets the polymorphism in that gene, or the selection acting on that gene, in terms of one or more ecological factors. On the other hand, we have evolutionary ecology, which has looked at the phenotype and asked – "What is important about the phenotype in relation to selection on survival and reproduction?" – then looked at various ways in which the phenotype has evolved to solve the problems posed by the environment. The quite different general problem here is: "How does natural selection design organisms?"

The two points of view have gradually expanded what we know about genotype and phenotype, but because the initial orientations and starting questions were quite different, it has not been easy to link them, especially since much of the intermediate mechanism has been left out. Genetics has produced only a small number of examples of ecological traits that can be traced directly to Mendelian mechanisms. Every time this kind of research has been explored, it has become more and more complex. One of the reasons graduate students were put off doing ecological genetics 10 or 15 years ago was that when they looked at the *Cepaea* work, or the stickleback work, or at the genetics underlying mimicry in African butterflies, they saw a can of worms. A few examples were well worked out, but when people looked deeper and deeper, the genetics underlying significant natural variation became more and more complex. Thus, it did not seem that this view, which initially seemed to hold great potential, was working. So they abandoned it and looked directly at the phenotype to find patterns in the components of fitness that are quite general. That has resulted in the second beachhead.

Neither of these two points of view could be logically sufficient because both discuss important mechanisms that are intimately connected with evolution. Only changes in gene frequencies record the history of selection. Only phenotypes survive and reproduce. We need creative research programs, not necessarily based on the traditional techniques of either field, that take as their organizing question the bridging of this gap.

Conclusion

I learned a great deal from the population geneticists at Oberjoch. The key area of agreement, to my surprise, was that selection acts on whole organisms, not on genes or on gene frequencies. My misinterpretation of population genetics resulted from my incorrect inference of its implicit assumptions from the abstractions of its models. Just because it is hard to find whole organisms in the models does not mean, it turns out, that they were not present in the minds of the modelers.

However, areas of disagreement remain, the biggest of which concerns fitness. I am willing to concede that population geneticists are well-justified in deciding that fitness shall be defined, within their own field, as a short-term measure that functions to help predict changes in gene frequency. To that extent, I can understand why they would say they know what it is. Nevertheless, their definition applies only within their field, and I am not willing to concede that their field and their definition encompass all of evolutionary biology. I also use fitness measures in my work, but for a different purpose – to predict the equilibrium properties of the phenotype. In these models, fitness measures have the function of ranking phenotypes in terms of probability of winning in competition with other phenotypes given certain ecological conditions. That strikes me as a thoroughly Darwinian interpretation of fitness, and so I am not willing to concede terminological priority to the population geneticists. Darwin discussed the concept before Wright, Haldane, or Fisher gave it their specific technical interpretations.

I would have no objection to coming to the understanding that phenotypic fitness, as used in evolutionary ecology, is a purely technical representation of natural selection that functions in predicting the equilibrium properties of the phenotype, whereas genotypic fitness, as used in population genetics, is a purely technical representation of natural selection that functions in predicting the rate of gene frequency change. Whereas the question, what is fitness?, may be settled in population genetics, it is not yet settled in evolutionary ecology. A proper answer to that question for phenotypic fitness, reflecting both life history evolution and population dynamics, will be of considerable interest to those population geneticists that, like the ones at Oberjoch, retain a clear image of the whole organism in their minds.

Subject Index

B.G. Mirkin, S.N. Rodin

Graphs and Genes

Translated from the Russian by H.L. Beus
1984. 46 figures. XIV, 197 pages
(Biomathematics, Volume 11)
ISBN 3-540-12657-0

Contents: Graphs in the analysis of gene structure. – Graphs in the analysis of gene semantics. – Graphs in the analysis of gene evolution. – Epilogue: Cryptographic Problems in genetics. – Appendix: Some notions about graphs. – References. – Index of genetics terms. – Index of mathematical terms.

Population Biology

Proceedings of the International Conference held at the University of Alberta, Edmonton, Canada, June 22–30, 1982
Editors: **H.I. Freedman, C. Strobeck**
1983. XVII, 440 pages
(Lecture Notes in Biomathematics, Volume 52)
ISBN 3-540-12677-5

From the contents: *C. Strobeck,* Invited Guest Lecture: The Use of Coefficients of Identity to Study Random Drift. – *A.R. Templeton,* Invited Guest Lecture: The Evolution of Life Histories Under Pleiotropic Constraints and K-Selection. – *N. Keyfitz,* Invited Guest Lecture: Forecasting the Human Population. – *P.L. Chesson,* Invited Guest Lecture: Coexistence of Competitors in a Stochastic Environment: The Storage Effect. – *P. Waltman,* Invited Guest Lecture: Models of Competition for a Single Resource. – *M.P. Hassell,* Invited Guest Lecture: The Dynamics of Host-Parasitoid Interactions with Heterogeneity Host Density Dependence and Variable Sex Ratios. – *S. Levin,* Invited Guest Lecture: Coevolution. – *C.W. Clark, M. Mangel,* Invited Guest Lecture: Search Theory in Ecology and Resource Management.

MIX
Papier aus verantwortungsvollen Quellen
Paper from responsible sources
FSC® C105338

If you have any concerns about our products,
you can contact us on
ProductSafety@springernature.com

In case Publisher is established outside the EU,
the EU authorized representative is:
Springer Nature Customer Service Center GmbH
Europaplatz 3, 69115 Heidelberg, Germany

Printed by Libri Plureos GmbH
in Hamburg, Germany